Perspectives in
Mathematical Sciences

INTERDISCIPLINARY MATHEMATICAL SCIENCES

Series Editor: Jinqiao Duan *(Illinois Inst. of Tech., USA)*

Editorial Board: Ludwig Arnold, Roberto Camassa, Peter Constantin, Charles Doering, Paul Fischer, Andrei V. Fursikov, Sergey V. Lototsky, Fred R. McMorris, Daniel Schertzer, Bjorn Schmalfuss, Yuefei Wang, Xiangdong Ye, and Jerzy Zabczyk

---

Published

Vol. 1: Global Attractors of Nonautonomous Dissipative Dynamical Systems
   David N. Cheban

Vol. 2: Stochastic Differential Equations: Theory and Applications
   A Volume in Honor of Professor Boris L. Rozovskii
   eds. Peter H. Baxendale & Sergey V. Lototsky

Vol. 3: Amplitude Equations for Stochastic Partial Differential Equations
   Dirk Blömker

Vol. 4: Mathematical Theory of Adaptive Control
   Vladimir G. Sragovich

Vol. 5: The Hilbert–Huang Transform and Its Applications
   Norden E. Huang & Samuel S. P. Shen

Vol. 6: Meshfree Approximation Methods with MATLAB
   Gregory E. Fasshauer

Vol. 7: Variational Methods for Strongly Indefinite Problems
   Yanheng Ding

Vol. 8: Recent Development in Stochastic Dynamics and Stochastic Analysis
   eds. Jinqiao Duan, Shunlong Luo & Caishi Wang

Vol. 9: Perspectives in Mathematical Sciences
   eds. Yisong Yang, Xinchu Fu & Jinqiao Duan

Interdisciplinary Mathematical Sciences – Vol. 9

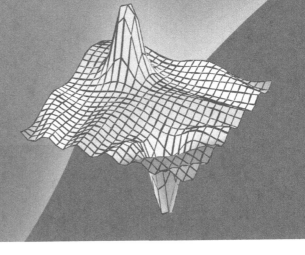

Perspectives in Mathematical Sciences

Editors

Yisong Yang
Polytechnic Institute of New York University, USA

Xinchu Fu
Shanghai University, China

Jinqiao Duan
Illinois Institute of Technology, USA

 World Scientific

NEW JERSEY · LONDON · SINGAPORE · BEIJING · SHANGHAI · HONG KONG · TAIPEI · CHENNAI

Published by

World Scientific Publishing Co. Pte. Ltd.

5 Toh Tuck Link, Singapore 596224

USA office: 27 Warren Street, Suite 401-402, Hackensack, NJ 07601

UK office: 57 Shelton Street, Covent Garden, London WC2H 9HE

British Library Cataloguing-in-Publication Data
A catalogue record for this book is available from the British Library.

ISBN-13 978-981-4289-30-6
ISBN-10 981-4289-30-2

Printed in Singapore.

Professor Youzhong Guo

Editorial Foreword

This volume of *Interdisciplinary Mathematical Sciences* collects invited contributions giving timely surveys on a diverse range of topics of contemporary research activities, including applications of analytic functions in periodic elastic mechanical problems, bifurcation and chaos in microelectromechanical systems, random dynamical systems and stochastic differential equations, epidemic disease propagation dynamics, inverse problems for equations of parabolic type, chaotic behavior of a two-component Bose–Einstein system, the p-Laplacian problems with a linking geometric structure, progress in chaos control, chaos synchronization and complex networks, mathematical modeling of cold plasma by partial differential equations of mixed type, gravitating Yang–Mills fields in arbitrary dimensions, the Hamiltonian constraint and Mandelstam identities in quantum gravity study, the lattice Boltzmann simulation and application to a nonlinear Schrödinger equation, stability of nonlocal time-delayed Burgers equation, bifurcation analysis of the Swift-Hohenberg equation, a new global-local computational method for differential equations, and a review on representing topological classes by harmonic quantities with applications in quantum field theory. The ordering of the chapters follows the alphabetic order of the last names of the first authors of the contributions.

We are grateful to all the authors for their contributions to this volume. Special thanks are due to Professor Zhenting Hou and Dr. Han-ping Chin, authors of Preface and Postscript, respectively, which provide vivid introductions to Professor Youzhong Guo for the reader. We would also like to thank Dr. Rajesh Babu, Production Specialist at World Scientific Publishing, for his professional editorial assistance.

We are privileged and honored to dedicate this volume to Professor Youzhong Guo, our teacher and life-long friend, on the occasion of his 75th birthday and 55 years of career in mathematical sciences.

Yisong Yang, Polytechnic Institute of New York University
Xinchu Fu, Shanghai University
Jinqiao Duan, Illinois Institute of Technology

August 10, 2009

Preface

This festschrift volume is dedicated to Professor Youzhong Guo on the occasion of his 75th birthday and 55 years of scientific career.

Born in Hangzhou, China, on October 25, 1935, and graduated from Nanjing University of Technology in 1955, Guo was recruited to Institute of Mathematical Sciences (Wuhan), Academia Sinica, as a research assistant to work with mathematician, Professor Guoping Li, academician of Academia Sinica. Since then, Guo has been engaged in mathematical sciences research and made many achievements. He published over one hundred articles ranging in pure and applied mathematics, especially in mathematical physics, system science, and mathematical economics. Guo's influence through his publications, lectures, and services has benefited people far beyond his fields.

In 1979, Guo was appointed Professor and Executive Vice-Director of Institute of Mathematical Sciences and Chairman of the Institute's Academic Committee. In 1987, Guo was appointed Chairman of Wuhan Municipal Commission of Science and Technology and became Director of Institute for Industry and Applied Mathematics. Afterwards he was elected as Vice-Chairman of Chinese Society of Industrial and Applied Mathematics, Chairman of Rational Mechanics and Modern Mathematics Commission of Chinese Society of Mechanics. Among his other past and present positions and responsibilities are: Editor-in-Chief/Editor of Lecture Notes for Mathematics, Acta Mathematica Scientia, Structural Analysis Systems and Engineering Analysis International, Director of System Engineering Society of China.

In the 1950's, Guo contributed in the approximation theory of functions, analytical theory of differential equations, variational principles, error estimates of solutions of differential equations in the Sobolev spaces. Guo also designed Wuhan Laboratory of Electrostatic Acceralator, Academia Sinica. Taking a cooperative research task with the Academy of Sciences of Soviet Union and working with Guoping Li, Guo studied the theory and applications of automorphic functions and the Minkowski–Denjoy functions for a long time. Their results were published as the first monograph of its kind in the world in 1978.

In the 1960's, Guo worked on field theory and the Three-Gorge Project. He developed the principles of transformations of fiber bundles for general relativistic quantum fields, known as the L-G method, which gives a new natural representation of exterior differential forms and allows much simplified calculations and analysis. He also put forward some new methods for the analysis of large scale systems, such as rheology theory of limiting equilibria, stability of side slopes, stresses in tunnels, and dynamics of blasts applied to the Three Gorges. In the so-called 'Cultural Revolution,' Academia Sinica stopped functioning, and Guo suffered ten years (1968–1978) of unjust imprisonment and stray.

In the 1970's, under extremely difficult circumstances, Guo was ordered to complete many designs in architecture and automatic system engineering, when he was behind bars. Guo also contributed in developing the energy band theory of semiconductors. In studying the anisotropic energy band theory and geo-dynamics based on the statistical characteristics of rock and soil, Guo established a statistical theory for rock and soil mechanics and pioneered mathematical seismology. After the Cultural Revolution, Guo was appointed to take charge of restoring and reconstructing Institute of Mathematical Sciences.

In the 1980's, focusing on nonlinear mechanism of earthquakes, Guo helped lay a mathematical foundation for structural analysis. He obtained a variational representation theory of plastic dislocation and proved an existence and uniqueness theorem in the theory of micro-elasto-visco-plasticity. He also found a mathematical formalism for saturated porous medium magneto-hydro-dynamics, corrected some major errors in electro-magnetic theory, and generated new models in seismology and bio-mechanics. Through studying asymptotic methods, Guo surprisingly proved the occurrence and propagation of global singularities in gradient problems. These achievements led to the election of Guo as a main-entry Author for Mathematical Physics, in the Great Chinese Encyclopedia.

In the 1990's, Guo was appointed Chairman of Wuhan Commission of Science and Technology and Director of Wuhan Institute for Industrial and Applied Mathematics. Then he was elected Vice Mayor of Wuhan City and exerted all his energy to the development of the local economy, education, and science. Being responsible for high-tech sectors, he always made his best effort in raising the living standard of the urban poor and developing the human resources in a scientific, practical, and persistent way. He helped to establish a new university, Jianghan University, through a decade-long tireless work. Within his tenure, the East-Lake High-Tech Development Zone was approved by the Central Government which was then listed as number one among its peers.

At the beginning of 2000, Guo was appointed to lead the national project, the Strategy Research for Central China Economic District Along Yangtze River, with a research group comprised of 12 academicians, 36 professors, and 38 young researchers. Jointly, they investigated this rapidly developing district and the river economy systematically, resulting in the publication of a series of 10 Evolution Economics research volumes to advise China's Central Government.

I would like to take this opportunity to wish Youzhong many more successes in his career and a long healthy life.

Zhenting Hou
Central South University
Changsha, China

February 12, 2009

Selected Books Authored by Youzhong Guo

1. *Theory of Conductors and Semi-Conductors* (with Guoping Li), Hubei Science and Technology Press, 1978.
2. *The Theory of Automorphic Functions and Minkowski Functions* (with Guoping Li), Academic Press, 1979.
3. *Mathematical Seismology* (with Guoping Li), Seismology Press, 1979.
4. *Mathematical Seismology* (with Guoping Li), Lecture Notes Inst. Math. Comp. Tech., Academia Sinica, 1979.
5. *The Theory on General Relativistic Quantum Fields* (with Guoping Li), Hubei Science and Technology Press, 1980.
6. *The Theory on General Relativistic Quantum Fields*, Seminar Notes Inst. Math. Comp. Tech., Academia Sinica, 1981.
7. *Differential Forms and Their Applications*, Hydro-Electro-Power Press, 1986.
8. *Theoretical Physics* (with Zengren Liu et al.), Central China University of Technology Press, 1986.
9. *Lie Theory and Fiber Bundles* (with Changgui Shao), *ibid*, 1986.
10. *The Operation Theory and its Applications* (with Jinzhong Mao et al.), Wuhan University of Technology Press, 1992.
11. *Mathematical Methods for Physical Sciences*, Wuhan University Press, 1993.
12. *Gazing at Central China - The High-Speed Developing Fourth Block in China*, Volumes 1, Central Literature Press, 2003.
13. *Gazing at Central China - The High-Speed Developing Fourth Block in China*, Volume 4 (with Zi Guo et al), *ibid*, 2003.
14. *Gazing at Central China - The High-Speed Developing Fourth Block in China*, Volume 8 (with Zhibin Han et al.), *ibid*, 2009.
15. *Gazing at Central China - The High-Speed Developing Fourth Block in China*, Volume 10 (with Zi Guo et al.), *ibid*, 2009.
16. *Modern Methods in Mathematical Physics*, China Science Press, to appear, 2009.
17. *Trudging Life – Selected Papers of Professor Youzhong Guo*, Writer's Publishing House, 2002.
18. *System Evolution and Evolution Economics*, Tin Ma Publishing Company Limited, 2008.

Contents

Perspectives in Mathematical Sciences
Interdisciplinary Mathematical Sciences, Volume 9, 2009
pp. 1–36

Chapter 1

Periodic Boundary Problems for Analytic Function Including Automorphic Functions

Haitao Cai[1] and Jian-Ke Lu[2]

[1] *Central South University, Changsha 410083, P.R.China,*
htcai34@yahoo.com.cn
[2] *Wuhan University, Wuhan 430074, China*

Two kinds of fundamental boundary problems for analytic functions, including automorphic functions, and their applications are discussed.

1. Periodic Boundary Value Problems for Analytic Functions

As a mathematical tool for the investigation of periodic problem in plane elasticity, two kinds of fundamental boundary value problems for analytic functions will be discussed in the present chapter, namely, the periodic Riemann boundary value problems, the periodic Riemann-Hilbert boundary value problems in the half plane. The more general formulations of these problems are those for automorphic functions which were studied by F. D. Gakhov and L.I. Chibrikova, Guoping Li and Youzhong Guo from the middle of the last century. However, the periodic problems discussed here are much more important in applications.

1.1. *Periodic Riemann Boundary Value Problems: Case of Closed Contours*

1) Formulation of the problems

Let L_k, $k = 0, \pm 1, \pm 2, \ldots$, be a set of smooth closed contours, non-intersecting to each other, oriented counter-clockwise, for the same shape and horizontally distributed with period $\pi a(a > 0)$.

The region interior to L_k is denoted by S_k^+, S_k^- is the complement of $S_k^+ + L_k$; and the region exterior to $L \equiv \sum_{k=-\infty}^{+\infty} L_k$ is denoted by S^-. We may always assume the origin $O \in S_0^+$ and all the points $\pm \frac{1}{2}\pi, \pm \frac{3}{2}\pi \ldots$ lie in S^-.

For briefness, we assume that L_k is a single closed smooth contour. In fact, the following discussions remain effective with slight modifications when L_k consists of a finite number of intersecting closed contours. Moreover, if L is an arbitrary smooth curve with period $a\pi$ (i.e., L_0 is a smooth curve from $-\frac{1}{2}a\pi + iy_0$ to $\frac{1}{2}a\pi + iy_0$ with the same slope at its end-points), the following are also valid in case of suitable modifications.

2) Periodic Riemann boundary value problem (problem P_1): find a function $\Phi(z)$ in the complex plane with period $a\pi$, satisfying

$$\Phi^+(t) = G(t)\Phi^-(t) + g(t), \, t \in L \tag{1.1}$$

where $\Phi^\pm(z)$ denote the boundary values (limiting value) of a sectionally holomorphic function $\Phi(z)$ (holomorphic in $S_k, k = 0, \pm 1, \pm 2, \cdots$, and S^-) from the positive (left) and the negative (right) sides of L respectively, while $G(t)$ and $g(t) \in H$ are given on L with period $a\pi$:

$$G(t + a\pi) = G(t), g(t + a\pi) = g(t), t \in L. \tag{1.2}$$

and $G(t) \neq 0$ (normal type).

Here, a functions $f(t) \in H(A, \mu)$ (function with Hölder Continuity) on L means

$$|f(t) - f(t')| \leq A|t - t'|^\mu, \quad \forall\, t, t' \in L$$

for certain positive constant A and $\mu(0 < \mu \leq 1)$. Note that ∞ is a limiting point of the points on L so that the solution $\Phi(z)$ of the problem (if any) could not have a definite limit as $z \to \infty$ in general. However, we may ask $\Phi(z)$ meeting certain requirements at $z = \pm\infty$ (that means , for $z = x + iy$, x may be arbitrary and $y \to \pm\infty$ respectively). We always require $\Phi(\pm\infty i)$ to be bounded (i.e. finite).

If $g(t) \equiv 0$, the problem is homogeneous, will be denoted by P_1^0, and otherwise, non-homogeneous.

Throughout this chapter, all periodic functions are always assumed with periodicity $\pi a(a > 0)$, for t an a definite limit.

Theorem 1 (Chibrikova) If $\Phi(\pm\infty i)$ are required to be bonded, then the homogeneous problem P_1^0 has $k + 1$ linearly independent solutions when its index $k \geq 0$ and has only the trivial solution when $k < 0$.

Theorem 2 (Chibrikova) If $\Phi(\pm\infty i)$ are required to be bonded, the general solution of the non- homogeneous problem P_1 has $k + 1$ arbitrary constants when $k \geq -1$; It is (uniquely) solvable when $k < -1$, iff $-k - 1$ conditions are valid as follows:

$$\int_{L_0} \frac{g(t)}{X^+(t)} \frac{\sin^{j-1}\frac{t}{a}}{\cos^{j+1}\frac{t}{a}} dt = 0, j = 1, \cdots, -k - 1;$$

In the case of index $k = 0$, then

$$\begin{aligned}
\Gamma(z) &= \frac{1}{2a\pi i} \int_{L_0} \log K \cdot \cot\frac{t - z}{a} dt \\
&= \frac{\log K}{2\pi i} [\log \sin\frac{t - z}{a}]_{L_0} \\
&= \begin{cases} \log K, & when\ z \in S^+, \\ 0, & when\ z \in S^-. \end{cases}
\end{aligned}$$

Where $\log K$ may be taken definitely and arbitrarily. Thus $X^+(z) = K$, $X^-(z) = 1$. Therefore, as $\Phi(\pm\infty i)$ are bounded, the general solution of the problem P_1 is

$$\Phi(z) = \begin{cases} \frac{1}{2a\pi i}\int_{L_0} g(t)\cot\frac{t-x}{a}dt + C, & z \in S^+, \\ \frac{1}{K}[\frac{1}{2a\pi i}\int_{L_0} g(t)\cot\frac{t-x}{a}dt + C], & z \in S^-. \end{cases}$$

1.2. Periodic Riemann Boundary Value Problems: Case of Open Ares or Discontinuous Coefficients

1) Case of open arcs

Consider the case $L = \sum_{k=-\infty}^{+\infty} L_k$ being periodic as before when L_0 consists of p non-intersecting smooth open arcs $a_r b_r$, $r = 1, \cdots, p$, oriented from a_r to b_r and both $G(t), g(t) \in H(A, \mu)$ on each arc (including its end-points) with $G(t) \neq 0$. We get the general solution of problem P_1^0:

$$\Phi(z) = X(z)P_k(\tan\frac{z}{a}),$$

where the characteristic function

$$X(z) = \prod(z)e^{\Gamma(z)},$$

in which

$$\prod(z) = \prod_{i=1}^{2p}(\tan\frac{z}{a} - \tan\frac{c_j}{a})^{\lambda_i}$$

$$\Gamma(z) = \frac{1}{2a\pi i}\int_{L_0}\log G(t)\tan\frac{t-z}{a}dt.$$

When a constant factor is merged into, then

$$\Phi(z) = \prod_{i=1}^{2p}\sin^{\lambda_i}\frac{z-c_j}{a}e^{\Gamma(z)}Q_k(\sin\frac{z}{a}, \cos\frac{z}{a}).$$

2) An important special case

For the need of application later, consider the special case where L_0 consist of a single- line-segment $\gamma_0 : -l \leq t \leq l$ $(l < \frac{1}{2}a\pi)$ along the real axis and $G(t) = -K$ is a negative real constant. We would seek for solutions of the corresponding problem P_1^0, i.e., solutions permitted to be integrably unbounded at not near $z = \pm l$. Now $c_1 = -l, c_2 = l$. The general solution is

$$\Phi(z) = \frac{1}{2a\pi i\sqrt{R(z)}}\int_{-l}^{l}g(t)\sqrt{R(t)}\cot\frac{t-z}{a}dt + \frac{C_0\tan\frac{z}{a} + C_1}{\sqrt{R(z)}}.$$

3) Case of discontinuous coefficients

Assume L_0 to be a closed contour as in Section 1, but $G(t)$ and $g(t)$ have a finite number of discontinuities $c_1, \cdots, c_p$ of the first kind on L_0, belong to $H(A, \mu)$ on each arc of L_0 (arc between two adjacent discontinuities) including its end-points (the values at these points are understood by the one sided limits) and $G(t) \neq 0$.

1.3. Periodic Riemann-Hilbert Boundary Value Problems: Case of the Half-plane

1) Formulation of the problems

Assume L_k is the segment from $-\frac{1}{2}a\pi + \frac{1}{2}ka\pi$ to $\frac{1}{2}a\pi + \frac{1}{2}ka\pi$ $(k = 0, \pm 1, \pm 2, \cdots)$ lying on the real axis with length $a\pi$ $(a > 0)$. Then $|L_k|$ is a set of periodic segments with period $a\pi$, the union of which is the real axis. Denote the lower half-plane by S^- and the upper, by S^+.

The Riemann-Hilbert boundary value problem of the half-plane is to find a function $w(z) \equiv u - iv$ holomorphic in S^- with period $a\pi$, satisfying the boundary conditions

$$a(x)u + b(x)v = F(x), \ x \in L_k, k = 0, \pm 1, \pm 2, \cdots,$$

where $a(x), b(x)$ and $F(x)$ are given functions with period $a\pi$, arcwise with Hölder-Lipschitz continuity, i.e., having a finite number of discontinuities on each $L_k \in H(A, \mu)$ on each continuous closed interval between two adjacent discontinuous points(called nodes), and $a(x), b(x)$ have no common zeros for any x. Without loss of generality, we may always assume that all of them are continuous at $x = \pm \frac{1}{2}a\pi$(and their congruent points).

2) An important particular case

We would give solutions for a particular but very important case which often occur in practice and would be met later in this chapter.

Let $\gamma_0 = [-l, l]$, $0 < l < \frac{a\pi}{2}$, and $\gamma_0' = L_0 - \gamma_0$. Denote the set of all the segments congruent to γ_0 and $\gamma_0' (\mathrm{mod}\ a\pi)$ by γ and γ' respectively. In the particular case considered hereby:

$$a(x) + ib(x) = \begin{cases} a_1 + ib_1 & x \in \gamma, \\ a_2 + ib_2 & x \in \gamma'. \end{cases} \tag{1.3}$$

where a_j, b_j are real constants with $a_j + ib_j \neq 0$ $(j = 1, 2)$ and $a_1 + ib_1 \neq a_1 + ib_2$.

A particular solution is well-known as

$$\Omega_1(z) = \frac{X(z)}{a\pi i} \int_{L_0} \frac{F(t)}{X^+(t)[a(t) - ib(t)]} \cot \frac{t - z}{a} dt, \mathrm{Im} z \neq 0.$$

2. Periodic Problems for Isotropic Material in Elastic Theory

There were many works in literature on the periodic problems in the theory of plane elasticity for isotropic material, e.g., researches on stress distribution in the neighborhoods of periodic holes by R. C. J. Howland, G. N. Savin, L. M. Tang, M. Isida, study on periodic contact problems by G. M. L. Gladwall and investigation for expression of periodic stress functions by S. Morigashi. All these works have limitations in either the shape of the holes or the boundary conditions, and the discussions are incomplete. Especially, there were few studies on the possibility of quasi-periodic displacements. see Ref.1-2,8,13,15-18,20-22.

2.1. *Stress Functions*

1) General expression of stress functions

Let the elastic plane possess a row of periodic holes with boundaries $L_j, j = 0, \pm 1, \pm 2, \cdots,$, where L_j consists of piecewise smooth closed contours $l_k^{(j)}, k = 1, \cdots, n$ and for the same k $l_k^{(j)}, k = 1, \cdots, n$ are periodically arranged. l_k^0 will be denoted briefly by l_k. All the contours are oriented counter-clockwise. The region occupied by the elastic body is denoted as S^-, the region bounded by $l_k^{(j)}$ as $S_k^{(j)+}$, and $S_k^{(0)+}$ briefly as S_k^+. Denote the strip region $|x| < \frac{1}{2}a\pi$ by S_0. Let $S_0^+ = \sum_{k=1}^{n} S_k^+$ and $S_0^+ = S_0 - S_0^+$.

Denote the stresses at any point $z = x + iy$ in S^- by $\sigma_x(z)$, $\sigma_y(z)$, $\tau_{xy}(z)$ and the (complex) displacement by $D(x) = u + iv$. It is well-known in Ref.23, that they may be expressed in terms of (complex) stress functions $\varphi(z)$ and $\psi(z)$ or their derivatives $\Phi(z) = \varphi'(z)$ and $\Psi(z) = \psi'(z)$ as follows:

$$\begin{cases} \sigma_x + \sigma_y = 2[\Phi(z) + \overline{\Phi(z)}], \\ \sigma_y - \sigma_x + 2i\tau_{xy} = 2[\bar{z}\Phi'(z) + \Psi(z)], \\ 2\mu d = 2\mu(u + iv) = \kappa\varphi(z) - z\overline{\varphi'(z)} - \overline{\psi(z)}, \end{cases} \tag{2.1}$$

where μ is the shearing modulus of the elastic medium, κ is a constant related to its Passion ratio $\sigma(1 < \kappa < 3)$ and $\varphi(z)$, $\psi(z)$ are functions analytic (in general, multi-valued) functions in S^- with their derivatives holomorphic, i.e., single-valued, in S^-.

We always assume that the stresses are periodic and bounded at $z = \pm\infty i$.

Theorem 3 In the isotropic infinite elastic plane weakened by a row of holes of any shape with period $a\pi(a > 0)$, if the stresses are known to be periodic and bounded at infinity , then the relative displacements must be quasi-periodic , i.e.,

$$\begin{cases} 2\mu[u(z + a\pi) - u(z)] = a\pi q \\ v(z + a\pi) - v(z) = 0 \end{cases} \tag{2.2}$$

where q is a certain real constant.

$a\pi q/2p$ is called the addendum of the quasi-periodic function $u(z)$. By the previous expression, we know that

$$\gamma = a\pi(\kappa\beta - q).$$

Theorem 4 Suppose there is a row of holes with period $a\pi$ $(a > 0)$ in the isotropic elastic plane, the boundary of which is $L_j, j = 0, \pm 1, \pm 2, \cdots$, where each L_j consists of n piecewise smooth closed contours $l_k^{(j)}, k = 1, 2, \cdots n$, and for a fixed $l_k^{(j)}, j = 0, \pm 1, \pm 2, \cdots$, are arranged periodically. Assume the stresses are periodic, bounded at $z = \pm\infty i$ and the principal vector of the external stresses on $l_k^{(0)}$ is $X_k + iY_k$. Then the stress functions and $\varphi(z)$ and $\psi(z)$ respectively have the expressions:

$$\varphi(z) = -\frac{1}{2\pi(\kappa+1)}(X_k + iY_k)\log\sin\frac{z-z_k}{a} + \beta z + \varphi_0(z) \qquad (2.3)$$

$$\psi(z) = \frac{\kappa}{2\pi(\kappa+1)}\sum_{k=1}^{n}(X_k - iY_k)\log\sin\frac{z-z_k}{a}$$

$$+ \frac{\kappa}{2a\pi(\kappa+1)}\sum_{k=1}^{n}(X_k - iY_k)\cot\frac{z-z_k}{a}$$

$$+ (\kappa\beta - \beta + q)z - z\varphi_0'(z) + \psi_0(z). \qquad (2.4)$$

where $\varphi_0(z)$ and $\psi_0(z)$ are functions holomorphic and $a\pi$-periodic in the elastic region S^-.

Corollary 4.1 Assume the elastic body as above. If the resultant of the principal vectors on the boundaries of the holes in a periodic strip is zero and the stresses at $z = \pm\infty\,i$ are $\sigma = \sigma_y(\pm\infty i)$ and $\tau = \tau_{xy}(\pm\infty i)$, then both the stress functions $\varphi(z)$and $\psi(z)$ are single-valued, and

$$\varphi(z) = \varphi_0(z) + \beta z,$$

$$\psi(z) = \psi_0(z) - z\varphi'(z) + \kappa\bar\beta z = (\kappa\bar\beta - \beta)z - z\varphi_0'(z) + \psi_0(z),$$

where $\beta = \frac{-\sigma + i\tau}{1+\kappa}$, while both $\varphi_0(z)$ and $\psi_0(z)$ are $a\pi$-periodic.

Theorem 5 If the displacements in an isotropic plane elastic body are quasi-periodic, then the stresses must be periodic.

2) Formulation of the fundamental problem
First fundamental problem Given the periodic stress function $X_n(t)+iY_n(t)$ on the boundary L of the elastic region and the stresses at $z = -\infty i$(or $z = +\infty i$), find the elastic equilibrium (i.e., the stress distribution in the elastic body).

This is the most general formulation of the problem under the assumption of stresses to be periodic and bounded at infinity. In this case, the stresses at $z = +\infty i$ (or$z = -\infty i$) is also known, and so do β and q.

The problem may be reduced to the following boundary value problem:

$$\varphi(t) + t\overline{\varphi'(t)} + \overline{\psi(t)} = f(t) + C(t), \ t \in L,$$

where $C(t)$ represents a step function on different boundary contour with different value and (we have assumed $z = 0 \in S^-$);

$$f(t) = -i\int_0^t (X_n + iY_n)ds, \quad t \in L,$$

where s is the arc-length parameter on boundary contour;

$$\varphi_0(t) + (t - \bar t)\overline{\varphi_0'(t)} + \overline{\psi_0(t)} = f_0(t) + C(t),$$

where we have put

$$
\begin{aligned}
f_0(t) =\ & \frac{1}{2\pi(\kappa+1)} \sum_{k=1}^{n} (X_k + iY_k) \log \sin \frac{t - z_k}{a} \\
& - \frac{\kappa}{2\pi(\kappa+1)} \sum_{k=1}^{n} (X_k + iY_k) \log \sin \frac{t - z_k}{a} \\
& + \frac{t - \bar{t}}{2a\pi(\kappa+1)} \sum_{k=1}^{n} (X_k + iY_k) \cot \frac{t - z_k}{a} \\
& - (\beta + \bar{\beta})t - (\kappa\beta - \bar{\beta} + q)\bar{t} + f(t),
\end{aligned}
$$

which is singer-valued on L obviously.

Second fundamental problem Given the relative displacements on the boundary contours of the periodic holes, the resultant principal vector $X + iY$ of the external stresses along the boundary contours in a periodic strip and the stresses at, $z = -\infty i$(or $z = +\infty i$), find the equilibrium.

Here, the relative displacements mean that they are quasi-periodic but the constant q is not given. As in the first fundamental problem, this is the most general formulation of the problem. It may be reduced to solve a Fredholm integral equation as well.

For convenience of discussion, in the sequel, we always assume that the displacements are also periodic.

3) Stress functions for elastic half-plane

Assume the elastic body occupies the lower half-plane S^-. In this case, the stresses and displacements may be expressed in terms of a single stress function. When z lies in the upper half-plane S^+, let

$$
\bar{\Phi}(z) = \overline{\Phi(\bar{z})}, \ \bar{\Psi}(z) = \overline{\Psi(\bar{z})}, \ z \in S^+,
$$

where $\Phi(z) = \varphi'(z), \Psi(z) = \psi'(z)$ are the stress functions.

Theorem 6 In the periodic problems for isotropic elastic half-plane, $\Phi(z)$ is $a\pi$-periodic for $z \in S^+$ and $z \in S^-$ including its boundary values; and is bounded at $z = \pm\infty i$.

2.2. *Periodic Fundamental Problems of Elastic Half-plane*

1) The first fundamental problem

Let the isotropic elastic body occupy the lower half-plane S^- in the z-plane. On the x-axis, denote $z = t$(t is real). Given the external stresses on the x-axis:

$$
\sigma_y(t) = -P(t), \quad \tau_{xy}(t) = T(t),
$$

being arcwise $\in H(A, \mu)$ and with period $a\pi$. Here P means the normal pressure distribution. Assume again both the stresses and displacements are periodic and the stresses at $z = -\infty i$ are bounded. Find the stress distribution and the displacements on the whole elastic body, called the periodic first fundamental problem of the half-plane.

We have

Theorem 7 Under the above assumptions, the solution of the periodic first fundamental problem uniquely exists.

The expression of $\Phi(z)$ can be written as

$$\Phi(z) = \frac{1}{2a\pi i} \int_{L_0} [P(t) + iT(t)] \cot \frac{t - z}{a} dt + \frac{1}{2}\frac{\kappa - 1}{\kappa + 1} P^* \tag{2.5}$$

P^* and T^* have clear mechanical meaning. and so

$$\sigma_y(-\infty \ i) = P^*, \tau_{xy}(-\infty \ i) = T^*, \tag{2.6}$$

$$\sigma_x(-\infty \ i) = -\frac{3 - \kappa}{1 + \kappa} P^*. \tag{2.7}$$

Corollary 7.1 For the periodic first fundamental problem of isotropic elastic half-plane, if the resultant of the external normal stresses on the boundary of a period is a pressure force:

$$\int_{L_0} P(t)dt = a\pi P^* > 0,$$

then σ_x is a compression at $z = -\infty i$, given by (2.7).

2) The second fundamental problem

Assume that, on the boundary L of the isotropic elastic half plane S^-, the complex displacement $u^- + iv^- = g(t)$ is given, where $g(t)$ is continuous and $a\pi$-periodic (up to an arbitrary constant term, corresponding to a rigid translation of the whole elastic body), $g'(t) \in H(A, \mu)$ arcwise, and the principal vector $X + iY$ of the external stresses on a period of the boundary is also given. Again we assume the stresses and the displacements are periodic and the stresses at $z = -\infty i$ are bounded. Find the elastic equilibrium here is the periodic second fundamental problem of the half-plane.

Theorem 8 Under the above assumptions, the solution of the periodic second fundamental problem of the half-plane uniquely exists.
We obtain

$$\Phi(z) = \begin{cases} \frac{\mu}{a\pi i} \int_{L_0} g'(t) \cot \frac{t-z}{a} dt - \frac{\kappa(Y-iY)}{(\kappa+1)a\pi}, & z \in S^+ \\ -\frac{\mu}{\kappa a\pi i} \int_{L_0} g'(t) \cot \frac{t-z}{a} dt + \frac{(Y-iX)}{(\kappa+1)a\pi}, & z \in S^- \end{cases} \tag{2.8}$$

The following two corollaries are evident.

Corollary 8.1 $\sigma_x(-\infty i) = 0$ iff $Y = 0$.

Corollary 8.2 The horizontal displacement u is bounded when and only when $X = 0$ and so does the vertical displacement v when and only when $Y = 0$.

3) The mixed fundamental problem

Let the elastic half-plane S^- as before. On the boundary L_0 in a period, given displacement (up to a constant term) on its sub-segment $\gamma_0(-l \le t \le l)$

$$u^- + iv^- = g(t), \ t \in \gamma_0$$

where $g(t)$ is continuous with $g'(t)$ and on $\gamma_0' = L_0 - \gamma_0$, given the external stresses for briefness, to be zero:

$$\sigma_y^-(t) = \tau_{xy}^-(t) = 0, \ l \le |t| \le \frac{1}{2}a\pi.$$

Besides, the principal vector $X + iY$ of the external stresses on γ_0 is also given. All the conditions given are $a\pi$-periodic. Again assume the stresses and the displacements are periodic and the stresses at $z = -\infty i$ are bounded. Find the elastic equilibrium. The problem is called the periodic mixed fundamental problem.

Theorem 9 Under the above conditions, the solution of the periodic mixed fundamental problem uniquely exists. We have

$$\Phi(z) = \frac{2\mu\varepsilon i}{\kappa + 1} \left\{ 1 - \frac{X(z)}{\cos\frac{l}{a}} \left(\cosh A \tan\frac{z}{a} - i \sinh A \right) \right\}. \tag{2.9}$$

2.3. *Periodic Contact Problems*

1) The case without friction

Assume a series of periodic stamps (of the same shape) are pressed on the boundary of an isotropic elastic half-plane S^-. In the present paragraph, it assumed that there exists no friction between the stamps and the half-plane. Out of the stamps, the condition subjected to the boundary of the half-plane is $\sigma_x = \tau_{xy} = 0$. Right beneath the bases of the stamps, only the periodic vertical displacement $v(t)$ is given while the horizontal displacement $u(t)$ is unknown. On the boundary L_0, let the interval pressed be $\gamma_0 : -l \le t \le l$. Since there is no friction $\tau_{xy} = 0$ also on γ_0 but σ_y is unknown . Besides, assume the load applied to each stamp is a positive pressure force P_0, i.e. $Y = -P_0$ and $X = 0$. Find the elastic equilibrium.

Let $y = f(x)$ be the equation of the vase of the stamp pressed on S^- and $f'(x) \in H(A, \mu)$. Thus the boundary conditions on L_0 are:

$$\tau_{xy}^-(t) = 0, \ t \in L_0; \ \sigma_y^-(t) = 0, \ t \in L_0 - \gamma_0$$
$$v(t) = f(t), \ t \in \gamma_0;$$

and the principal vector of the external stresses on γ_0 is $X + iY = -P_0 i$.

Under these boundary conditions, by the principle of equilibrium, we know that

$$\sigma = \sigma_y(-\infty i) = -\frac{P_0}{a\pi}, \ \tau = \tau_{xy}(-\infty i) = 0. \tag{2.10}$$

Note that $\Phi(z)$ is holomorphic in the plane cut by γ (γ_0 and its periodic congruent),

$$\Phi(z) = \frac{2\mu}{(\kappa+1)a\pi\sqrt{R(z)}}\int_{\gamma_0} f'(t)\sqrt{R(t)}\cot\frac{t-z}{a}dt + \frac{\beta_0\tan\frac{z}{z}+\beta_1}{\sqrt{R(z)}} + \beta_2, \quad (2.11)$$

where we have written $X(z)$ as

$$X(z) = \frac{1}{i\sqrt{R(z)}}, \quad R(z) = \tan^2\frac{l}{a} - \tan^2\frac{x}{a} \quad (2.12)$$

where $\sqrt{R(z)}$ takes positive value $\sqrt{R(t)}$, when z tends to $t \in \gamma_0$ from S^+. It is also seen that by (2.12),

$$\lim_{z\to\pm\infty i} (z-\bar{z})\Phi'(z) = 0 \quad (2.13)$$

remains valid.

From the periodic condition of the displacements and the condition of equilibrium of the stresses at $z = \pm\infty i$, we get

$$\beta_0 = -\frac{2\mu}{(\kappa+1)a\pi}\int_{\gamma_0} f'(t)\sqrt{R(t)}dt,$$

$$\beta_1 = \frac{P_0}{2a\pi\cos\frac{l}{a}},$$

$$\beta_1 = \frac{\kappa-1}{\kappa+1}\frac{P_0}{2a\pi}.$$

The required unique solution is finally obtained

$$\Phi(z) = \frac{2\mu}{(\kappa+1)a\pi\sqrt{R(z)}}\int_{\gamma_0} f'(t)\sqrt{R(t)}(\cot\frac{t-z}{a} - \tan\frac{z}{a})dt$$

$$+ \frac{P_0}{2a\pi\cos\frac{l}{a}\sqrt{R(z)}} + \frac{\kappa-1}{\kappa+1}\frac{P_0}{2a\pi} \quad (2.14)$$

The pressure distribution right beneath the bases of the stamps could be easily evaluated by

$$P(t_0) = \frac{4\mu}{(\kappa+1)a\pi\sqrt{R(t_0)}}\int_{-l}^{l} f'(t)\sqrt{R(t)}(\cot\frac{t-t_0}{a} - \tan\frac{t_0}{a})dt$$

$$+ \frac{P_0}{a\pi\cos\frac{l}{a}\sqrt{R(t_0)}} \quad (2.15)$$

Example Periodic stamps with horizontal rectilinear base.
Here $f'(t) = 0$, by (2.14) and (2.15):

$$\Phi(z) = \frac{P_0}{2a\pi\cos\frac{l}{a}\sqrt{\tan^2\frac{l}{a}-\tan^2\frac{z}{a}}} + \frac{\kappa-1}{\kappa+1}\frac{P_0}{2a\pi}$$

$$= \frac{P_0\cos\frac{l}{a}}{2a\pi\sqrt{\sin\frac{l+z}{a}\sin\frac{l-z}{a}}} + \frac{\kappa-1}{\kappa+1}\frac{P_0}{2a\pi},$$

$$P(t) = \frac{P_0}{a\pi \cos \frac{l}{a} \sqrt{\tan^2 \frac{l}{a} - \tan^2 \frac{t}{a}}}$$

$$= \frac{P_0}{a\pi \sqrt{\sin \frac{l+t}{a} \sin \frac{l-t}{a}}},$$

where the radical involved is taken as the branch, when the plane is cut by γ taking positive value as $z \to t \in \gamma_0$ from S^+.

2) The case with friction

Now assume the friction coefficient $k \neq 0$ between the periodic stamps and eh elastic half-plane, that means, beneath the stamps, between the shearing stress $T(t) = \tau_{xy}(t)$ and the normal pressure $P(t) = -\sigma_y(t)$, there exist the relation

$$T(t) = kP(t), \ t \in \gamma_0. \tag{2.16}$$

Again assume $v^-(t) = f(t)$ on γ_0 with $f(t) \in H(A, \mu)$ and the external pressure force P_0 are given on γ_0. The principal vector of the external stresses on γ_0 is known as $X + iY = T_0 - iP_0 = (k - i)P$. On $\gamma_0' = L_0 - \gamma_0$, $T(t) = P(t) = 0$.

We obtain the general solution

$$\Phi(z) = \frac{2\mu(1 + ik)e^{\pi ai} \cos \pi a X(z)}{a\pi(\kappa + 1)} \int_{\gamma_0} \frac{f'(t)}{X^+(t)} \cot \frac{t - z}{a} dt$$

$$+ X(z)(1 + ik)i(\beta_0 \tan \frac{z}{a} + \beta_1) + \beta_2. \tag{2.17}$$

From the condition of periodicity of the displacements and the condition of equilibrium of the stresses at $z = -\infty i$, we get

$$\begin{cases} \beta_0 = -\frac{2\mu \cos \pi a}{(\kappa + 1)a\pi} \int_{\gamma_0} f'(t)Q(t)dt + \frac{P_0 \sin(1 - \lambda)\pi a}{2a\pi \cos \pi a \cos \frac{l}{a} \sqrt{R(z)}} \\ \beta_1 = \frac{P_0 \cos(1 - \lambda)\pi a}{2a\pi \cos \pi a \cos \frac{l}{a}} \\ \beta_2 = \frac{\kappa - 1}{\kappa + 1} \frac{(k^2 + 1)P_0}{2a\pi} \end{cases} \tag{2.18}$$

and

$$P(t_0) = \frac{2\mu \sin 2\pi a}{(\kappa + 1)} f'(t_0) + \frac{4\mu \cos^2 \pi a}{a\pi(\kappa + 1)Q(t_0)} \int_{-l}^{l} f'(t)Q(t) \cot \frac{t - t_0}{a} dt$$

$$+ \frac{2 \cos \pi a}{Q(t_0)} (\beta_0 \tan \frac{t_0}{a} + \beta_1), \ t \in \gamma_0. \tag{2.19}$$

Example Periodic stamps with horizontal rectilinear base.
In this case, $f'(t) = 0$.

$$\Phi(z) = \frac{P_0(1 + ik)e^{\pi ai} \cos[\frac{z}{a}(1 - \lambda)\pi a]}{2a\pi \cos \pi a \sin^{\frac{1}{2}+a} \frac{l+z}{a} \sin^{\frac{1}{2}-a} \frac{l-z}{a}} + \frac{\kappa - 1}{\kappa + 1} \frac{(k^2 + 1)P_0}{2a\pi} \tag{2.20}$$

$$P(t) = \frac{P_0 \cos[\frac{t}{a}(1-\lambda)\pi a]}{a\pi \sin^{\frac{1}{2}+a}\frac{l+t}{a} \sin^{\frac{1}{2}-a}\frac{l-t}{a}}, \ t \in \gamma_0. \tag{2.21}$$

3. Periodic Problems for Anisotropic Medium

3.1. *The stress Functions*

1) Basic assumptions

All the discussion below are under the following assumption, the medium of the $a\pi$-periodic region is anisotropic, the stresses and displacements are periodic and the stresses at infinity are bounded, and the involved boundary conditions are periodic. Thus, we need only restrict our discussions in a period part of the elastic body. Moreover, we assume the elastic body occupies the lower half-plane in the z-plane ($z = x + iy$) and so there is only one point at infinity $z = -\infty i$.

The principal vector of $X(-\infty i) + iY(-\infty i)$ the external stresses at $z = -\infty i$ is understood by the limit of the principal vector of the stresses along a line-segment from z to $z + a\pi$ in S^- as $z \to -\infty i$ i.e.

$$X(-\infty i) = a\pi\tau_{xy}(-\infty i), \ Y(-\infty i) = a\pi\sigma_y(-\infty i).$$

If the principal vector of the external stresses on the boundary in a period is $X+iY$, then by the condition of equilibrium,

$$x + iY = X(-\infty i) + iY(-\infty i)$$

we have

$$\sigma_y(-\infty i) = \frac{Y}{a\pi}, \tau_{xy}(-\infty i) = \frac{X}{a\pi}. \tag{3.1}$$

2) Periodicity of the stress functions for anisotropic medium

For anisotropic elastic body, the stress components $\sigma_x, \sigma_y, \tau_{xy}$ and displacement components u, v may be expressed by means of $\varphi(z_1)$ and $\psi(z_2)$ or their derivatives $\Phi(z_1) = \varphi'(z_1)$ and $\Psi(z_2) = \psi'(z_2)$ (stress functions):

$$\sigma_x = \mu_1^2\Phi(z_1) + \overline{\mu}_1^2\overline{\Phi(z_1)} + \mu_2^2\Psi(z_2) + \overline{\mu}_2^2\overline{\Psi(z_2)} \tag{3.2}$$

$$\sigma_y = \Phi(z_1) + \overline{\Phi(z_1)} + \Psi(z_2) + \overline{\Psi(z_2)} \tag{3.3}$$

$$\tau_{xy} = -[\mu_1\Phi(z_1) + \overline{\mu}_1\overline{\Phi(z_1)} + \mu_2\Psi(z_2) + \overline{\mu}_2\overline{\Psi(z_2)}] \tag{3.4}$$

$$u = p_1\varphi(z_1) + \overline{p}_1\overline{\varphi(z_1)} + p_2\psi(z_2) + \overline{p}_2\overline{\psi(z_2)} \tag{3.5}$$

$$u = q_1\varphi(z_1) + \overline{q}_1\overline{\varphi(z_1)} + q_2\psi(z_2) + \overline{q}_2\overline{\psi(z_2)} \tag{3.6}$$

where $\varphi(z_1)$ and $\psi(z_2)$ are functions holomorphic in z_1 and z_2 respectively, in which,

$$z_1 = x + \mu_1 y, \ z_2 = x + \mu_2 y,$$

and

$$
\begin{cases}
p_1 = \beta_{11}\mu_1^2 + \beta_{12} - \beta_{16}\mu_1, \\
p_2 = \beta_{11}\mu_2^2 + \beta_{12} - \beta_{16}\mu_2, \\
q_1 = \dfrac{\beta_{12}\mu_1^2 + \beta_{22} - \beta_{26}\mu_1}{\mu_1}, \\
q_2 = \dfrac{\beta_{12}\mu_2^2 + \beta_{22} - \beta_{26}\mu_2}{\mu_2},
\end{cases}
\tag{3.7}
$$

and $\mu_1, \overline{\mu}_1, \mu_2, \overline{\mu}_2$ are the root of the equation

$$
\beta_{11}s^4 - 2\beta_{16}s^3 + (2\beta_{11} + \beta_{66})s^2 - 2\beta_{26}s + \beta_{22} = 0.
\tag{3.8}
$$

while

$$
\begin{array}{ccc}
\beta_{11} & \beta_{12} & \beta_{16} \\
\beta_{12} & \beta_{22} & \beta_{26} \\
\beta_{16} & \beta_{26} & \beta_{66}
\end{array}
$$

are the elastic coefficients of the anisotropic elastic body.

Lemma 1 Under the basic assumptions, the stress functions $\Phi(z_1)$ and $\Psi(z_2)$ are $a\pi$- periodic functions.

3.2. *Periodic Fundamental Problems of Anisotropic Half-plane*

1) The first fundamental problem

Assume the anisotropic body occupies the lower half-plane S^- of the z-plane. Denote $z = t$ (real) on the x-axis. Given the external stresses on the x-axis:

$$
\sigma_y(t) = -P(t), \quad \tau_{xy}(t) = T(t),
\tag{3.9}
$$

which are arcwise Hölder continuous and periodic. Under the basic assumptions, find the stress distribution and displacements, called the periodic first fundamental problem.

Theorem 10 Under the above assumptions, the first fundamental problem of the anisotropic half-plane is uniquely solvable.
We get stress functions

$$
\Phi(z_1) = \frac{-1}{(\mu_1 - \mu_2)2a\pi i} \int_{L_0} [\mu_2 P(t) - T(t)] \cot \frac{t - z_1}{a} dt + \gamma_1,
\tag{3.10}
$$

$$
\Psi(z_2) = \frac{1}{(\mu_1 - \mu_2)2a\pi i} \int_{L_0} [\mu_1 P(t) - T(t)] \cot \frac{t - z_2}{a} dt + \gamma_2,
\tag{3.11}
$$

where

$$
\begin{aligned}
\gamma_1 = -\frac{1}{2}\frac{1}{\mu_1 - \mu_2}[(\mu_2 - \mu_1)\Phi(-\infty i) \\
+ (\mu_2 - \mu_1)\overline{\Phi(-\infty i)} + (\mu_2 - \overline{\mu_2})\Psi(-\infty i)],
\end{aligned}
\tag{3.12}
$$

$$\gamma_2 = \frac{1}{2}\frac{1}{\mu_1 - \mu_2}[(\mu_1 - \mu_2)\Psi(-\infty i)\Phi(-\infty i)$$
$$+ (\mu_1 - \overline{\mu_1})\overline{\Phi(-\infty i)} + (\mu_1 - \overline{\mu_2})\overline{\Psi(-\infty i)}]. \tag{3.13}$$

In order to determine γ_1 and γ_2 by noting (3.12) and (3.13), it is easily seen that

$$Re|\gamma_1 + \gamma_2| = 0, \tag{3.14}$$

$$Re|\mu_1\gamma_1 + \mu_2\gamma_2| = 0. \tag{3.15}$$

2) The second fundamental problem

Assume that on the boundary x-axis of the anisotropic elastic half-plane S^-, the displacement

$$u^- + iv^- = g_1(t) + ig_2(t) \tag{3.16}$$

is given, where $g_1(t) + ig_2(t)$ is continuous and $g_1'(t) + ig_2'(t)$ is Hölder continuous arcwise. Moreover, on the segment L_0 of the boundary in a period, the principal vector $X + iY$ of the external stresses is also given. Under these basic boundary conditions, find the equilibrium, called the periodic second fundamental problem.

Theorem 11 Under the above assumptions, the solution of the second fundamental problem of the uniquely exists.

We get stress functions

$$\Phi(z_1) = \frac{1}{q_1p_2 - p_1q_2}\frac{1}{2a\pi i}\int_{L_0}[q_2g_1' - p_2g_2')]\cot\frac{t - z_1}{a}dt + \gamma_1 \tag{3.17}$$

$$\Psi(z_2) = \frac{-1}{(q_1p_2 - p_1q_2)}\frac{1}{2a\pi i}\int_{L_0}[q_1g_1' - p_1g_2')]\cot\frac{t - z_2}{a}dt + \gamma_2 \tag{3.18}$$

where γ_1 and γ_2 may be obtained by solving

$$Re[p_1\gamma_1 + p_2\gamma_2] = 0,$$

$$Re|q_1\gamma_1 + q_2\gamma_2| = 0,$$

$$Re[\gamma_1 + \gamma_2] = \frac{Y}{2a\pi},$$

$$Re[\mu_1\gamma_1 + \mu_2\gamma_2] = -\frac{X}{2a\pi}.$$

3.3. *Periodic Contact Problem for Anisotropic Medium*

1) Formulation of the problem and the solutions

Formulation of the periodic contact problem in anisotropic half-plane S^- is as follows.

Assume that a row of periodic stamps (with bases of the same shape) pressed on S^- and there exists friction between the stamps and with coefficient of friction ρ, that is, beneath the stamps, the shearing stress $T(x) = \tau_{xy}(x)$ and normal pressure $P(x) = -\sigma_y(x)$ obey the Coulomb's law:

$$T(x) = \rho P(x), \ x \in \gamma_0,$$

or

$$\tau_{xy}(x) + \sigma_y(x) = 0, \ x \in \gamma_0,$$

where $\gamma_0 : -l_0 \le x \le l_0$, is the contact line segment in $L_0 : -\frac{1}{2}a\pi < x < \frac{1}{2}a\pi$. Assume the stamps are in the limiting situation of equilibrium. On the free interval, there is no load, i.e.

$$\sigma_y = 0, \ \tau_{xy} = 0.$$

Besides, assume the vertical displacement

$$v^-(\dot{x}) = f(x) \tag{3.19}$$

is given, where $y = f(x)$ is the equation of the bases of the stamps, which is $a\pi$-periodic with $f'(x) \in H(A, \mu)$. Moreover, the external pressure force P_0 applied on each stamp is also given and so the principal; vector of the external stresses is $X + iY = T_0 - iP_0 = (\rho - i)P_0$. Find the elastic equilibrium under these assumptions.

$$\begin{cases} \sigma_y(x) = 0, \ \tau_{xy}(x) = 0, \ x \in L_0 - \gamma_0, \\ \tau_{xy}(x) + \rho\sigma_y(x) = 0, \ v^-(x) = f(x), \ x \in \gamma_0. \end{cases} \tag{3.20}$$

In order to solve the problem, we have to transform (3.1). Introduce two functions represented by integrals with Hilbert kernel:

$$w_1(z) = u_1 - iv_1 = \int_{L_0} \sigma_y(t) \cot \frac{t - z}{a} dt, \tag{3.21}$$

$$w_2(z) = u_2 - iv_2 = \int_{L_0} \tau_{xy}(t) \cot \frac{t - z}{a} dt + \beta. \tag{3.22}$$

Therefore, we obtain

$$w_1(z) = \frac{\pm ie^{\pm\theta\pi i} \cos\theta\pi E(z)}{A_3 - \rho A_4} \int_{\gamma_0} \frac{f'(t)}{E(t)} \cot \frac{t - z}{a}$$
$$\pm ie^{\pm\omega_1 i}(C_1 \tan \frac{z}{a} + C_2)E(z), \quad z \in S^\pm, \tag{3.23}$$

and our solution is $w_1(z)$ when $z \in S^-$.

As for $w_2(z)$, by boundary condition, it is evident that

$$w_2(z) = -\rho w_1(z) + \beta. \tag{3.24}$$

From periodic condition for displacements and equilibrium condition at $z = -\infty\, i$, the real conditions C_1, C_2 and β may be obtained by solving

$$\left\{ (\varepsilon_1 + i\delta_1)e^{-i\frac{2l\theta}{a}} + (\varepsilon_2 + i\delta_2)e^{i\frac{2l\theta}{a}} \right\} C_1$$

$$+ i\left\{ (\varepsilon_1 + i\delta_1)e^{-i\frac{2l\theta}{a}} - (\varepsilon_2 + i\delta_2)e^{i\frac{2l\theta}{a}} \right\} C_2 + (\varepsilon_3 + i\delta_3)\beta a\pi$$

$$= \frac{\cos \pi\theta}{A_3 - \rho A_4} \left\{ (\varepsilon_1 + i\delta_1)e^{-i\frac{2l\theta}{a}} + (\varepsilon_2 + i\delta_2)e^{i\frac{2l\theta}{a}} \right\} \int_{\gamma_0} \frac{f'(t)}{E(t)} dt$$

$$\frac{P_0}{\cos\frac{l}{a}} - \frac{\cos\pi\theta \sin\frac{2l\theta}{a}}{A_3 - \rho A_4} \int_{\gamma_0} \frac{f'(t)}{E(t)} dt = C_2 \cos\frac{2l\theta}{a} - C_1 \sin\frac{2l\theta}{a}.$$

2) The pressure beneath the stamps

$$p(x) = \frac{1}{2(A_3 - \rho A_4)} \left\{ -\sin 2\theta\pi f'(x) + \frac{2\cos^2\theta\pi E(x)}{a\pi} \int_{\gamma_0} \frac{f'(t)}{E(t)} \cot\frac{t-x}{a} dt \right\}$$

$$+ \frac{\cos\theta\pi}{a\pi} E(x)(C_1 \tan\frac{x}{a} + C_2), x \in \gamma_0.$$

Example Consider the case where the stamps possess periodic horizontal rectilinear bases, here, $f'(x) = 0$. Then

$$p(x) = \frac{\cos\theta\pi}{a\pi} E(x)(C_1 \tan\frac{x}{a} + C_2), \ x \in \gamma_0,$$

where C_1 and C_2 are determined by

$$\left\{ (\varepsilon_1 + i\delta_1)e^{-i\frac{2l\theta}{a}} + (\varepsilon_2 + i\delta_2)e^{i\frac{2l\theta}{a}} \right\} C_1$$

$$+ i\left\{ (\varepsilon_1 + i\delta_1)e^{-i\frac{2l\theta}{a}} - (\varepsilon_2 + i\delta_2)e^{i\frac{2l\theta}{a}} \right\} C_2 + (\varepsilon_3 + i\delta_3)\beta a\pi = 0$$

$$C_2 \cos\frac{2l\theta}{a} - C_1 \sin\frac{2l\theta}{a} = \frac{P_0}{\cos\frac{l}{a}}.$$

See Ref.3-7,14,19.

4. Periodic Crack Problems in Plane Elasticity

For isotropic elastic plane weakened by a periodic row of cracks, W.T.Koiter had studied the first fundamental problems by complex variable methods for the cases where the cracks are rectilinear and collinear (in the direction of period), or parallel and perpendicular to the direction of period, under very special assumptions for the external stresses subjected on the cracks as well as those at infinity. See Ref.6-7,14,17,19,28,30.

4.1. *Fundamental Problems of Isotropic Plane with Periodic Collinear Cracks*

1) General comments

Consider the isotropic elastic infinite plane weakened by periodic rectilinear cracks with the same direction as the period $a\pi$. without loss of generality, we may ask that they are situated on the real axis.

Assume that there are n cracks in the periodic strip $|x| < \frac{1}{2}a\pi$, namely, l_k : $a_k \leq t \leq b_k (a_{k+1} > b_k)$, $k = 1, ..., n-1$, positively oriented from a_k to b_k. denote $l_0 = \sum_{k=1}^{n} l_k$, and the principal vector of the external stresses on l_k by $X_k + iY_k$, The elastic region is denoted by S. Other notations are the same as in previous section.

The following discussions are made under the assumptions that the stresses are periodic and bounded at $z = \pm\infty i$ while the displacements are quasi-periodic.

Introduce functions

$$\left. \begin{aligned} w(z) &= z\bar{\Phi}(z) + \bar{\Psi}(z), \\ \Omega(z) &= w'(z) = \bar{\Phi}(z) + z\bar{\Phi}'(z) + \bar{\Psi}'(z), \end{aligned} \right\} z \in S, \qquad (4.1)$$

where $\Phi(z), \Psi(z)$ are complex stress functions and $\bar{\Phi}(z) = \overline{\Phi(\bar{z})}, \bar{\Psi}(z) = \overline{\Psi(\bar{z})}$. It is easily seen

$$\sigma_y - i\tau_{xy} = \Phi(z) + \Omega(\bar{z}) + (z - \bar{z})\overline{\Phi'(z)}, \quad z \in S. \qquad (4.2)$$

By periodicity of $\Phi(z)$We known that $\Omega(z)$ is also periodic.

Assume both $\Phi(z)$ and $\Omega(z)$ are at most integrably unbounded at the tips of l_k and

$$\lim_{z \to t} y\Phi'(z) = 0 (z = x + iy \in S, t \in L_0). \qquad (4.3)$$

Obviously,

$$\left. \begin{aligned} X_k &= \int_{l_k} [\tau_{xy}^-(t) - \tau_{xy}^+(t)]dt, \\ Y_k &= \int_{l_k} [\sigma_y^-(t) - \sigma_y^+(t)]dt, \end{aligned} \right\} k = 1, ..., n, \qquad (4.4)$$

where $\sigma_y^\pm + i\sigma_{xy}^\pm$ are the external stresses on the upper bank and the lower bank of l_k respectively. Thereby, their resultant $X + iY$ is given by

$$\begin{cases} X = \int_{l_0} [\tau_{xy}^-(t) - \tau_{xy}^+(t)]dt, \\ Y = \int_{l_0} [\sigma_y^-(t) - \sigma_y^+(t)]dt, \end{cases} \qquad (4.5)$$

Let $\Phi_0(z)$ and $\Psi_0(z)$ be as before, then $\Phi_0(\pm\infty i) = \Psi_0(\pm\infty i) = 0$. We can easily verify that

$$
\begin{cases}
\Phi(\pm\infty i) = \mp\dfrac{Y - ix}{2a\pi(k+1)} + \beta, \\[3mm]
\Psi(\pm\infty i) = \mp\dfrac{(k-1)Y + i(k+1)X}{2a\pi(k+1)} - \beta + k\bar{\beta} - q;
\end{cases}
\tag{4.6}
$$

and then, by (4.1)

$$
\Omega(\pm\infty i) = \pm\frac{k(Y - iX)}{2a\pi(k+1)} + k\beta - q.
\tag{4.7}
$$

2 The first fundamental problem

Assume $\sigma_{xy}^{\pm}(t), \tau_{xy}^{\pm}(t) \in H$ are given, and σ_-, τ_-, h_- (consequently $\sigma_+, \tau_+, h_+, \beta, q$) are also given. Find the equilibrium.

According to the boundary condition, we assure:

$$
\Phi^+(t) + \Omega^-(t) = \sigma_y^+ - i\tau_{xy}^+,
$$
$$
\Phi^-(t) + \Omega^+(t) = \sigma_y^- - i\tau_{xy}^-.
$$

By addition and subtraction, our problem is easily transferred to the following two boundary value problems:

$$
[\Phi(t) + \Omega(t)]^+ + [\Phi(t) + \Omega(t)]^- = 2p(t),
\tag{4.8}
$$

$$
[\Phi(t) - \Omega(t)]^+ - [\Phi(t) - \Omega(t)]^- = 2q(t),
\tag{4.9}
$$

where we have put

$$
p(t) = \frac{1}{2}[\sigma_y^+(t) + \sigma_y^-(t)] - \frac{i}{2}[\tau_{xy}^+(t) + \tau_{xy}^-(t)],
\tag{4.10}
$$

$$
q(t) = \frac{1}{2}[\sigma_y^+(t) - \sigma_y^-(t)] - \frac{i}{2}[\tau_{xy}^+(t) - \tau_{xy}^-(t)].
\tag{4.11}
$$

We have

$$
\begin{cases}
\Phi(z) = \dfrac{X(z)}{2a\pi i}\displaystyle\int_{l_0} \frac{p(t)}{X^+(t)}\cot\frac{t-z}{a}dt + \dfrac{1}{2a\pi i}\displaystyle\int_{l_0} q(t)\cot\frac{t-z}{a}dt \\[3mm]
\qquad + X(z)P_n(\tan\dfrac{z}{a}) + C, \\[3mm]
\Omega(z) = \dfrac{X(z)}{2a\pi i}\displaystyle\int_{l_0} \frac{p(t)}{X^+(t)}\cot\frac{t-z}{a}dt - \dfrac{1}{2a\pi i}\displaystyle\int_{l_0} q(t)\cot\frac{t-z}{a}dt \\[3mm]
\qquad + X(z)P_n(\tan\dfrac{z}{a}) - C,
\end{cases}
\tag{4.12}
$$

where

$$X(z) = \prod_{k=1}^{n} (\tan \frac{z}{a} - \tan \frac{a_k}{a})^{-\frac{1}{2}} (\tan \frac{z}{a} - \tan \frac{b_k}{a})^{-\frac{1}{2}} \qquad (4.13)$$

The radicals in which may be arbitrarily taken as a continuous branch in the z-plane cut by the periodic cracks, for instance, that branch fulfilling

$$\lim_{z \to \frac{a\pi}{2}} \tan^n \frac{z}{a} X(z) = 1.$$

And $C_0, ..., C_m$ can be easily determined by the following equations:

$$k \int_{l_k} [\Phi^+(t) - \Omega^-(t)]dt + \int_{l_k} [\Phi^+(t) - \Omega^-(t)]dt = 0, k = 1, ..., n, \qquad (4.14)$$

$$k \int_{\Lambda_t} \Phi(z)dz - \int_{\Lambda_t} \Omega(z)dz = a\pi q. \qquad (4.15)$$

3) The second fundamental problem

For simplicity, we assume that there occurs only one single crack in a period, that is, $\gamma_0 : [-l, l]$.

Given the periodic displacements $u^{\pm}(t) + iv^{\pm}(t)$ respectively on the upper and the lower banks of the cracks, with $u^{\pm'}(t) + iv^{\pm'}(t) \in H(A, \mu)$, the resultant of the principal vectors of the external stresses on $\gamma_0 : X + iY$, and the stresses at $z = -\infty i$ (and hence at $z = +\infty i$), find the equilibrium.

We get

$$\begin{cases} \Phi(z) = \dfrac{1}{2a\pi k i \sqrt{R(z)}} \displaystyle\int_{-l}^{l} f(t)\sqrt{R(t)} \cot \dfrac{t-z}{a} dt \\ \quad + \dfrac{1}{2a\pi k i} \displaystyle\int_{-l}^{l} g(t) \cot \dfrac{t-z}{a} dt + \dfrac{C_0 \tan \frac{z}{a} + C_1}{k\sqrt{R(z)}} + \beta - \dfrac{q}{2k}, \\ \Omega(z) = -\dfrac{1}{2a\pi i \sqrt{R(z)}} \displaystyle\int_{-l}^{l} f(t)\sqrt{R(t)} \cot \dfrac{t-z}{a} dt \\ \quad + \dfrac{1}{2a\pi i} \displaystyle\int_{-l}^{l} f(t) \cot \dfrac{t-z}{a} dt - \dfrac{C_0 \tan \frac{z}{a} + C_1}{\sqrt{R(z)}}, \end{cases} \qquad (4.16)$$

where

$$\begin{cases} C_0 = -\dfrac{1}{2a\pi i} \displaystyle\int_{-l}^{l} f(t)\sqrt{R(t)}dt - \dfrac{iq}{2\cos \frac{l}{a}}, \\ C_1 = -\dfrac{k(Y - iX)}{2(k+1)a\pi \cos \frac{l}{a}}. \end{cases} \qquad (4.17)$$

This problem was incomplete discussion by H.F.Bueckner.

Remark The corresponding mixed problem may be studied by method similar to that used here.

4.2. *Fundamental Problems of Anisotropic Elastic Plane with Periodic Collinear Cracks*

1) General comments

Assume that, in the anisotropic infinite elastic plane, there are periodically arranged rectilinear cracks $L_j, j = 0, \pm 1, \pm 2, \ldots$ Lying on the $x - axis$, each of which has the length $2l(l < \frac{1}{2}a\pi)$, L_0 being the interval $[-l, l]$, as shown in Fig. 1.1 Denote $L = \sum\limits_{j=-\infty}^{\infty} L_j$ and its complement L'.

Assume that there exist periodic external loads on both sides of the cracks but no external stresses at infinity. We would study respectively the cases where the loads are symmetric or anti-symmetric on L_0.

The following discussions are made under the basic assumptions that both the stresses and displacements are periodic and the stresses at infinity are bounded. Both the stress $\Phi(z_1)$ and $\Psi(z_2)$ are periodic.

At $z = \pm\infty i$, the principal vectors $X(\pm\infty i) + iY(\pm\infty i)$ of external stresses, by assumption, are zeros:

$$X(\pm\infty i) + iY(\pm\infty i) = a\pi\left[\tau_{xy}(\pm\infty i) + i\sigma_y(\pm\infty i))\right] = 0. \qquad (4.18)$$

Our discussions may be restricted in the periodic strip $|Rez| < \frac{1}{2}a\pi$.

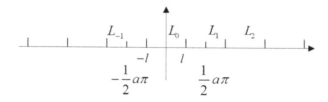

Fig. 1.1.

2) Symmetric loads

Assume $\Phi_1(z_1)$ and $\Psi_1(z_2)$ are the stress functions corresponding to the case where periodic loads (in the plane) applied on the cracks are symmetric to the x-axis, i.e., tension loads only.

On the real axis, let $z_1 = z_2 = \tau$, on account of symmetry, by(3.4)

$$\mu_1\Phi_1(\tau) + \mu_2\Psi_1(\tau) = 0, \tau \in L'. \qquad (4.19)$$

Assume $\sigma_y^{\pm}(\tau), \tau \in L$, are given (periodic). By (3.5) and (4.19) the boundary condition can be expressed in $\Phi_1(z)$ only:

$$\left.\begin{aligned}
\sigma_y^{+}(\tau) &= \frac{\mu_2 - \mu_1}{\mu_2}\Phi_1^{+}(\tau) + \frac{\bar{\mu}_2 - \bar{\mu}_1}{\bar{\mu}_2}\bar{\Phi}_1^{-}(\tau), \\
\sigma_y^{-}(\tau) &= \frac{\mu_2 - \mu_1}{\mu_2}\Phi_1^{-}(\tau) + \frac{\bar{\mu}_2 - \bar{\mu}_1}{\bar{\mu}_2}\bar{\Phi}_1^{+}(\tau),
\end{aligned}\right\} \quad \tau \in L_0, \qquad (4.20)$$

$$[\frac{\mu_2 - \mu_1}{\mu_2}\Phi_1(\tau) + \frac{\bar{\mu}_2 - \bar{\mu}_1}{\bar{\mu}_2}\bar{\Phi}_1(\tau)]^+ + [\frac{\mu_2 - \mu_1}{\mu_2}\Phi_1(\tau) + \frac{\bar{\mu}_2 - \bar{\mu}_1}{\bar{\mu}_2}\bar{\Phi}_1(\tau)]^-$$

$$= 2f_1(\tau), \tau \in L_0, \tag{4.21}$$

$$[\frac{\mu_2 - \mu_1}{\mu_2}\Phi_1(\tau) - \frac{\bar{\mu}_2 - \bar{\mu}_1}{\bar{\mu}_2}\bar{\Phi}_1(\tau)]^+ - [\frac{\mu_2 - \mu_1}{\mu_2}\Phi_1(\tau) - \frac{\bar{\mu}_2 - \bar{\mu}_1}{\bar{\mu}_2}\bar{\Phi}_1(\tau)]^-$$

$$= 2g_1(\tau), \tau \in L_0 \tag{4.22}$$

where

$$f_1(\tau) = \frac{1}{2}(\sigma_y^+ + \sigma_y^-), g_1(\tau) = \frac{1}{2}(\sigma_y^+ - \sigma_y^-), \tau \in L.$$

Assume both $f_1(\tau)$ and $g_1(\tau) \in H(A, \mu)$. Evidently, they are periodic. $\Phi_1(\pm\infty i)$ and $\Psi_1(\pm\infty i)$ are finite by the assumption that the stresses at infinity are bounded.

The general solution of the boundary value problem is

$$\Phi_1(z_1) = \Phi_1^*(z_1) + \frac{\mu_2}{\mu_2 - \mu_1}\frac{c_0 \tan\frac{z_1}{a} + c_1}{\sqrt{R(z_1)}} + \frac{\mu_2\beta}{\mu_2 - \mu_1}, \tag{4.23}$$

where

$$\frac{\mu_2 - \mu_1}{\mu_2}\Phi_1^*(z_1) = \frac{1}{2a\pi i}\int_{-l}^{l} g_1(\tau)\cot\frac{\tau - z_1}{a}d\tau$$

$$+\frac{1}{2a\pi i\sqrt{R(z_1)}}\int_{-l}^{l} f_1(\tau)\sqrt{R(\tau)}\cot\frac{\tau - z_1}{a}d\tau. \tag{4.24}$$

By the similar way, we find that

$$\Psi_1(z_2) = \Psi_1^*(z_1) + \frac{\mu_1}{\mu_1 - \mu_2}\frac{c_0^* \tan\frac{z_1}{a} + c_1^*}{\sqrt{R(z_2)}} + \frac{\mu_1\beta^*}{\mu_1 - \mu_2}, \tag{4.25}$$

where

$$\frac{\mu_1 - \mu_2}{\mu_1}\Psi_1^*(z_2) = \frac{1}{2a\pi i}\int_{-l}^{l} g_1(\tau)\cot\frac{\tau - z_2}{a}d\tau$$

$$+\frac{1}{2a\pi i\sqrt{R(z_2)}}\int_{-l}^{l} f_1(\tau)\sqrt{R(\tau)}\cot\frac{\tau - z_2}{a}d\tau, \tag{4.26}$$

in which c_0^*, c_1^* are undetermined real constants and β^* complex.

In order to determine $c_0, c_1, c_0^*, c_1^*, \beta$ and β^*, it is sufficient to consider the periodicity of displacements and the stresses at $z = \pm\infty i$.

3) Anti-symmetric loads

Now, let us consider the problem of anti-symmetric loads (in the plane), i.e., $\tau_{xy}^{\pm}$ are given on the cracks. denote the stress function in the case by $\Phi_2(z_1)$ and $\Psi_2(z_2)$. By the condition of anti-symmetry,

$$\Phi_2(\tau) + \Psi_2(\tau) = 0, \quad \tau \in L'.$$

Then by (3.4), the boundary conditions become

$$[(\mu_2 - \mu_1)\Phi_2(\tau) + (\bar\mu_2 - \bar\mu_1)\bar\Phi_2(\tau)]^+ + [(\mu_2 - \mu_1)\Phi_2(\tau)$$

$$+(\bar\mu_2 - \bar\mu_1)\bar\Phi_2(\tau)]^- = 2f_2(\tau), \quad \tau \in L_0. \tag{4.27}$$

$$[(\mu_2 - \mu_1)\Phi_2(\tau) - (\bar\mu_2 - \bar\mu_1)\bar\Phi_2(\tau)]^+ - [(\mu_2 - \mu_1)\Phi_2(\tau)$$

$$-(\bar\mu_2 - \bar\mu_1)\bar\Phi_2(\tau)]^- = 2g_2(\tau), \tau \in L_0 \tag{4.28}$$

where

$$f_2(\tau) = \frac{1}{2}(\tau_{xy}^+ + \tau_{xy}^-), g_2(\tau) = \frac{1}{2}(\tau_{xy}^+ - \tau_{xy}^-)$$

are periodic, assumed $\in H$.

Thus, we may write

$$\Phi_2(z_1) = \bar\Phi_2^*(z_1) + \frac{1}{\mu_2 - \mu_1}\frac{d_0 \tan \frac{z_1}{a} + d_1}{\sqrt{R(z_1)}} + \frac{\gamma}{\mu_2 - \mu_1} \tag{4.29}$$

where we have

$$(\mu_2 - \mu_1)\Phi_2^*(z_1) = \frac{1}{2a\pi i}\int_{-l}^{l} g_2(\tau)\cot\frac{\tau - z_1}{a}d\tau$$

$$+\frac{1}{2a\pi i\sqrt{R(z_1)}}\int_{-l}^{l} f_2(\tau)\cot\frac{\tau - z_1}{a}d\tau.$$

And

$$\Psi_2(z_2) = \Psi_2^*(z_2) + \frac{1}{\mu_1 - \mu_2}\frac{d_0^* \tan \frac{z_1}{a} + d_1^*}{\sqrt{R(z_2)}} + \frac{\gamma^*}{\mu_1 - \mu_2} \tag{4.30}$$

where d_0^*, d_1^* are undetermined real constants and γ^*, complex, and at the same time, we have defined

$$(\mu_1 - \mu_2)\Psi_2^*(z_2) = \frac{1}{2a\pi i}\int_{-l}^{l} g_2(\tau)\cot\frac{\tau - z_2}{a}d\tau$$

$$+\frac{1}{2a\pi i\sqrt{R(z_2)}}\int_{-l}^{l} f_2(\tau)\sqrt{R(\tau)}\cot\frac{\tau - z_2}{a}d\tau,$$

$d_0^*, d_1^*\gamma, \gamma^*$ can be determined by the periodicity of displacements and the condition at $z = \pm\infty i$.

In particular, consider the subcase $\tau_{xy}^+(\tau) = \tau_{xy}^-(\tau)$ or $g_2(\tau) = 0$. Then, (4.29) and (4.30) are simplified respectively to

$$(\mu_2 - \mu_1)\Phi_2(z_1) = \frac{1}{2a\pi i\sqrt{R(z_1)}}\int_{-l}^{l} f_2(\tau)\sqrt{R(\tau)}\cot\frac{\tau - z_1}{a}d\tau$$

$$+\frac{d_0\tan\frac{z_1}{a}+d_1}{\sqrt{R(z_1)}}+\gamma, \tag{4.31}$$

$$(\mu_2-\mu_1)\Psi_2(z_2)=\frac{1}{2a\pi i\sqrt{R(z_2)}}\int_{-l}^{l}f_2(\tau)\sqrt{R(\tau)}\cot\frac{\tau-z_2}{a}d\tau$$

$$+\frac{d_0^*\tan\frac{z_1}{a}+d_1^*}{\sqrt{R(z_1)}}+\gamma^*. \tag{4.32}$$

Next, consider the more special case where uniform shearing forces are applied on the cracks: $g_2(\tau)=0$ and $f_2(\tau)=-q$, $\tau\in L$. The solution becomes:

$$(\mu_2-\mu_1)\Phi_2(z_1)=\frac{-q}{2a\pi i\sqrt{R(z_1)}}\int_{-l}^{l}\sqrt{R(\tau)}\cot\frac{\tau-z_1}{a}d\tau$$

$$+\frac{d_0\tan\frac{z_1}{a}+d_1}{\sqrt{R(z_1)}}+\gamma \tag{4.33}$$

$$(\mu_1-\mu_2)\Psi_2(z_2)=\frac{-q}{2a\pi i\sqrt{R(z_2)}}\int_{-l}^{l}\sqrt{R(\tau)}\cot\frac{\tau-z_2}{a}d\tau$$

$$+\frac{d_0^*\tan\frac{z_2}{a}+d_1^*}{\sqrt{R(z_2)}}+\gamma^* \tag{4.34}$$

4) Stress intensity factors

In order to analyze the stress distribution around the tips of L_0, put, in the neighborhood of $t=l$,

$$x=l+r\cos\theta,y=l+r\sin\theta$$

where r/l is assumed to be sufficiently small and polar coordinates r,θ represent respectively the radial distance of the point $z=x+iy$ to the tip $t=l$ of the crack and the angle of inclination of the radial ray to the crack.

Note that when $z_j=l$,

$$(\tan^2\frac{z_j}{a}-\tan^2\frac{l}{a})^{\frac{1}{2}}\approx\sec^2\frac{l}{a}[2r(\cos\theta+\mu_j\sin\theta)]^{\frac{1}{2}},\quad j=1,2.$$

Thus, the stress functions, either for the case of symmetric loads or anti-symmetric loads, can be written as

$$\left.\begin{array}{l}\Phi_j(z_1)=\dfrac{F_j}{[r(\cos\theta+\mu_j\sin\theta)]^{\frac{1}{2}}}+O(1)\\[2mm]\Psi_j(z_2)=\dfrac{G_j}{[r(\cos\theta+\mu_j\sin\theta)]^{\frac{1}{2}}}+O(1)\end{array}\right\}\quad j=1,2 \tag{4.35}$$

where

$$F_1=\frac{k_1\mu_2\cos\frac{l}{a}}{2\sqrt{2}(\mu_2-\mu_1)},F_2=\frac{k_2\cos\frac{l}{a}}{2\sqrt{2}(\mu_2-\mu_1)},$$

$$G_1 = \frac{k_1\mu_1 \cos \frac{l}{a}}{2\sqrt{2}(\mu_1 - \mu_2)}, G_2 = \frac{k_2 \cos \frac{l}{a}}{2\sqrt{2}(\mu_1 - \mu_2)} = -F_2.$$

where $k_j, j = 1, 2$, are called the stress intensity factors, and can be evaluated directly from the stress function $\Phi_j(z_1)$ or $\Psi_j(z_2)$, namely

$$\begin{cases} k_1 = 2\sqrt{2}\dfrac{\mu_2 - \mu_1}{\mu_2} \lim_{z_1 \to t_0} (\tan \dfrac{z_1}{a} - \tan \dfrac{t_0}{a})^{\frac{1}{2}} \Phi_1(z_1), \\ k_2 = 2\sqrt{2}(\mu_2 - \mu_1) \lim_{z_1 \to t_0} (\tan \dfrac{z_1}{a} - \tan \dfrac{t_0}{a})^{\frac{1}{2}} \Phi_2(z_1), \end{cases} \tag{4.36}$$

where $t_0 \in L_0$, or

$$\begin{cases} k_1 = 2\sqrt{2}\dfrac{\mu_1 - \mu_2}{\mu_1} \lim_{z_2 \to t_0} (\tan \dfrac{z_2}{a} - \tan \dfrac{t_0}{a})^{\frac{1}{2}} \Psi_1(z_1), \\ k_2 = 2\sqrt{2}(\mu_1 - \mu_2) \lim_{z_2 \to t_0} (\tan \dfrac{z_2}{a} - \tan \dfrac{t_0}{a})^{\frac{1}{2}} \Psi_2(z_2). \end{cases} \tag{4.37}$$

When $a \to \infty$, the stress intensity factors are identical to the result due to G. C. Sih and H. Liebowitz, see Ref.27, for the case only one single crack on the x-axis.

5. Generalized Symmetric Boundary Value Problems for Automorphic Functions

Various types of periodic problems often encountered in continuum mechanics are the concrete performances of some kind of generalized invariances, symmetric or conservations, which can be mathematically written in a unified way:

$$f[T(z)] \equiv f(z), \quad T \in \mathbf{G}, z \in \Sigma. \tag{5.1}$$

where $\mathbf{G}$ is a function group with Ω as its invariance domain, $f(z)$ is a meromorphic function within Ω, then $f(z)$ is a automorphic function in regard to $\mathbf{G}$. Therefore, most of the foregoing results can be generalized to corresponding boundary value problems for automorphic functions, and their conditions could be also relaxed more widely. See Ref.10-11,29.

5.1. *Relaxing Conditions*

1) Condition $\wedge(1, 1)$

Let $s \in L$ be the coordinate, the arc length l, measuring from a fixed-point along L; $z = z(s), 0 \le s \le l$, be the equation of L; λ, δ and h be non-zero constants. Let $\varphi(z)$ be a continuous function with the modulus of continuity $\overset{*}{\omega}(\delta, \varphi)$ or $\omega(\delta, \varphi)$, defined separately by

$$\begin{cases} \overset{*}{\omega}(\delta, \varphi) \equiv \sup_{|s-s'| \le \delta} |\varphi(z(s)) - \varphi(z(s'))|, \\ \omega(\delta, \varphi) \equiv \sup_{|z_1-z_2| \le \delta} |\varphi(z_1) - \varphi(z_2)|; \end{cases} \tag{5.2}$$

They are both non-decreasing continuous functions, positive valued. Moreover,

$$\overset{*}{\omega}(\delta, \varphi) \le \omega(\delta, \varphi) \le (1 + \frac{1}{\alpha}) \overset{*}{\omega}(\delta, \varphi),$$

$$\omega(\lambda\delta, \varphi) \le (\lambda + 1)(1 + \frac{1}{\alpha})\omega(\delta, \varphi).$$

Corresponding to Hölder-Lipschitz Condition $\mathbf{H}(H, \alpha)$, we know

$$\omega(\delta, \varphi) \le H\delta^{\alpha}. \tag{5.3}$$

Let $\varphi(z)$ be a function satisfied Condition $\wedge(1,1)$:

$$\int_0^1 \frac{\omega(t, \phi)}{t} dt < \infty; \quad \wedge(1,0)$$

$$\int_0^1 \frac{1}{t} dt \int_0^t \frac{\omega(\tau, \phi)}{\tau} d\tau < \infty, \int_0^1 dt \int_0^t \frac{\omega(\tau, \phi)}{\tau^2} d\tau < \infty. \quad \wedge(0,1).$$

Let $\omega(\delta, \varphi)$ be substituted by $\overset{*}{\omega}(\delta, \varphi)$, accordingly we have $\overset{*}{\wedge}(1,0)$, $\overset{*}{\wedge}(0,1)$ and $\overset{*}{\wedge}(1,1)$.

The total functions satisfied Condition $\boldsymbol{H}(\mathrm{H},\alpha)$, Condition $\overset{*}{\mathbf{H}}(H, \alpha)$, Condition $\wedge(1,1)$ or Condition $\overset{*}{\wedge}(1,1)$, composed Group $\boldsymbol{H}(\mathrm{H},\alpha)$, Group $\overset{*}{\mathbf{H}}(H, \alpha)$, Group $\wedge(1,1)$ or Group $\overset{*}{\wedge}(1,1)$ separately.

2) Sectionally Automorphic Functions

Let $L_k, k \equiv 0, 1, 2, \ldots$ be a set of rectifiable closed contours, non-intersecting to each other, oriented counter-clockwise, the region interior to L_k is denoted by S_k^+ and S^- is the complement of $S_k^+ + L_k$; for $L = \sum\limits_{k=-\infty}^{+\infty} L_k$ correspondingly we have S^+ and S^-, as usual. Let G be a function group (Fuchs group or Elementary group) with its elements,

$$\boldsymbol{T(z)} \equiv \boldsymbol{z}, \ T\,(\boldsymbol{z}), \boldsymbol{T(z)}, \ldots,$$

where $\boldsymbol{T(z)}$ is a linear transformation, and $L_k \equiv \boldsymbol{T}(L_0)$ is a equivalence curve of L_0. $\varphi(z)$ is a sectionally automorphic function; (5.1) is valid if $\varphi(z)$ is a automorphic function for G; $\varphi(z)$ is holomorphic except the points in L and can be extended to L continuously.

5.2. *Generalized Plemelj Formulae and Singular Integral Equations*

1) Generalized Plemelj Formula

Let $\varphi(z)$ be a function satisfied Condition$\wedge(1,1)$ in $\boldsymbol{L}$ and consider the Cauchy Type Integral:

$$\Phi(z) \equiv \frac{1}{2\pi i} \int_L \frac{\varphi(t)}{t - z} dt, \quad z \notin L. \tag{5.4}$$

The Generalized Plemelj Formulae can easily be proved, see Ref.11, 29.

Theorem 12 (Generalized Plemelj Formulae) Let $\varphi(z)$ be a function satisfied Condition $\wedge(1,0)$ in L, then the Cauchy Type Integral $\Phi(z)$ is holomorphic both on S^+ and S^-, vanishes at infinity; when z approaches to z_0, $\Phi(z)$ tends to definite limit $\Phi^+(z_0)$ or $\Phi^-(z_0)$ according to z in S^+ or S^-, they satisfy the following relations:

$$\begin{cases} \Phi^+(z_0) + \Phi^-(z_0) = \dfrac{1}{\pi i} \displaystyle\int_L \dfrac{\varphi(t)}{t - z_0} dt, \\ \Phi^+(z_0) - \Phi^-(z_0) = \varphi(z_0). \end{cases} \tag{5.5}$$

or

$$\begin{cases} \Phi^+(z_0) = \dfrac{1}{2\pi i} \displaystyle\int_L \dfrac{\varphi(t)}{t - z_0} dt + \dfrac{1}{2}\varphi(z_0), \\ \Phi^-(z_0) = \dfrac{1}{2\pi i} \displaystyle\int_L \dfrac{\varphi(t)}{t - z_0} dt - \dfrac{1}{2}\varphi(z_0) \end{cases} \tag{5.6}$$

where the meaning of integrals should be understand for the principal value integral.

Therefore, $\Phi(z)$ is sectionally holomorphic, and $\Phi^+(z)$, $\Phi^-(z)$ are continuous in L. When z_0 is an angular point laying in L with inner angle α, the coefficient $\pm 1/2$ of $\Phi(z)$ should be changed into $+(1 + \alpha/2)$ and $-(\alpha/2)$; when z_0 is a exterior cusp, $\alpha = 0$; when z_0 is a inner cusp, $\alpha = 2\pi$.

2) Singular Integral Equation

Because $\Phi^+(z)$ and $\Phi^-(z)$ need not to satisfy the same condition, so the above results could not be applied to singular integral equation. But we have

Theorem 13   Let $\varphi(z)$ be a function satisfied Condition $\wedge(1,1)$, then $\Phi^+(z)$ and $\Phi^-(z)$ will satisfy the same condition $\wedge(1,0)$:

$$\int_0^1 \frac{\omega(t, \Phi^+)}{t} dt < +\infty; \quad \int_0^1 \frac{\omega(t, \Phi^-)}{t} dt < +\infty. \tag{5.7}$$

By rewriting (5.6), we obtain

$$\Phi^-(z_0) = \frac{1}{2\pi i} \int_L \frac{\varphi(t) - \varphi(z_0)}{t - z_0} dt.$$

Let $z_0 + h = z(s_0 + k)$, then

$$\Phi^-(z_0 + h) - \Phi^-(z_0)$$

$$= \frac{1}{2\pi i} \int_L \left[\frac{\varphi(t) - \varphi(z_0 + h)}{t - (z_0 + h)} - \frac{\varphi(t) - \varphi(z_0)}{t - z_0} \right] dt$$

$$\equiv \frac{1}{2\pi i} \int_{L_\varepsilon} [*] dt + \frac{1}{2\pi i} \int_{\Delta_\varepsilon} [*] dt$$

$$\equiv \frac{1}{2\pi i} (I_\varepsilon + J_\varepsilon),$$

where $[\cdot]$ is a simple notation of integrand; let $2|k|\equiv\varepsilon$, we have

$$|J_\varepsilon| \leq \int_{s_0-\varepsilon}^{s_0+\varepsilon} \frac{\overset{*}{\omega}\,(|s-(s_0+k)|,\varphi)}{\alpha|s-(s_0+k)|}ds + \int_{s_0-\varepsilon}^{s_0+\varepsilon} \frac{\overset{*}{\omega}\,(|s-s_0|,\varphi)}{\alpha|s-s_0|}ds$$

$$\leq \frac{1}{\alpha}\int_{-s-k}^{s-k} \frac{\overset{*}{\omega}\,(|t|,\varphi)}{|t|}dt + \frac{1}{\alpha}\int_{-\varepsilon}^{\varepsilon} \frac{\overset{*}{\omega}\,(|t|,\varphi)}{|t|}dt$$

$$\leq \frac{4}{\alpha}\int_0^{4|k|} \frac{\overset{*}{\omega}\,(t,\varphi)}{t}dt;$$

and

$$|I_\varepsilon| \leq \int_{L_\varepsilon}\left|\frac{\varphi(t)-\varphi(z_0+h)}{t-(z_0+h)}\right|\left|\frac{h}{t-z_0}\right|dt + \pi|\varphi(z_0+h)-\varphi(z_0)|$$

$$\leq \frac{|K|}{\alpha^2}\int_{L_\varepsilon} \frac{\overset{*}{\omega}\,(|s-(s_0+K)|,\varphi)}{|s-(s_0+K)||s-s_0|}ds + \pi\,\overset{*}{\omega}\,(|K|,\varphi)$$

$$\leq 2K|K|\int_{L_\varepsilon} \frac{\overset{*}{\omega}\,(|s-(s_0+K)|,\varphi)}{|s-(s_0+K)|^2}ds + \pi\,\overset{*}{\omega}\,(|K|,\varphi)$$

$$\leq 2K|K|\int_{|K|}^1 \frac{\overset{*}{\omega}\,(t)}{t^2}dt + \pi\,\overset{*}{\omega}\,(|K|,\varphi).$$

Where h, k, K are different constants, notice the above inequalities regarding to $|I_\varepsilon|$ and $|J_\varepsilon|$, then

$$\overset{*}{\omega}\,(t,\Phi) \leq K_1 t\int_0^1 \frac{\overset{*}{\omega}\,(t)}{t^2}dt + K_2\int_0^1 \frac{\overset{*}{\omega}\,(t)}{t}dt + \frac{1}{2}\,\overset{*}{\omega}\,(t).$$

Using Condition $\wedge(1,0)$, we have proved the second inequality in formula (5.7); to the following Cauchy Type Integral

$$\overset{*}{\Phi}\,(z) \equiv \frac{1}{2\pi i}\int_{L^-} \frac{\varphi(t)}{t-z}dt, \quad z\in S^-;$$

here $\overset{*}{\Phi^-}\,(z) = -\Phi^+(z)$, so

$$\omega(t,\Phi^+) = \omega(t,\overset{*}{\Phi^-}).$$

Because of the above result, we have the first inequality in formula (5.7), Under the condition of $\varphi^\pm(z)\in\wedge(1,1)$, the following corollaries are much useful and easy to prove.

Corollary 13.1 For any given continuous function $\varphi^+(z_0)$ being the boundary value of a function $\Phi(z)$ continuous on S^++L, holomorphic on S^+, the sufficient and necessary condition is

$$\frac{1}{2\pi i}\int_L \frac{\varphi^+(t)}{t-z}dt = 0, z\in S^-. \tag{5.8}$$

Corollary 13.2 For any given continuous function $\varphi^-(z_0)$ being the boundary value of a function $\Phi(z)$ continuous on $S^- + L$ holomorphic on S^-, with a given principle part $\mathbf{r(z)}$ at ∞ point, the sufficient and necessary condition is

$$\frac{1}{2\pi i}\int_L \frac{\varphi^-(t)}{t-z}dt - \Gamma(z) = 0, z \in S^+. \tag{5.9}$$

5.3. *Generalized Plemelj Formula and Automorphic Functions*

1) Automorphic Functions of Finite Group
For finite group $\mathbf{G} \equiv \{\mathbf{T_0(z)} \equiv \mathbf{z}, \mathbf{T_1(z)}, \cdots, \mathbf{T_{n-1}(z)}\}$, consider the function

$$\Phi(z) \equiv \frac{1}{2\pi i}\int_{L_0}\sum_{k=0}^{n-1}\Big[\frac{1}{\tau - T_k(z)} - \frac{1}{\tau - T_k(\infty)}\Big]\varphi(\tau)d\tau. \tag{5.10}$$

where $\varphi(\tau) \in \wedge(1,1), \tau \in L_0$. Easy to know that $\varphi(\infty) = 0$, and analytic everywhere, except $L = L_0 + L_1 + \ldots + L_{n-1}$ (L is total of equivalence curves of L_k with orientational invariance, its equation is $\tau - \boldsymbol{T}(\boldsymbol{z}) = 0$ or $\boldsymbol{z} = \boldsymbol{T}(\tau), \tau \in \boldsymbol{L}, \boldsymbol{k} = 1, \ldots, 2n-1$). Evidently, formula (5.1) valid, namely $\Phi(\boldsymbol{z})$ is an automorphic functions of finite group $\mathbf{G}$. Taking into account the character of $\Phi(\boldsymbol{z})$ in $\boldsymbol{L}$, with z replaced by $T_k^{-1}(z)$ in formula (5.10), notice that the items under the summation varying continuously; except the $(\boldsymbol{k}+1)$th item transforming into an Cauchy Type Integral, with L as its discontinuous line. Therefore, let $z \to T_k^{-1}(z_0) \in L_k, z_0 \in L_0$, we have

$$\Phi^\pm(T_k^{-1}(z_0))$$

$$= \frac{1}{\pi i}\int_{L_0}\sum_{j=0}^{n-1}[\frac{1}{\tau - T_j(T_k^{-1}(z_0))} - \frac{1}{\tau - T_j(\infty)}]\varphi(\tau)d\tau \pm \frac{1}{2}\varphi(z_0)$$

$$= \frac{1}{\pi i}\int_{L_0}\sum_{j=0}^{n-1}[\frac{1}{\tau - T_k(z_0)} - \frac{1}{\tau - T_k(\infty)}]\varphi(\tau)d\tau \pm \frac{1}{2}\varphi(z_0). \tag{5.11}$$

Similar to Plemelj Formula, we have
Corollary 12.1 For an automorphic function defined by (5.10), its limit value $\Phi^\pm(z_0), z_0 \in L_0$ exist, and

$$\left.\begin{array}{l}\Phi^+(T_k^{-1}(z_0)) + \Phi^{-1}(T_k^{-1}(z_0)) \\[2mm] = \frac{1}{\pi i}\int_{L_0}\sum_{j=0}^{n-1}[\frac{1}{\tau - T_k(z_0)} - \frac{1}{\tau - T_k(\infty)}]\varphi(\tau)d\tau, \\[3mm] \Phi^+(T_k^{-1}(z_0)) - \Phi^{-1}(T_k^{-1}(z_0)) = \varphi(z_0), \quad k = 0, 1, \ldots, n-1.\end{array}\right\} \tag{5.12}$$

2) Automorphic Functions of Infinite Group
Let $\mathbf{F(z)}$ be a simple automorphic function to infinite group $\mathbf{G} \equiv \{\mathbf{T_0(z)}, \mathbf{T_1(z)}, \cdots\}$, with a simple pole z_0 in elementary region; $g(z_0) \in \wedge(1,1)$

be given in L_0. Consider the function

$$\Phi(z) \equiv \frac{1}{2\pi i} \int_{L_0} g(\tau) \frac{F'(\tau)}{F(\tau) - F(z)} d\tau. \tag{5.13}$$

It is a sectional automorphic function, with L_0 as it's discontinuous line, vanishing at each generalized equivalent point of z_0. The integral kernel can be expressed as follows:

$$\frac{F'(\tau)}{F(\tau) - F(z)} \equiv \frac{1}{\tau - z} + \Omega(\tau, t),$$

where Ω is a function continuous in L_0. Then, we have

Corollary 12.2 For a sectional automorphic function $\Phi(z)$ defined by (5.13), it's limit value $\Phi^{\pm}(z_0)(z_0 \in L_0)$ exist, and

$$\begin{cases} \Phi^+(z_0) + \Phi^-(z_0) = \dfrac{1}{\pi i} \int_{L_0} g(\tau) \dfrac{F'(\tau)}{F(\tau) - F(z)} d\tau, \\ \Phi^+(z_0) - \Phi^-(z_0) = g(z_0). \end{cases} \tag{5.14}$$

5.4. *Singular Integral Equations and Boundary Value Problems*

In the ordinary circumstances, the exact solution of a singular integral equation can only be approached by an approximating solution. However as the equation kernel is an analytic function to the main variable, by use of analytic continuation, introducing an auxiliary (analytic) function, we can transform successfully the singular integral question to a boundary value problem,and can obtain the closed solution. The basic property of singular integral equation considered here is the automorphic behavior of the kernel.

1) Automorphic Functions of Finite Group

Under the same symbols as above,consider the following singular integral question

$$a(t)\varphi(t) + \frac{b(t)}{\pi i} \int_{L_0} \sum_{k=0}^{n-1} \left[\frac{1}{\tau - T_k(t)} - \frac{1}{\tau - T_k(\infty)} \right] \varphi(\tau) d\tau = f(t), \tag{5.15}$$

where $a(t), b(t), c(t)$ are given functions satisfied Condition $\wedge(1,1)$, and $a^2(t) - b^2(t) = 1$.

Substituting the function $\Phi(z)$ in (5.10) as an auxiliary function into singular integral question (5.15), using relation (5.12), we know that $\Phi(z)$ satisfies boundary condition

$$\Phi^+(T_k^{-1}(t)) = \frac{a(t) - b(t)}{a(t) + b(t)} \Phi^-(T_k^{-1}(t)) + \frac{f(t)}{a(t) + b(t)}.$$

It can be rewritten as follows

$$\Phi^+(t) = G(T_k(t))\Phi^-(t) + g(T_k(t)), k = 0, 1, \ldots, n - 1, \tag{5.16}$$

where

$$G(t) \equiv \frac{a(t) - b(t)}{a(t) + b(t)}, \quad g(t) \equiv \frac{f(t)}{a(t) + b(t)}. \tag{5.17}$$

This is a kind of problems called Riemann Boundary Value Problem, due to the nature of its boundary condition. But it is different from the original definition of Riemann Boundary Value Problem; the essential difference between them lies in the function $\Phi(z)$ which is a given expression of sectional automorphic function to finite group **G**.

2) Automorphic Functions of Infinite Group

Under the same symbols as above,consider the singular integral question (5.15), denoting by

$$a(t)\varphi(t) + \frac{b(t)}{\pi i} \int_{L_0} \frac{F'(\tau)}{F(\tau) - F(t)} \varphi(\tau) d\tau = f(t). \tag{5.18}$$

using $\Phi(z)$ in (5.13) as its auxiliary function with substituting function $\varphi(\tau)$ for $g(\tau)$, from Corollary 12.2, the above singular integral question (5.18) can be transformed immediately to Riemann Boundary Value Problem

$$\Phi^+(t) = G(t)\Phi^-(t) + g(t). \tag{5.19}$$

According to (5.13), function $\mathbf{F(z)}$ has a single pole at z_0, so $\Phi(\infty) = 0$.

Now, let function $\mathbf{F(z)}$ in (5.18) be a sectional automorphic function with a finite number of poles, the closed solution of generalized Riemann Boundary Value Problem (5.19) can still work out in the same way.

5.5. *Boundary Value Problems and Automorphic Functions*

1) Automorphic Functions of Finite Group

The main purpose of this paragraph is to solve Riemann Boundary Problem (5.16) for finite group **G**, where L_0 is closed, so do its equivalences L_k and $L \equiv L_0 + L_1 + \cdots + L_{n-1}$. The solution of homogeneous Riemann Boundary Value Problem with the condition on L is known as canonical function $X(Z)$, is equal to the product of those canonical functions $X_k(Z)$ with various conditions $L_k, k = 1, \cdots, 2n - 1$.

Under the present case, we need not to structure the canonical function $X_k(Z)$ for each L_k, by the nature of canonical function, showing obviously:

$$X_k(z) = X_0(T_k(z)). \tag{5.20}$$

And the canonical function $X_0(Z)$ for each L_0 for was easily be solved, what discovered it to come first was Hilbert, therefore, this kind of questions is called the Hilbert Problem:

$$X_0(z) = (z - t_0)^{-k} e^{\Gamma_0(t)} \quad (t_0 \in L_0), \tag{5.21}$$

where κ is the index of $\mathbf{G}(t)$ for L_0:

$$\begin{cases} k \equiv \dfrac{1}{2\pi i} \displaystyle\int_{L_0} d\ln G(\tau), \\ \Gamma_0(z) \equiv \dfrac{1}{2\pi i} \displaystyle\int_{L_0} \dfrac{\ln G(\tau)}{\tau - z} d\tau. \end{cases} \tag{5.22}$$

Therefore, the canonical function $X(Z)$ for L can be written as follows:

$$X(z) = \Pi_{k=0}^{n-1} X_0(T_k(z)) = \Pi_{k=0}^{n-1}(T_k(t) - t_0)^{-k} e^{\sum\limits_{k=0}^{n-1} \Gamma_0(T_k(z))}, \tag{5.23}$$

It never vanishes everywhere, takes the generalized equivalent points of ∞-point as poles of k-order, and

$$X(z) = X(T_k(z)), \quad k = 1, 2, \ldots, n-1. \tag{5.24}$$

To find a solution $\Phi(z)$ for the homogeneous problem,

$$\Phi^+(t) = G(T_k(t))\Phi^-(t), \quad t \in L_k, k = 1, 2, \ldots, n-1, \tag{5.25}$$

so as to meet the condition $\Phi(\infty) = 0$. Eliminated $\mathbf{G}(T_k(t))$ from equations (5.24) and (5.25), we know

$$\Phi^+(t)/X^+(t) = \Phi^-(t)/X^-(t),$$

and by the principle of analytically continuation, $\Phi(\boldsymbol{z})/\boldsymbol{X}(\boldsymbol{z})$ is meromorphic on the whole plan with generalized equivalent points of ∞-point, denoted by z_∞, as its poles of $(\kappa\text{-}1)$-order. Moreover it is automorphic due to the invariance under transforming $T_k(z) \in \mathbf{G}$ and can be expressed as one of the follows:

$$\begin{cases} \Phi(z) = X(z)P_{k-1}(F^*)/[F^*(z) - F^*(z_\infty)]^{k-1}, \\ \Phi(z) = X(z)P_{k-1}(F). \end{cases} \tag{5.26}$$

where $F(z) \equiv \sum\limits_{i=0}^{n-1} T_i(z)$ and $F(z) \equiv \sum\limits_{i=0}^{n-1}[T_i(z-a)]^{-1}$ are elemental automorphic functions, a is a constant; $\mathbf{P}(\mathbf{F})$ is an arbitrary polynomial of order not larger than $(k-1)$.

(5.26) is the expression of the κ-solutions with linear independence. When $k \leq 0$, the solution for the homogeneous problem is not exist.

Now we consider non-homogeneous problem (5.16), and change the form into

$$\Phi^+(t)/X^+(t) - \Phi^-(t)/X^-(t) = g(T_k(t))/X^+(t), \quad t \in L_k, \quad k = 0, 1, \cdots, 2n-1.$$

In this way, the problem is turned to find an automorphic function according to the discontinuities in boundary L_k or $g(t) \in \wedge(1,1)$. Easy to verify that the function

$$\Psi(z) = \frac{1}{2\pi i} \sum_{k=0}^{n-1} \int_{L_0} \frac{g(\tau)}{X^+(\tau)(\tau - T_k(z))} d\tau \tag{5.27}$$

is a sectional automorphic function satisfied the above boundary condition.

Therefore, for equation (5.16), $X(z)\,\Psi(z)$ is a special solution, and its general solution is

$$\Phi(z) = X(z)\{\Psi(z) + P_{k-1}(F^*)/[F^*(z) - F^*(z_\infty)]^{k-1}\}, \qquad (5.28)$$

where $X(z)$ is a canonical function determined by (5.16), and $\Psi(z)$ by (5.27).

When $k < 0$, P_{k-1} must be zero, point infinity must be the zero-point of order $(-k+1)$. In order to achieve these requirements, we expand $\Psi(z)$ in the neighborhood of infinity $V(\infty)$ and obtain

$$\Psi(z) = -\frac{1}{2\pi i} \sum_{j=1}^{\infty} \Big[\sum_{k=0}^{n-1} \int_{L_0} \frac{T_k'(\tau) T_k^{j-1}(\tau) g(\tau)}{X^+(\tau)} d\tau \Big] z^{-j}.$$

Let all the coefficients of z^{-j} be vanished, we have the conditions:

$$\int_{L_0} \frac{g(\tau)}{X^+(\tau)} \sum_{k=0}^{n-1} T_k'(\tau) T_k^{j-1}(\tau) d\tau = 0, \quad j = 1, 2, \ldots, -k. \qquad (5.29)$$

Similar to usual Riemann Boundary Value Problem, we have

Theorem 14 For homogeneous problem (5.25), there are solutions of k-linear independence, when $k > 0$; no solution, when $k \le 0$. For non-homogeneous problem (5.16), there is solution unconditionally, when $k > 0$; there is solution, only when the conditions (5.29) are satisfied.

2) Automorphic Functions of Infinite Group

Riemann Boundary Value Problem discussed here is to find a sectional automorphic function $\Phi(z)$ satisfied the boundary condition (5.19), where $G\ (\ t\ ) \neq 0$ and $g\ (\ t\ )$ are given in L_0 and satisfied the condition $\wedge(1,1)$.

Paramount considering the jump question

$$\Phi^+(t) - \Phi^-(t) = g(t), \quad t \in L. \qquad (5.30)$$

The usual Cauchy Type Integration cannot satisfy the request, because it is not a automorphic function. From Corollary 12.2, we know that the function $\Phi(z)$ defined by (5.13) is uniquely the solution of (5.30); due to vanishing at z_0. The canonical function $X(z)$ of question (5.30) satisfied the boundary condition (5.13) in the homogeneous case is a sectional automorphic function with the index k of $G\ (\ t\ )$.

Easy to know

$$X(z) = [F(z) - F(t_0)]^{-k} e^{\Gamma(z)}, \quad t_0 \in L_0, \qquad (5.31)$$

where $F(z)$ is a simple automorphic function used in (5.30) with z as its simple pole,

$$\Gamma(z) \equiv \frac{1}{2\pi i} \int_{L_0} \ln G(\tau) \frac{F'(\tau)}{F(\tau) - F(z)} d\tau. \qquad (5.32)$$

The canonical function defined such a way can be accurate to a constant.

So the general solution of (5.30) is

$$\Phi(z) = X(z)[\Psi(z) + P_k(F)], \tag{5.33}$$

where

$$\Psi(z) \equiv \frac{1}{2\pi i} \int_{L_0} \frac{g(\tau)F'(\tau)}{X^+(\tau)[F(\tau) - F(z)]} d\tau, \tag{5.34}$$

$X(z)$ is defined by (5.31), P_k is an arbitrary polynomial of order k.

From the application's point of view, to fine a solution vanishing at point z_0 has special importance; for example in the case of solving a problem of singular integral equation. Therefore we have to consider P_{k-1} substituting for P_k; if $k \geq 0$, $P_{k-1} \equiv 0$ is essential; if $k < 0$, then we have the conditions of solvability:

$$\int_{L_0} \frac{g(\tau)}{X^+(\tau)} [F'(\tau)]^{j-1} F'(\tau) d\tau = 0, j = 1, 2, \ldots, -k. \tag{5.35}$$

Summarizing the above statements, we have

Theorem 15 (5.33) is the solution of (5.30). If an additional condition $\Phi(z_0) = 0$ must be satisfied by the solution $\Phi(z)$, P_{k-1} has to substitute for P_k; if $k \geq 0$, $P_{k-1} \equiv 0$ is essential; if $k < 0$, for the existence of solution, then a set of solvable conditions (5.35) must be satisfied.

6. Some Closed Formulae

Two kinds of singular integral equations proposed in section-4 would be solved by use of theorems 15 and 16.

1) Singular Integral Equation (5.15)

A closed solution for singular integral equation (5.15) would be found by using the solution (5.28) of (5.16), and generalized Plemelj formulae

$$\varphi(t) = \Phi^+(t) - \Phi^-(t). \tag{6.1}$$

From (6.1) and (5.28), easy to prove:

$$\varphi(t) = \frac{1}{2}[1 + 1/G(t)]g(t) + X^+(t)[1 - 1/G(t)]$$
$$\{\Psi(t) - \frac{1}{2}[P_{k-1}(F^*)/[F^*(t) - F^*(z_\infty)]^{k-1}]\}. \tag{6.2}$$

In the above equation, let $X(t)$, $\Psi(t)$ and $G(t)$, $g(t)$ be replaced by (5.23), (5.27) and (5.20) separately, and pay attention to condition $a^2(t) - b^2(t) = 1$, we obtain

$$\varphi(t) = a(t)f(t) - b(t)Z(t) \sum_{k=0}^{n-1} \frac{1}{\pi i} \int_{L_0} f(\tau) d\tau / Z(\tau)[\tau - T_k(t)]$$
$$+ b(t)Z(t)P_{k-1}(F^*)/[F^*(t) - F^*(z_\infty)]^{k-1}, \tag{6.3}$$

where

$$Z(t) \equiv [a(t) + b(t)]X^*(t) = [a(t) - b(t)]X^-(t)$$

$$= \Pi_{k=0}^{n-1}[T_k(t) - t_0]^{-k} \exp \sum_{k=0}^{n-1} \Gamma_0(T_k(t)),$$

$$\Gamma_0(t) \equiv \frac{1}{2\pi i} \int_{L_0} \ln G(\tau) d\tau/(\tau - t), \quad G(t) \equiv [a(t) - b(t)]/[a(t) + b(t)].$$

If $k \geq 0$, then let $P_{k-1} = 0$; if $k < 0$, for the existence of the solution, a set of solvable conditions

$$\int_{L_0} [\sum_{k=0}^{n-1} T_k'(\tau) T_k^{j-1}(\tau)] f(\tau) d\tau/Z(\tau) = 0, \quad j = 1, 2, \ldots, -k \qquad (6.4)$$

must be satisfied.

2) Singular Integral Equation (5.19)

Using the method similar to the previous section, and sectional automorphic function (5.13), applying Corollary 12.2, according to the Plemelj formula, based on the formula similar to (5.33), we can obtain the solution of Riemann Boundary Problem, namely, the solution of Singular Integral Equation (5.19):

$$\varphi(t) = a(t)f(t) - \frac{b(t)Z(t)}{\pi i} \int_{L_0} \frac{f(t)F'(\tau)d\tau}{Z(\tau)[F(\tau) - F(t)]}$$

$$+ b(t)Z(t)P_{k-1}(F), \qquad (6.5)$$

where

$$Z(t) = [a(t) + b(t)]X^+(t) = [a(t) - b(t)]X^-(t)$$

$$= [F(t) - F(t_0)]^{-k} e^{\Gamma(t)}.$$

The result here is much like what stated in the previous section. If $k > 0$, then let $P_{k-1} \equiv 0$; if $k \leq 0$, for the existence of the solution, a set of solvable conditions

$$\int_{L_0} \frac{f(\tau)}{Z(\tau)} [F(\tau)]^{j-1} F'(\tau) d\tau = 0, \quad j = 1, 2, \ldots, -k, \qquad (6.6)$$

must be satisfied.

7. Some Remarks

There were although already the rich literatures in the field of automorphic function boundary value problems, singular integral equations and its applications in mechanics, but also there are many meaningful works awaiting to solve; as space is limited, we proposed certain remarks take the end of this chapter.

1) Trigonometric function, hyperbolic function, elliptical function, modular function and so on are primary automorphic functions, and suitable to characterize the phenomenon of different periodic phenomena, or some type conservation laws in nature. Many achievements stated before, under the same controlled conditions, may be generalized to the corresponding results of automorphic functions.

2) The condition $\wedge(1,1)$ described by modulus of continuity is more general than the Hölder-Lipschitz condition $H(H,\alpha)$. There are many results here in condition $H(H,\alpha)$, can be conditionally generalized to those on the condition $\wedge(1,1)$.

3) Using the methods here, certain type of singular integral equations can easily be solved in a closed way, most commonly with a Cauchy kernel, a logarithm kernel or an exponent kernel, including different combination of these three kind of kernels.

4) These results have widespread and important applications to the elasticity theory and the fluid mechanics. No matter in the theory or application, there are still much more works awaiting to do.

References

1. Cai Hai-tao, Problem of the periodic contact of elastic theory for the isotropic elastic plane, J. of Central-South Inst. Mining and Metallurgy, 1979, 113-123.
2. Cai Hai-tao, The problem of periodic contact in the plane theory of elasticity, Acta Math. Appl. Sinica, 2 (1979), 181-195.
3. Cai Hai-tao, On the first and the second periodic fundamental problems of semi-infinite medium with anisotropic elasticity, Acta Mech. Sci., (1979), 240-247.
4. Cai Hai-tao, Periodic crack problem of plane anisotropic medium, Acta Math Sci., 2 (1982), 35-44.
5. Cai Hai-tao, The periodic cracks of an infinite anisotropic medium for plane skew-symmetric loadings, Acta Mech. Solida Sinica. 2(1986) 155-159
6. Cai Hai-tao, A periodic array of cracks in an infinite anisotropic medium, Engineering Fracture Mechanics, 1993, 6, 46, 127-131.
7. Cai Hai-tao, Yuan Xiu-gui, The crack problem of two bounded half orthotropic planes materials, Engineering Fracture Mech, 1995, 52, 898-900.
8. Cai Hai-tao, Complex analysis and application to compound material, Complex Analysis and its Applications, 1994, Longman, 292-297.
9. Cai Hai-tao, An application of complex analysis to periodic movable loading problems, Complex Variables, 1996, 30, 145-151.
10. Gakhov F. D., Boundary Value Problems, Nauka, Moscow, 1977.
11. Guo You-Zhong, Boundary value problems and singular integral equations with automorphic functions, Lecture Notes, Inst. of Math. And Comp. Tech., Academia Sinica, 1965.
12. Lu Jian-ke, Periodic Riemann boundary value problem and its application to elasticity, Acta Math. Sinica, 13(1963), 343-388.
13. Lu Jian-ke, On fundamental problems for the infinite elastic plane with crack, J. Wuhan Univ.(math. specialed.), 1963, No.2, 37-49.
14. Lu Jian-ke, On the plane welding problems of different materials, J. Wuhan Univ. (Math. Specialized), 1963, No.2, 50-66.
15. Lu Jian-ke, On mathematical problems of elastic plane with cyclic symmetry, J. Wuhan Univ., 1964, No.2, 1-13.
16. Lu Jian-ke, On fundamental problems of plane elasticity with periodic stresses, Acta Mech. Sinica, 7(1964), 316-327

17. Lu Jian-ke, On problems of an infinite elastic plane with arbitrary periodic cracks, J.of Central-South Inst. of Mining and Metallury, 1980, No.2, 9-19.

18. Lu Jian-ke, A remark on plane elasticity with periodic stresses, J. Wuhan Univ., 1980, No.2, 9-10.

19. Lu Jian-ke, Circular welding problems with a crack, Appl. Math. Mech. (English), 4 (1983). 751-763.

20. Lu Jian-ke, New formulations of the second fundamental problem in plane elasticity, Appl. Math. Mech., 6 (1985), 223-230.

21. Lu Jian-ke, Boundary Value Problems for Analytic Functions. World Scientific, Singapore, 1993.

22. Lu Jian-ke, Complex Variable Methods in Plane Elasticity, World Scientific, Singapore, 1995.

23. Muskhelishvili, N. I., Some Basic problems of the Mathematical Theory of Elasticity, Noordhoff Groningen, 1953.

24. Muskhelishvili, N. I., Singular Integral Equations, Noordhoff, Leyden, 1977.

25. Sarin, G. N., Stress Concentration around holes, Sci. Press, Beijing, 1958.

26. Sarin, G. N., Stresses in elastic plane with an infinite row of equal cuts, Dokl. USSR, 23 (1999), 515-519.

27. Sih, G. C., Liebowitz, H., Mathematical theories of brittle fracture, in "Fracture" (edited by H. Liebowitz) 2 (1968), 67-190. Academic Press.

28. Cai Hai Tao, Lu Jianke, Mathematical Theory in periodic plane elasticity, Taylor & Francis, England, 2000.

29. Li Guo-Ping, Guo You-Zhong, The theory of automorphic functions and Minkowski functions, Academic Press, 1979.

30. Lu Jian-ke, Cai Hai-tao, Introduction to periodic plane elastic problems, Wuhan Univ., Press. Wuhan, 2008.

Perspectives in Mathematical Sciences
Interdisciplinary Mathematical Sciences, Volume 9, 2009
pp. 37–52

Chapter 2

Subharmonic Bifurcations and Chaos for a Model of Micro-Cantilever in MEMS*

Yushu Chen[1,2], Liangqiang Zhou[1,2]** and Fangqi Chen[3,2]

[1] *Research Center of Nonlinear Dynamics,*
Tianjin University, Tianjin 300072, PR China
[2] *State Key Laboratory of Engines,*
Tianjin University, Tianjin 300072, PR China
[3] *Department of Mathematics,*
Nanjing University of Aeronautics and Astronautics,
Nanjing 210016,PR China

Dedicated to Professor Youzhong Guo on the occasion of his 75th birthday

The dynamical system of an idealized micro-cantilever induced the cubic nonlinear spring stiffness term and excited by periodic voltages in micro-electro-mechanical system (MEMS) is considered. With Taylor expansion, the system can be reduced approximately as one with forcing and quadratic excitations. The global bifurcations and chaotic behaviors of the system are studied through both analytical and numerical methods. With Melnikov method, the conditions which the system parameters satisfy for chaos are obtained. The critical curves which separate the chaotic regions and non-chaotic regions of the system are plotted. There is a "chaotic band" for this system, whose area changes as the coefficient of the cubic spring stiffness term. Subharmonic bifurcations are studied via subharmonic Melnikov method, and the critical curves for subharmonic bifurcations are also plotted. With numerical methods, the phase portraits of the system are obtained, which demonstrate some new interesting dynamical phenomena for this system.

1. Introduction

The dynamical behavior of micro-electro-mechanical system (MEMS) has attracted much attention in the past a few years. There exist intrinsic nonlinearities and exterior nonlinearities arising from coupling of different domains, creep phenomena and nonlinear damping effects, etc.[1-3,13]. Many micro-cantilever based MEMS

---

* This research was supported by the National Natural Science Foundation of China (No.10632040), the Natural Science Foundation of Tianjin, China (No. 09JCZDJC26800), China Postdoctoral Science Foundation (No. 20090450765).
** Corresponding author, Email: zlqrex@tom.com.

sensors have been utilized for high precision chemical detection and small force detection. Some micro-cantilever mass sensors have been developed and applied as chemical sensors, biosensors and other sensors as well [4-5].

A lot of researches have been done on the nonlinear dynamical behavior of micro-cantilever based in MEMS. Passiana et al.[6] presented a micro-cantilever of an atomic force microscope (AFM) dynamic system to display the useful resonance behavior at kilohertz frequencies. Zook and Burns [7] calculated the natural frequencies of a micro-beam using finite element method. Choi and Lovell [8] computed the static deflection of a micro-beam numerically by the shooting method. Ahn et al.[9] modeled a micro-beam under electrostatic actuation as a single degree-of-freedom spring-mass-damper system. Zhang [10] presented the nonlinear dynamical system of micro-cantilever under combined parametric and forcing excitations in MEMS, and studied the dynamical behavior with numerical methods.

In this paper, using Melnikov method, we study the chaotic behavior and subharmonic bifurcations rigorously and analytically for this class of system. The critical curves which separate the chaotic regions and non-chaotic regions are plot. There is a "chaotic band" for this system. There exist subharmonic bifurcations of odd or even orders for this system under certain conditions. The critical curves for subharmonic bifurcations are also plot. With numerical methods, the phase portraits of the system are obtained, and some new interesting phenomena are found. It is a complementarity of the research in [10].

2. Formulation of the problem

Consider the dynamical model shown in Fig. 1. For this simplified mass-spring-damping system, the governing equation of motion for the system in MEMS is [10]

$$m\ddot{y} + c\dot{y} + ky = F_E(t) \tag{1}$$

where y is the vertical displacement of the micro-cantilever relative to the origin of the fixed place, m is the mass, k and c are the effective spring stiffness and the spring coefficient of the simplified system, respectively, and

$$F_E(t) = \frac{\varepsilon_0 A}{2} \frac{V^2(t)}{(d-y)^2} \tag{2}$$

where ε_0 is the absolute dielectric constant of vacuum, A is the overlapping area between the two plates, and d is the gap between them.

Introducing the following dimensionless variables,

$$x = \frac{y}{d}, \quad \xi = \frac{c}{\sqrt{km}}, \quad \omega_0 = \sqrt{\frac{k}{m}}, \quad \tau = \omega_0 t, \quad T = \frac{\varepsilon_0 A}{2m\omega_0^2 d^3},$$

therefore (1) and (2) give

$$\ddot{x} + \xi\dot{x} + x = T\frac{V^2(\tau)}{(1-x)^2} \tag{3}$$

Using the approximation $1/(1-x)^2 \approx 1 + 2x + 3x^2 + O(x^3)$, inducing the cubic nonlinear spring stiffness term $k_1 x^3$, system (3) can be written as follows under the applied voltage $V = (V_0 + V_P)\cos(\omega t + \phi)$ [10]

$$\ddot{x} = -\xi\dot{x} - (x + k_1 x^3) + (2x + 3x^2)T(V_0 + V_P)^2\cos^2(\Omega t) + T(V_0 + V_P)^2\cos^2(\Omega t) \quad (4)$$

where $\Omega = \omega/\omega_0$.

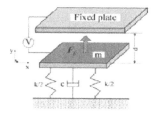

Fig. 1. A simplified dynamical model of the micro-cantilever in MEMS.

3. Melnikov analysis

When $k_1 < 0$, using the transformation $x = X/\sqrt{-1/k_1}$, (4) can be written as

$$\ddot{X} + \bar{\xi}\dot{X} - (X - X^3) + (2\sqrt{-k_1}X - 3k_1 X^2)T(V_0 + V_P)^2\cos^2\Omega t + T(V_0 + V_P)^2\cos^2\Omega t \quad (5)$$

where $\bar{\xi} = \sqrt{-k_1}\xi$.

When $k_1 > 0$, using the transformation $x = X/\sqrt{1/k_1}$, (4) is written as follows

$$\ddot{X} + \bar{\xi}\dot{X} - (X + X^3) + (2\sqrt{k_1}X + 3k_1 X^2)T(V_0 + V_P)^2\cos^2\Omega t + T(V_0 + V_P)^2\cos^2\Omega t \quad (6)$$

where $\bar{\xi} = \sqrt{k_1}\xi$.

3.1 The case $k_1 < 0$

Denoting $x_1 = X, x_2 = \dot{X}, \bar{\xi} = \varepsilon\xi, T(V_0 + V_P)^2 \equiv f = \varepsilon f$, Eq.(5) can be written as

$$\begin{cases} \dot{x_1} = x_2 \\ \dot{x_2} = -\varepsilon\xi x_2 - (x_1 - x_1^3) + \varepsilon f(2\sqrt{-k_1}x_1 - 3k_1 x_1^2)\cos^2\Omega t + \varepsilon f\cos^2\Omega t \end{cases} \quad (7)$$

When $\varepsilon = 0$, the unperturbed system is

$$\begin{cases} \dot{x_1} = x_2 \\ \dot{x_2} = -(x_1 - x_1^3) \end{cases} \quad (8)$$

System (8) is a Hamilton system with Hamiltonian $H(x_1, x_2) = x_2^2/2 + x_1^2/2 - x_1^4/4$. $(0,0)$ is the center of system (8), $(1,0)$ and $(-1,0)$ are saddles of system (8).

There exist two heteroclinic orbits connecting $(\pm 1, 0)$ when $H(x_1, x_2) = h = 1/4$. The expressions of the heteroclinic orbits are

$$\begin{cases} x_{1\pm}(t) = \pm\tanh(\frac{\sqrt{2}}{2}t) \\ \\ x_{2\pm}(t) = \pm\frac{\sqrt{2}}{2}\operatorname{sech}(\frac{\sqrt{2}}{2}t) \end{cases} \tag{9}$$

When $0 < H(x_1, x_2) < 1/4$ there exist closed periodic orbits around $(0, 0)$, whose expressions are

$$\begin{cases} x_{1k}(t) = \frac{\sqrt{2}k}{\sqrt{1+k^2}}\operatorname{sn}(\frac{t}{\sqrt{1+k^2}}, k) \\ \\ x_{2k}(t) = \frac{\sqrt{2}k}{\sqrt{1+k^2}}\operatorname{cn}(\frac{t}{\sqrt{1+k^2}}, k)\operatorname{dn}(\frac{t}{\sqrt{1+k^2}}, k) \end{cases} \tag{10}$$

where sn, cn, dn are Jacobi elliptic functions, $0 < k < 1$ is the modulus of the Jacobi ellipse functions. The period of the orbit is $T_k = 4\sqrt{1+k^2}K(k)$, where $K(k)$ is the complete elliptic integral of the first kind. It is easy to verify that $dT_k/dk > 0$.

Here we compute the Melnikov integrals of system (7) along the heteroclinic orbits (9). The result computing along the orbit $(x_{1+}(t), x_{2+}(t))$ is the same as that computing along the orbit $(x_{1-}(t), x_{2-}(t))$, so we just give the result computing along the orbit $(x_{1+}(t), x_{2+}(t))$ as follows:

$$M(t_0) = -\xi \int_{-\infty}^{+\infty} x_{2+}^2(t)dt + f[2\sqrt{-k_1} \int_{-\infty}^{+\infty} x_{1+}(t)x_{2+}(t)\cos^2\Omega(t+t_0)dt$$

$$-3k_1 \int_{-\infty}^{+\infty} x_{1+}^2(t)x_{2+}(t)\cos^2\Omega(t+t_0)dt + \int_{-\infty}^{+\infty} x_{2+}(t)\cos^2\Omega(t+t_0)dt] \tag{11}$$

$$\equiv -\frac{\sqrt{2}}{2}\xi I_0 + f(I_1 - I_2\sin\omega t_0 + I_3 + I_4\cos\omega t_0 + I_5 + I_6\cos\omega t_0)$$

where

$$I_0 = \int_{-\infty}^{+\infty} \operatorname{sech}^4 t\, dt = \frac{4}{3},$$

$$I_1 = \sqrt{-k_1} \int_{-\infty}^{+\infty} \tanh t\, \operatorname{sech}^2 t\, dt = \frac{2\sqrt{-k_1}}{3},$$

$$I_2 = \sqrt{-k_1} \int_{-\infty}^{+\infty} \tanh t\, \operatorname{sech}^2 t\, \cos\omega t\, dt = \frac{\sqrt{-k_1}\pi\omega^2}{\sinh(\pi\omega/2)},$$

$$I_3 = -\frac{3}{2}k_1 \int_{-\infty}^{+\infty} \tanh^2 t\, \operatorname{sech}^2 t\, dt = -\frac{3}{5}k_1,$$

$$I_4 = -\frac{3}{2}k_1 \int_{-\infty}^{+\infty} \tanh^2 t\, \operatorname{sech}^2 t\, \cos\omega t\, dt = -\frac{3k_1\pi(\omega^3-2\omega)}{2\sinh(\pi\omega/2)},$$

$$I_5 = \frac{1}{2} \int_{-\infty}^{+\infty} \operatorname{sech}^2 t\, dt = 1,$$

$$I_6 = \frac{1}{2} \int_{-\infty}^{+\infty} \operatorname{sech}^2 t\, \cos\omega t\, dt = \frac{\pi\omega}{2\sinh(\pi\omega/2)},$$

$$\omega \equiv 2\Omega.$$

Let

$$R_{\min} \equiv \max\{\frac{I_1+I_3+I_5-\sqrt{(I_4+I_6)^2+I_2^2}}{\frac{\sqrt{2}}{2I_0}}, 0\}, \quad R_{\max} \equiv \frac{I_1+I_3+I_5+\sqrt{(I_4+I_6)^2+I_2^2}}{\frac{\sqrt{2}}{2I_0}},$$

by (11), we estimate that the Melnikov function $M(t_0)$ has simple zeros when $R_{\min} < \xi/f \equiv R_{het}(\omega) < R_{\max}$, and Smale horseshoe chaos occurs.

The critical curves which separate the chaotic zones and non-chaotic zones for different values of k_1 are plot as Fig. 2.

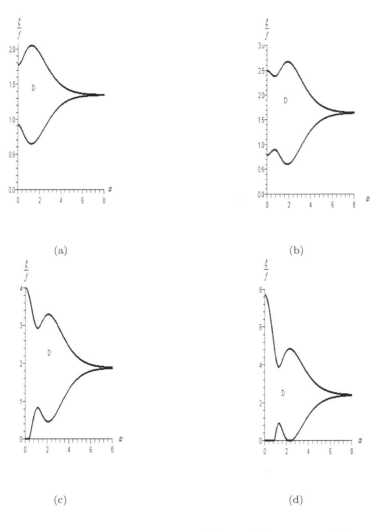

(a)

(b)

(c)

(d)

Fig. 2. The critical curves for chaos of system (7) with (a)$k_1 = -0.1$, (b)$k_1 = -0.3$, (c)$k_1 = -0.5$, (d)$k_1 = -1$.

From Fig. 2 we can draw the following conclusions: Chaos occurs in the region D; the area of D increases as the increases of the absolute value of k_1; chaos won't take place when ω increases to a positive value which is near 8.

Next subharmonic bifurcation for system (7) is considered. It can be computed that the subharmonic Melnikov function for the periodic orbits (10) is

$$M^{m/n}(t_0) = -\xi \int_0^{mT} x_{2k}^2(t)dt + f[\sqrt{-k_1} \int_0^{mT} x_{1k}(t)x_{2k}(t)\cos^2\Omega(t+t_0)dt$$

$$-3k_1 \int_0^{mT} x_{1k}^2(t)x_{2k}(t)\cos^2\Omega(t+t_0)dt + \int_0^{mT} x_{2k}(t)\cos^2\Omega(t+t_0)dt]$$

$$\equiv \xi I_1(m,n) + f[I_2(m,n)\sin\omega t_0 + I_3(m,n)\cos\omega t_0 + I_4(m,n)\cos\omega t_0]$$

where

$$\omega \equiv 2\Omega,$$

$$T = 4\sqrt{1+k^2}K(k) = \frac{2\pi m}{\omega n},$$

$$I_1(m,n) = \frac{8m}{3(1+k^2)}[(k^2-1)K(k) + (1+k^2)E(k)],$$

$$I_2(m,n) = \begin{cases} 0, n \neq 1\text{ or } m \text{ is odd} ; \\ \frac{\sqrt{-k_1}\pi^3 m^2}{(1+k^2)K^2(k)}\text{cosech}\frac{\pi mK'(k)}{2K(k)}, & n=1 \text{ and } m \text{ is even} \end{cases}$$

$$I_3(m,n) = \begin{cases} 0, \ n \neq 1 \text{ or } m \text{ is even}; \\ -\frac{\sqrt{2}k_1\pi^2 m^2}{8(1+k^2)K^3(k)}[4K^2(k)(1+k^2) \\ -\pi^2 m^2]\text{cosech}\frac{\pi mK'(k)}{2K(k)}, \ n=1 \text{ and } m \text{ is odd} \end{cases}$$

$$I_4(m,n) = \begin{cases} 0, \ n \neq 1 \text{ or } m \text{ is even}; \\ -\frac{\pi^2 m^2}{2kK(k)}\text{cosech}\frac{\pi mK'(k)}{4K(k)}, \ n=1 \text{ and } m \text{ is odd} \end{cases}$$

By Melnikov analysis, when

$$\frac{\xi}{f} < \frac{I_3(m,n) + I_4(m,n)}{I_1(m,n)} = \frac{-3\sqrt{2}k_1\pi(1+k^2)^{3/2}(1-\omega^2)}{8\omega[(k^2-1)K(k)+(1+k^2)E(k)]} \cdot \frac{1}{\sinh(\sqrt{1+k^2}K'(k)\omega)}$$

$$+\frac{3\pi(1+k^2)^2\omega}{8k[(k^2-1)K(k)+(1+k^2)E(k)]} \cdot \frac{1}{\sinh(\sqrt{1+k^2}K'(k)\omega/2)} \equiv R_1^m(\omega),$$

subharmonic bifurcation of odd orders will occur. Numerical calculations demonstrate that subharmonic bifurcation of first order occurs while $\omega \in (0.1, 0.25)$ for $m = 1$; subharmonic bifurcation of m-order occurs while $\omega \in (0.15m, 0.5 + 0.15m)$ for an odd number m $(m > 1)$. The critical curves are shown as Fig. 3.

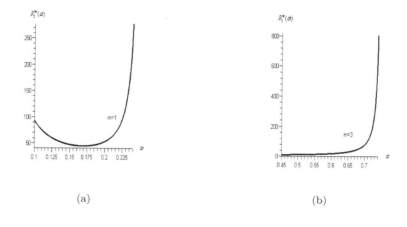

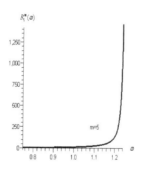

(c)

Fig. 3. The critical curves for subharmonic bifurcations of odd orders for system (7).

Melnikov analysis suggests that when

$$\frac{\xi}{f} < \frac{I_2(m,n)}{I_1(m,n)} = \frac{3\sqrt{-k_1}\pi(1+k^2)^{3/2}}{2\sqrt{2}[(1+k^2)E(k) - (1-k^2)K(k)]} \cdot \frac{1}{\sinh(\sqrt{1+k^2}K'(k)\omega)} \equiv R_2^m(\omega)$$

subharmonic bifurcation of even orders will occur. Numerical calculations demonstrate that subharmonic bifurcation of m-order occurs while $\omega \in (0.15m, 0.25m)$ for an even number m. The critical curves are shown as Fig. 4.

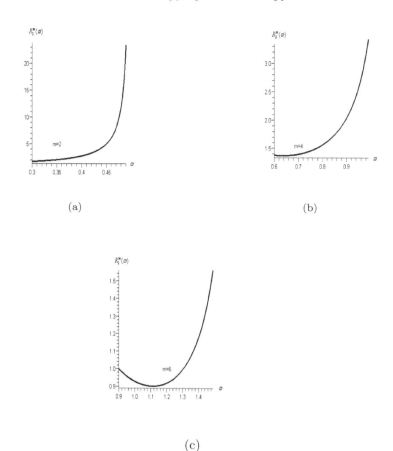

Fig. 4. The critical curves for subharmonic bifurcations of even orders for system (7).

3.2 The case $k_1 > 0$

Letting $x_1 = X, x_2 = \dot{X}, \bar{\xi} = \varepsilon\xi, T(V_0 + V_P)^2 \equiv f = \varepsilon f$, (8) is changed into the following system

$$\begin{cases} \dot{x}_1 = x_2 \\ \dot{x}_2 = -\varepsilon\xi x_2 - (x_1 + x_1^3) + \varepsilon f(2\sqrt{k_1}x_1 + 3k_1 x_1^2)\cos^2\Omega t + \varepsilon f\cos^2\Omega t \end{cases} \tag{12}$$

When $\varepsilon = 0$, the unperturbed system is

$$\begin{cases} \dot{x}_1 = x_2 \\ \dot{x}_2 = -(x_1 + x_1^3) \end{cases} \tag{13}$$

System (13) is also a Hamilton system with hamiltonian $H(x_1, x_2) = x_2^2/2 + x_1^2/2 + x_1^4/4$. $(0,0)$ is the center of system (13).There exist closed periodic orbits around $(0,0)$, whose expressions are

$$\begin{cases} x_{1k}(t) = \sqrt{\frac{2k^2}{1-2k^2}}\,\text{cn}(\frac{t}{\sqrt{1-2k^2}}, k) \\ x_{2k}(t) = -\frac{\sqrt{2k^2}}{\sqrt{1-2k^2}}\,\text{sn}(\frac{t}{\sqrt{1-2k^2}}, k)\text{dn}(\frac{t}{\sqrt{1-2k^2}}, k) \end{cases} \tag{14}$$

The periods of the orbits are $T_k = 4\sqrt{1-2k^2}K(k)$.

By complicated calculations, the subharmonic Melnikov function for the periodic orbit with period $T = 4\sqrt{1-2k^2}K(k) = 2\pi m/(\omega n)$ can be written as

$$M^{m/n}(t_0) = -\xi \int_0^{mT} x_{2k}^2(t)\mathrm{d}t + f[2\sqrt{k_1}\int_0^{mT} x_{1k}(t)x_{2k}(t)\cos^2\Omega(t+t_0)\mathrm{d}t$$

$$+3k_1\int_0^{mT} x_{1k}^2(t)x_{2k}(t)\cos^2\Omega(t+t_0)\mathrm{d}t + \int_0^{mT} x_{2k}(t)\cos^2\Omega(t+t_0)\mathrm{d}t]$$

$$\equiv \xi J_1(m,n) + f[J_2(m,n) + J_3(m,n) + J_4(m,n)]\sin\omega t_0.$$

where

$$\omega \equiv 2\Omega,$$

$$J_1(m,n) = \frac{8m}{3\sqrt{1-2k^2}}[\frac{1-k^2}{1-2k^2}K(k) - E(k)],$$

$$J_2(m,n) = \begin{cases} 0, n \neq 1 \text{or } m \text{ is odd}; \\ -4\sqrt{k_1}mK(k)\text{cosech}\frac{\pi mK'(k)}{2K(k)}, n = 1 \text{ and } m \text{ is even} \end{cases}$$

$$J_3(m,n) = \begin{cases} 0, \ n \neq 1 \text{or } m \text{ is even}; \\ -\frac{-\sqrt{2}k_1\pi^2 m^2}{6(1-2k^2)^{3/2}K^3(k)}[\frac{\pi^2 m^2}{4} - 4K^2(k)(1-2k^2)]\text{cosech}\frac{\pi mK'(k)}{2K(k)}, \\ n = 1 \text{ and } m \text{ is odd} \end{cases}$$

$$J_4(m,n) = \begin{cases} 0, \ n \neq 1 \text{ or } m \text{ is even}; \\ -\frac{\sqrt{2}\pi^2 m^2}{\sqrt{1-2k^2}K(k)}\text{sech}\frac{\pi mK'(k)}{4K(k)}, \ n = 1 \text{ and } m \text{ is odd} \end{cases}$$

By Melnikov analysis, when

$$\frac{\xi}{f} < \frac{J_3(m,n) + J_4(m,n)}{J_1(m,n)} = \frac{\sqrt{2}k_1(1-2k^2)^{3/2}\omega(\omega^2-4)}{8[(1-k^2)K(k) - (1-2k^2)E(k)]} \cdot \frac{1}{\cosh(\sqrt{1-2k^2}K'(k)\omega)}$$

$$+\frac{3\sqrt{2}(1-2k^2)^{3/2}\omega}{4[(1-k^2)K(k) - (1-2k^2)E(k)]} \cdot \frac{1}{\cosh(\sqrt{1-2k^2}K'(k)\omega/2)} \equiv R_3^m(\omega)$$

subharmonic bifurcations of odd orders will occur. Numerical calculations demonstrate that subharmonic bifurcation of m-order occurs while $\omega > m$ for an odd number m. The critical curves are shown as Fig. 5.

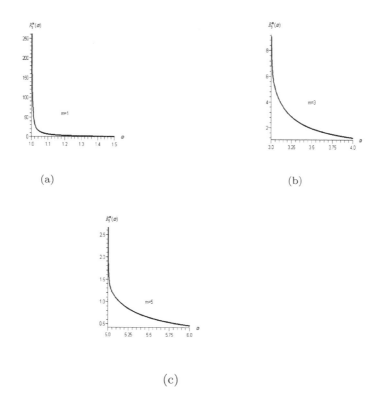

Fig. 5. The critical curves for subharmonic bifurcations of odd orders for system (12).

By Melnikov analysis, we can also conclude that subharmonic bifurcation of even orders will occur when

$$\frac{\xi}{f} < \frac{J_2(m,n)}{J_1(m,n)} = \frac{3\sqrt{k_1}K(k)(1-2k^2)^{3/2}}{2[(1-k^2)K(k)-(1-2k^2)E(k)]} \cdot \frac{1}{\sinh(\sqrt{1-2k^2}K'(k)\omega)} \equiv R_4^m(\omega).$$

Numerical calculations demonstrate that subharmonic bifurcation of m-order occurs while $\omega > m$ for an even number m. The critical curves are shown as Fig. 6.

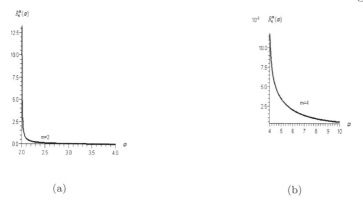

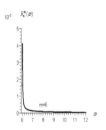

(c)

Fig. 6. The critical curves for subharmonic bifurcations of even orders for system (12).

4. Numerical simulations

Numerical results for system (7) and (12) have been obtained by using forth-order Runge-Kutta method in this section.

Taking $\varepsilon = 0.1, \xi = 0.1, f = 0.1, \Omega = 1$, varying k_1, the phase portraits are obtained as Fig. 7 and Fig. 8.

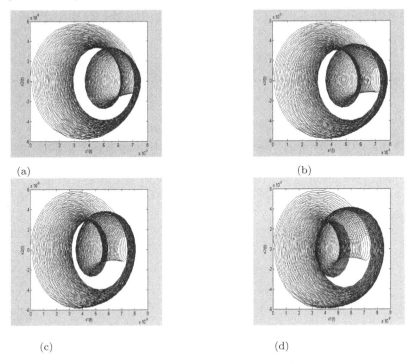

(a)　　　　　　　　　　　　　　　(b)

(c)　　　　　　　　　　　　　　　(d)

Fig. 7. The phase portraits of system (7) for (a) $k_1 = -0.1$, (b) $k_1 = -0.3$, (c) $k_1 = -0.5$, (d) $k_1 = -1$.

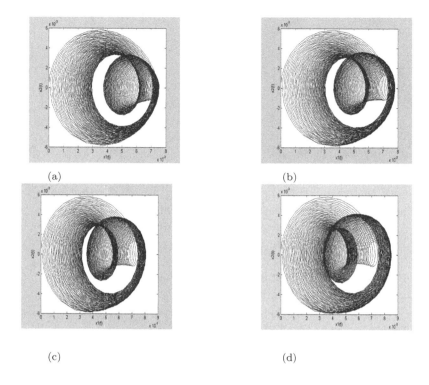

(a) (b)

(c) (d)

Fig. 8. The phase portraits of system (12) for (a) $k_1 = 0.1$, (b) $k_1 = 0.3$, (c) $k_1 = 0.5$, (d) $k_1 = 1$.

From Fig. 7 and Fig. 8 we can see that the phase portraits of both system (7) and (12) change little as k_1 varies, and there is little difference in the phase portraits between the two systems.

Next we fix $k_1 = -1$ (for system (7)) or $k_1 = 1$ (for system (12)), $\varepsilon = 0.1, \xi = 0.1, \Omega = 1$, vary f, and obtain the phase portraits as Fig. 9 and Fig. 10.

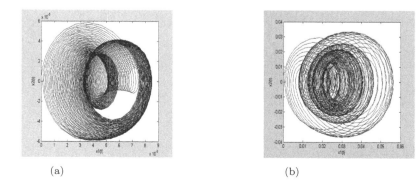

(a) (b)

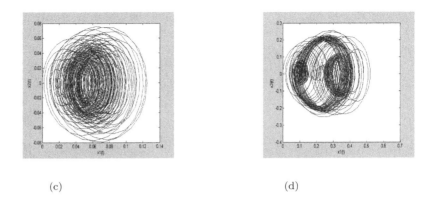

(c)   (d)

Fig. 9. The phase portraits of system (7) for (a) $f = 0.1$, (b) $f = 0.5$, (c) $f = 1$, (d) $f = 2.7$.

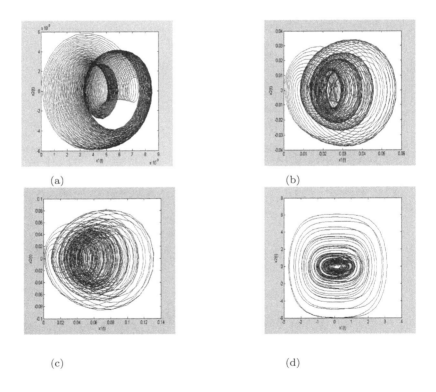

Fig. 10. The phase portraits of system (12) for (a) $f = 0.1$, (b) $f = 0.5$, (c) $f = 1$, (d) $f = 2.7$.

From Fig. 9 and Fig. 10 we can see that the phase portraits of both system (7) and (12) change markedly as f changes. There is some difference in the phase portraits between the two systems.

We also fix $k_1 = -1$ (for system (7)) or $k_1 = 1$ (for system (12)), $\varepsilon = 0.1, \xi = 0.1, f = 0.1$, vary Ω, and obtain the phase portraits as Fig. 11 and Fig. 12.

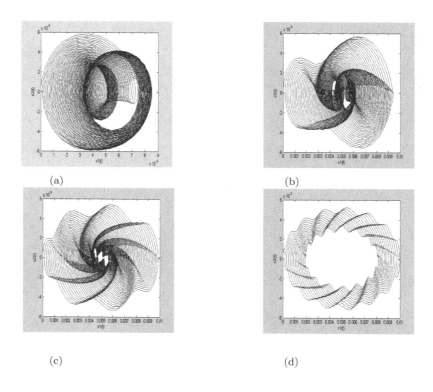

Fig. 11. The phase portraits of system (7) for (a) $\Omega = 1$, (b) $\Omega = 2$, (c) $\Omega = 4$. (d) $\Omega = 8$.

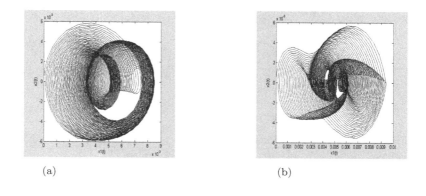

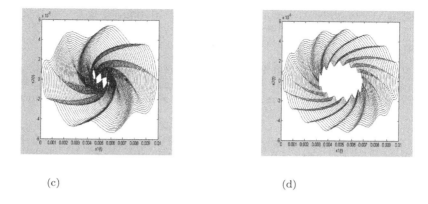

(c)   (d)

Fig. 12. The phase portraits of system (12) for (a) $\Omega = 1$, (b) $\Omega = 2$, (c) $\Omega = 4$. (d)
$\Omega = 8$.

From Fig. 11 and Fig. 12 we can see that the phase portraits of both system (7) and (12) change markedly as Ω changes, but there is little difference in the phase portraits between the two systems.

5. Conclusions

The dynamical behavior of an idealized micro-cantilever subjected to combined nonlinear parametric and forcing excitations in MEMS is studied. The chaotic motions arising from the transverse intersections of the stable and unstable manifolds of the heteroclinic orbits are analyzed by means of a version of Melnikov's method. The critical curves which separate the chaotic zones and non-chaotic zones of the system are plot. There is a "chaotic band" for this class of system, whose area changes as the coefficient of the cubic spring stiffness term. The subharmonic bifurcation is studied via subharmonic Melnikov method. There exist subharmonic bifurcations of even or odd orders. Numerical simulations demonstrate some new interesting dynamical phenomena. These results provide some inspiration and guidance for the analysis and dynamic design for this class of system.

References

[1] S.D. Senturia, Simulation and design of Microsystems: a 10-year preserve, *Sens. Actuators A* 67 (1998) 1-7.
[2] N.R. Aluru, J. White, An efficient numerical technique for electormechanical simulation of complicated microelectromechanical structures, *Sens. Actuators A* 58 (1997) 1-11.
[3] M. Ashhab, M.V. Salapaka, M. Dahleh, I. Mezic, Dynamic analysis and control of microcantilevers, *Automatica* 35 (1999) 1663-1670.

[4] T.D. Stowe, K. Yasumura, T.W. Kenny, D. Botkin, K. Wago, D. Rugar, Attonewton force detection using ultrathin silicon cantilevers, Appl. *Phys. Lett* 71 (1997) 288-290.

[5] T. Kenny, Nanometer-scale force sensing with MEMS devices, *IEEE Seas. J* 1 (2001) 148-157.

[6] A. Passiana, G. Muralidharana, A. Mehtaa, H. Simpsonb, T.L. Ferrella, T. Thundant, Manipulation of microcantilever oscillations, *Ultramicroscopy* 97 (2003) 391-399.

[7] J.D. Zook, D.W. Burns, Characteristics of polysilicon resonant microbeams, *Sens. Actuators A* 35 (1992) 51-59.

[8] B. Choi, E.G. Lovell, Improved analysis of microbeams under mechanical and electrostatic loads, *J. Micromech. Microeng* 7 (1997) 24-29.

[9] Y. Ahn, H. Guckel, J.D. Zook, Capacitive microbeam resonator design, *J. Micromech. Microeng* 11 (2001) 70-80.

[10] W. Zhang, G. Meng, Nonlinear dynamical system of micro-cantilever under combined parametric and forcing excitations in MEMS, *Sens. Actuators A* 119 (2005) 291-299.

[11] S.Wiggins, *Introduction to Applied Non-linear Dynamical Systems and Chaos*, Springer, New York 1990.

[12] Y. Chen, Y. Leung, *Bifurcation and Chaos in Engineering*, Springer, New York 1998.

[13] Y. Guo, Z. Liu, X. Jiang, Z. Han, Higher-order Melnikov method, *Applied Mathematics and Mechanics* 12 (1991) 21-32.

Perspectives in Mathematical Sciences
Interdisciplinary Mathematical Sciences, Volume 9, 2009
pp. 53–70

Chapter 3

Canonical Sample Spaces
for Random Dynamical Systems

Jinqiao Duan *, Xingye Kan † and Bjorn Schmalfuss

*Department of Applied Mathematics, Illinois Institute of Technology,
Chicago, IL 60616, USA*
duan@iit.edu

*Department of Applied Mathematics, Illinois Institute of Technology,
Chicago, IL 60616, USA*
xkan@iit.edu

*Institut für Mathematik
Fakultät EIM, Warburger Strasse 100, 33098
Paderborn, Germany
schmalfu@math.upb.de*

This is an overview about natural sample spaces for differential equations driven by various noises. Appropriate sample spaces are needed in order to facilitate a random dynamical systems approach for stochastic differential equations. The noise could be white or colored, Gaussian or non-Gaussian, Markov or non-Markov, and semimartingale or non-semimartingale. Typical noises are defined in terms of Brownian motion, Lévy motion and fractional Brownian motion. In each of these cases, a canonical sample space with an appropriate metric (or topology that gives convergence concept) is introduced. Basic properties of canonical sample spaces, such as separability and completeness, are then discussed.

Moreover, a flow defined by shifts, is introduced on these canonical sample spaces. This flow has an invariant measure which is the probability distribution for Brownian motion, or Lévy motion or fractional Brownian motion. Thus canonical sample spaces are much richer in mathematical structures than the usual sample spaces in probability theory, as they have metric or topological structures, together with a shift flow (or driving flow) defined on it. This facilitates dynamical systems approaches for studying stochastic differential equations.

---

*Partially supported by NSF grants 0620539 and 0731201, the Cheung Kong Scholars Program and the K. C. Wong Education Foundation
†Partially supported by a Graduate Research Fellowship from Illinois Institute of Technology

1. Random dynamical systems

Stochastic differential equations (SDEs) or stochastic partial differential equations (SPDEs) arise as mathematical models for complex systems under various random influences in engineering and science. Here we only consider random dynamical systems defined by SDEs. Such SDEs define random dynamical systems (RDS) with appropriate sample spaces, much as ordinary differential equations define deterministic dynamical systems.

Theory of random dynamical systems allows to discuss the qualitative behavior of stochastic systems that are not only driven by a white noise, Markov processes and semimartingales, but also driven by non-Markov processes or by non-semimartingales (e.g., fractional Brownian motion). To analyze these more general noise cases, appropriate sample spaces and ergodic theory play an important role. In this article, we discuss canonical or natural sample spaces for SDEs with various noises.

We recall the definition of a random dynamical system (RDS) in the state space $H = \mathbb{R}^n$, with the underlying probability space $(\Omega, \mathcal{F}, \mathbb{P})$, and with time t varying in $\mathbb{T} = \mathbb{R} = (-\infty, \infty)$ or $\mathbb{T} = \mathbb{R}^+ = [0, \infty)$, as in Arnold.[1] The state space $\mathbb{R}^n$ is equipped with the Euclidean norm (or length) $|x| = \sqrt{x_1^2 + \cdots + x_n^2}$ and the usual scalar product $< x, y > = x_1 y_1 + \cdots + x_n y_n$.

Note that a deterministic dynamical system on the state space H is a mapping $\psi : \mathbb{T} \times H \to H$, $(t, x) \mapsto \psi(t, x)$, such that the flow property is satisfied:

$$\psi(0, x) = x, \quad \psi(t + s, x) = \psi(t, \psi(s, x)),$$

for all $t, s \in \mathbb{T}$ and $x \in H$.

For a random dynamical system, we need an extra ingredient, namely, a model for the noise. Moreover, the flow property has a twist (thus called cocycle property) due to the effect of noise.

Definition 3.1. (Random dynamical system)
A random dynamical system (RDS), denoted by φ, consists of two ingredients:
(i) **Model for the noise**: A driving flow on a probability space $(\Omega, \mathcal{F}, \mathbb{P})$, i.e., a flow $(\theta_t)_{t \in \mathbb{T}}$ on the sample space Ω, such that $\mathbb{P}$ is invariant, namely $\theta_t \mathbb{P} = \mathbb{P}$ for all $t \in \mathbb{T}$, and $(t, \omega) \mapsto \theta_t \omega$ is measurable from $\mathbb{T} \times \Omega$ to Ω.
(ii) **Model for the evolution**: A cocycle φ over θ, i.e. a measurable mapping $\varphi : \mathbb{T} \times \Omega \times H \to H$, $(t, \omega, x) \mapsto \varphi(t, \omega, x)$, such that the family $\varphi(t, \omega, \cdot) = \varphi(t, \omega) : H \to H$ of random mappings satisfies the cocycle property:

$$\varphi(0, \omega) = id_H, \varphi(t + s, \omega) = \varphi(t, \theta_s \omega) \circ \varphi(s, \omega) \text{ for all } t, s \in \mathbb{T}, \omega \in \Omega. \quad (1.1)$$

Here the driving flow θ_t describes stationary dynamics of noise in an appropriately chosen sample space (see below). The mathematical model for noises in engineering and science is usually a stationary generalized stochastic process.[21]

When $(t, x) \mapsto \varphi(t, \omega, x)$ is continuous for all $\omega \in \Omega$, we say that φ is a continuous RDS. Since the continuity in space x is quite common for RDS generated by stochastic differential equations, we usually do not specifically mention this spatial continuity. We often call φ a continuous-time or discrete-time RDS when it is continuous or discrete in time t.

It follows from[1] that $\varphi(t, \omega)$, $t \in \mathbb{R}$, is a homeomorphism of H and

$$\varphi(t, \omega)^{-1} = \varphi(-t, \theta_t \omega).$$

2. Dynamical systems driven by white noises

2.1. *Brownian Motion*

The physical phenomenon *Brownian motion*[a] is due to the incessant hitting of pollen by the much smaller molecules of the liquid. The hits occur a large number of times in any small time interval, independently of each other and the effect of a particular hit is small compared to the total effect.[43] The physical theory of this motion, set up by Albert Einstein in 1905, suggests that the motion is random, and has the following properties:

i) the motion is continuous;
ii) it has independent increments;
iii) the increments are stationary and Gaussian random variables.

Figure 3.1 shows a sample path of the Brownian motion.

Intuitively speaking, property i) says that the sample path of the Brownian motion is continuous. Property ii) means that the displacements of a pollen particle over disjoint time intervals are independent random variables. Property iii) is natural considering the *Central Limit Theorem*.

We now describe the Brownian motion in the mathematical language, i.e., introduce the first definition of the Brownian motion.[2,9,21]

Definition A: A stochastic process $\{B_t(\omega) : t \geq 0\}$ defined on a probability space $(\Omega, \mathcal{F}, P)$ is called a *Brownian motion* or a *Wiener process* if the following conditions hold:
1) $B_0(\omega) = 0$ a.s.;
2) the sample paths $t \to B_t(\omega)$ are a.s. continuous;
3) $B_t(\omega)$ has stationary independent increments;
4) the increments $B_t(\omega) - B_s(\omega)$ has the normal distribution with mean 0 and variance $t - s$, i.e. $B_t(\omega) - B_s(\omega) \sim N(0, t - s)$ for any $0 \leq s < t$.

---

[a]The phenomenon was first observed by Jan Ingenhouz in 1785, but was subsequently rediscovered by Brown in 1828, according to sources used by Eric Weisstein's *World of Physics*, which can be found on the Internet at `http://scienceworld.wolfram.com/physics/BrownianMotion.html`

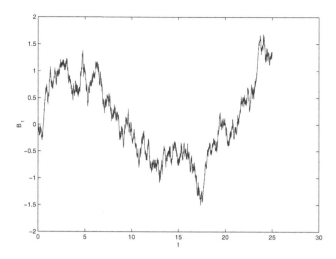

Fig. 3.1. *A sample path of Brownian motion* $B(t)$

Since the stochastic process $B_t(\omega)$ is a mapping from a probability space to a metric space and is governed by its law, i.e. the probability measure on the metric space induced by $B_t(\omega)$, we want to find it in order to have a better understanding of the Brownian motion. In fact, we can find the finite dimensional distribution of the Brownian motion using condition 3) and 4) in Definition A, and this gives another definition of the Brownian motion as we shall see. We first write down the second definition[9] and then prove it is equivalent to Definition A.

Definition B: A stochastic process $\{B_t(\omega) : t \geq 0\}$ defined on a probability space $(\Omega, \mathcal{F}, P)$ is called a *Brownian motion* or a *Wiener process* if the following conditions hold:

1') $B_0(\omega) = 0$ a.s.;

2') the sample paths $t \to B_t(\omega)$ are a.s. continuous;

3') for any finite sequence of times $0 = t_0 < t_1 < t_2 < \cdots < t_n$ and Borel sets $B_1, \cdots, B_n \subset \mathbb{R}$

$$P\{B_{t_1}(\omega) \in B_1, \cdots, B_{t_n}(\omega) \in B_n\}$$

$$= \int_{B_1} \cdots \int_{B_n} p(t_1, 0, x_1) p(t_2 - t_1, x_1, x_2) \cdots$$

$$\cdots p(t_n - t_{n-1}, x_{n-1}, x_n) dx_1 \cdots dx_n \tag{2.1}$$

where

$$p(t; x, y) = \frac{1}{\sqrt{2\pi t}} e^{-\frac{(x-y)^2}{2t}} \tag{2.2}$$

defined for any $x, y \in \mathbb{R}$ and $t > 0$ is called the *transition density*.

We can see the conditions 1'),2') in Definition B are completely the same as 1),2) in Definition A, respectively, the only thing we need to do in the proof of the equivalence of the two is to show 3),4) $\Leftrightarrow$ 3').

Proof. 3),4)$\Rightarrow$ 3')
Firstly, we show that Brownian motion B_t has Markov property, by proving that the conditional distribution of B_{t+s} given $\mathcal{F}_t$ is the same as that given B_t, in terms of moment generating function.[32] In fact,

$$
\begin{aligned}
E(e^{uB_{t+s}}|\mathcal{F}_t) &= E(e^{u[(B_{t+s}-B_t)+B_t]}|\mathcal{F}_t) \\
&= e^{uB_t} E(e^{u(B_{t+s}-B_t)}|\mathcal{F}_t) \\
&= e^{uB_t} E(e^{u(B_{t+s}-B_t)}) \\
&= e^{uB_t} e^{u^2 s/2} \\
&= e^{uB_t} E(e^{u(B_{t+s}-B_t)}|B_t) \\
&= E(e^{uB_{t+s}}|B_t).
\end{aligned}
$$

Secondly, we compute the joint distribution of the Brownian motion.

$$
\begin{aligned}
&P\{B(t_{k+1}) \le x_{k+1}, B(t_k) \le x_k\} \\
&= P\{B(t_k) \le x_k, [B(t_{k+1}) - B(t_k)] + B(t_k) \le x_{k+1}\} \\
&= P\{B(t_k) \le x_k, [B(t_{k+1}) - B(t_k)] \le x_{k+1} - B(t_k)\} \\
&= P\{B(t_{k+1}) - B(t_k) \le x_{k+1} - B(t_k)|B(t_k) \le x_k\} P\{B(t_k) \le x_k\}
\end{aligned}
$$

(Conditional probability)

$$
\begin{aligned}
&= \int_{-\infty}^{x_k} \int_{-\infty}^{x_{k+1}-x} \frac{1}{\sqrt{2\pi(t_{k+1}-t_k)}} e^{-\frac{y^2}{2(t_{k+1}-t_k)}} dy \frac{1}{\sqrt{2\pi t_k}} e^{-\frac{x^2}{2t_k}} dx \\
&= \int_{-\infty}^{x_k} \frac{1}{\sqrt{2\pi t_k}} e^{-\frac{x^2}{2t_k}} dx \int_{-\infty}^{x_{k+1}} \frac{1}{\sqrt{2\pi(t_{k+1}-t_k)}} e^{-\frac{(y-x)^2}{2(t_{k+1}-t_k)}} dy \\
&= \int_{-\infty}^{x_k} p(t_k; 0, x) dx \int_{-\infty}^{x_{k+1}} p(t_{k+1} - t_k; x, y) dy.
\end{aligned}
$$

Finally, we have the finite dimensional distribution.

$$P\{B(t_1) \le x_1, B(t_2) \le x_2, \cdots, B(t_n) \le x_n\}$$

$$= P\{B(t_1) \le x_1 | B(t_0) = 0\} P\{B(t_2) \le x_2 | B(t_1) \le x_1\}$$

$$P\{B(t_3) \le x_3 | B(t_1) \le x_1, B(t_2) \le x_2\} \cdots P\{B(t_n) \le x_n | B(t_1) \le x_1 B(t_{n-1}) \le x_{n-1}\}$$

$$= P\{B(t_1) \le x_1 | B(t_0) = 0\} P\{B(t_2) \le x_2 | B(t_1) \le x_1\}$$

$$P\{B(t_3) \le x_3 | B(t_2) \le x_2\} \cdots P\{B(t_n) \le x_n | B(t_{n-1}) \le x_{n-1}\} \text{ (Markov property)}$$

$$= \frac{P\{B(t_1) \le x_1\} P\{B(t_2) \le x_2, B(t_1) \le x_1\} \cdots P\{B(t_n) \le x_n, B(t_{n-1}) \le x_{n-1}\}}{P\{B(t_1) \le x_1\} P\{B(t_2) \le x_2\} \cdots P\{B(t_{n-1}) \le x_{n-1}\}}$$

$$= \int_{-\infty}^{x_1} p(t_1; 0, y_1) dy_1 \int_{-\infty}^{x_2} p(t_2 - t_1; y_1, y_2) dy_2 \cdots \int_{-\infty}^{x_n} p(t_n - t_{n-1}; y_{n-1}, y_n) dy_n$$

which is obviously the same as 3').

Conversely, we need to show 3')$\Rightarrow$ 3),4)

However, this part of work is completely done in[9](see page 153-155).   □

2.2. Wiener Measure

In Section 1, we treat the Brownian motion as a stochastic process, i.e., a collection of time-parameterized random variables. Since a stochastic process ξ is governed by its law, i.e., the probability measure $\mu := P\xi^{-1}$ on the space it maps to, one may ask the question why the Brownian motion can be determined only by its finite dimensional distributions although we have proved the so-defined Brownian motion (Definition B) coincides with the definition describing the phenomenon (Definition A). Motivated by this question, we shall treat the Brownian motion as a "random variable", which we call *random function*,[6] and this point of view introduces the *Wiener measure*, the probability measure defined on an appropriate space which gives us the Brownian motion. We will sketch the ideas showing the existence and uniqueness of this special probability measure, starting from three different spaces, i.e, $\mathbb{R}[0, \infty), C[0, \infty), C_0[0, \infty)$, the spaces of arbitrary functions, continuous functions, and continuous functions passing zero at time 0, defined on $[0, \infty)$, respectively.

First approach This way of showing the existence and uniqueness of the Wiener measure is the one most often used to construct a Markov Process and is rather technical. The main idea is the following: First define a set function on the algebra generated by the cylinder sets in $\mathbb{R}[0, \infty)$ according to the finite-dimensional distribution of Brownian motion. Note that it is a set function rather than a probability measure since it is just finitely additive (not countably additive); then by the celebrated *Kolmogorov extension theorem*,[21] this set function is uniquely extended to a probability measure on the $\sigma-$algebra generated by the algebra mentioned above.

More precisely, we now give the definitions and theorems.

- $\mathbb{R}[0, \infty)$ denotes the set of all real-valued functions on $[0, \infty)$.

- **Cylinder set**[21] A subset of $\mathbb{R}[0, \infty)$ of the form

$$A = \{\omega \in \mathbb{R}[0, \infty) : (\omega(t_1), \ldots, \omega(t_n)) \in B_n\} \qquad (2.3)$$

where B_n is a Borel subset of $\mathbb{R}^n$ is called a *cylinder set*.
Fixing $t_1, \ldots t_n$ but varying B_n over the entire Borel subsets of $\mathbb{R}^n$, the class of such all cylinder sets forms a $\sigma-$algebra $\mathfrak{B}^{(t_1, \ldots, t_n)}$.

- **Set function** A set function $\Phi_{t_1, \ldots, t_n}$ is defined on the measurable space $(\mathbb{R}[0, \infty), \mathfrak{B}^{(t_1, \ldots, t_n)})$, given by

$$\Phi_{t_1, \ldots, t_n}(A) := P\{(B_{t_1}, \ldots, B_{t_n}) \in B_n\}$$

By varying the choice of the finite time points $\{t_1, \ldots, t_n\} \subset [0, \infty)$, we get a class of such set functions $\Phi := \{\Phi_{t_1, \ldots, t_n}\}$, and this class is independent of the cylindrical expression of the functions in $\mathbb{R}[0, \infty)$, i.e. if the cylinder set A of (2.3) has another expression, say

$$A = \{\omega \in \mathbb{R}[0, \infty) : (\omega(s_1), \ldots, \omega(s_m)) \in B_m\},$$

then

$$\Phi_{t_1, \ldots, t_n}(A) = \Phi_{s_1, \ldots, s_m}(A).$$

- **Algebra and σ-algebra**
· $\mathfrak{U}[0, \infty)$ denotes the algebra of subsets of $\mathbb{R}[0, \infty)$ consisting all cylinder sets
· $\mathfrak{B}[0, \infty)$ denotes the smallest $\sigma-$algebra containing $\mathfrak{U}[0, \infty)$
· Φ is a finitely additive measure on $(\mathbb{R}[0, \infty), \mathfrak{U}[0, \infty))$ such that the restriction of Φ to $\mathfrak{B}^{(t_1, \ldots, t_n)}$ coincides with $\Phi_{t_1, \ldots, t_n}$

Theorem 2.1 (Kolmogorov extension theorem). *The set function* Φ *on* $(\mathbb{R}[0, \infty), \mathfrak{U}[0, \infty))$ *is uniquely extendible to a probability measure* $\tilde{\Phi}$ *on* $(\mathbb{R}[0, \infty), \mathfrak{B}[0, \infty))$.

It has been proved[21,31] that there exist a unique probability measure P on $(\mathbb{R}[0, \infty), \mathfrak{B}[0, \infty))$, under which the coordinate mapping process

$$B_t(\omega) := \omega(t); \ \omega \in \mathbb{R}[0, \infty), t \geq 0$$

satisfies condition 3) and 4) of Definition A in section 2.1, so it does not introduce a "Brownian motion" as we defined. Fortunately, by another famous theorem of Kolmogorov,[41] the continuity problem has been solved.

Theorem 2.2 (Kolmogorov continuity theorem). *Suppose that the process* $X = X\{t\}_{t \geq 0}$ *satisfies the following condition: for all* $T > 0$ *there exist positive constants* α, β, D *such that*

$$E[|X_t - X_s|^\alpha] \leq D \cdot |t - s|^{1+\beta}; \ 0 \leq s, t \leq T$$

then there exists a continuous modification *of* X.

Modification[41] Suppose that $\{X_t\}, \{Y_t\}$ are stochastic processes on $(\Omega, \mathcal{F}, P)$, then we say that $\{X_t\}$ is a *modification* of $\{Y_t\}$ if

$$P(\{\omega; X_t(\omega) = Y_t(\omega)\}) = 1 \ \forall t$$

Note that if X_t is a modification of Y_t, then they have the same finite-dimensional distributions.

Now there is one problem needs to be solved: Why is $B_0(\omega) = \omega(0) = 0$ a.s.?

Definition[6] *Wiener measure*, denoted by μ_W, is a probability measure on a measurable space $(C, \mathcal{C})$ having the following two properties:
i) each $\omega(t) \in C$ is normally distributed under μ_w with mean 0 and variance t, i.e,

$$\mu_w\{\omega(t) \leq x\} = \frac{1}{\sqrt{2\pi t}} \int_{-\infty}^{x} e^{-\frac{u^2}{2t}} du, \ x \in \mathbb{R}.$$

For $t = 0$ this is interpreted to mean that $\mu_w\{\omega(0) = 0\} = 1$.
ii) the stochastic process $\{\omega(t) : t \geq 0\}$ has independent increments under μ_w, i.e., $\forall 0 \leq t_0 \leq t_1 \leq \cdots \leq t_n$, $\omega(t_1) - \omega(t_0), \omega(t_2) - \omega(t_1), \cdots \omega(t_n) - \omega(t_{n-1})$ are independent under μ_w.

Remark It may not seem so obvious that this approach in fact proves the existence of the Wiener measure. Also, we cannot obtain the Wiener measure simply by assigning measure one to $C[0, \infty)$; see Karatzas and Shreve.[31] One might hope to construct the measure directly on $C[0, \infty)$. Indeed, this is the main idea of the second approach we are going to present.

Second approach There are some advantages if we start from $C[0, \infty)$ since we can make it *Polish* (complete and separable) by assigning an appropriate *metric*. It has been proved that the Borel σ−algebra generated by the open sets in $C[0, \infty)$ is equal to the σ−algebra generated by all the cylinder sets.[31] Thus there is a totally different way of constructing Wiener measure.

Metric on $C[0, \infty)$

$$\rho(\omega_a, \omega_b) := \sum_{n=1}^{\infty} \frac{1}{2^n} \max_{0 \leq t \leq n} (|\omega_a(t) - \omega_b(t)| \wedge 1) \tag{2.4}$$

This metric introduces the topology of uniform convergence on compact intervals. For convenience, we denote $C[0, \infty)$ by C and its Borel σ-algebra by $\mathcal{C}$.

The main idea is to construct a sequence of probability measure $\{P_n\}$ on $(C, \mathcal{C})$ such that $P_n \Rightarrow \mu_w$. Since (2.1) gives the f.d.d of the Wiener measure μ_w, we must let the f.d.d of $\{P_n\}$ weakly converges to those of μ_w. Although weak convergence in C need not follow the weak convergence of the f.d.d alone in general, it does if we add the condition that $\{P_n\}$ is tight.[6,31] In fact, if a metric space is separable and complete, then each probability measure on the measurable space is tight (see Theorem 1.3 in[6]). It can be shown that equipped with the metric defined by (2.4), the canonical space is separable and complete.[6] For detail, see Billingsley.[6]

Remark What is still not natural is that $B_0(\omega) = \omega(0) = 0$ a.s. So next we will talk about constructing Wiener measure on $C_0[0, \infty)$. It appears that the above two different approaches may be adapted to this case.

We can define the Brownian motion B_t for $t \in \mathbb{R}$ as follows: Taking two independent Brownian motions $\hat{B}_t$ and $\tilde{B}_t$, we define

$$B_t = \begin{cases} \hat{B}_t, & \text{if } t \geq 0; \\ \tilde{B}_{-t}, & \text{if } t < 0. \end{cases} \tag{2.5}$$

We can work on the space $C_0(\mathbb{R}, \mathbb{R}^n)$ for two-sided Brownian motion. The Wiener measure defined above should be similar defined in this space.[1]

2.3. *Canonical sample space*

We consider a SDE

$$dX_t = b(X_t)dt + \sigma(X_t)dB(t). \tag{2.6}$$

The canonical sample space is $\Omega := C_0(\mathbb{R}, \mathbb{R}^n)$, space of continuous functions that are zero at time zero, equipped with the compact open topology, the Borel σ-field $\mathcal{F} := \mathcal{B}(C_0(\mathbb{R}^+, \mathbb{R}^n))$, and Wiener (probability) measure μ_W.

We introduce a driving flow θ_t on this canonical sample space Ω is given by the Wiener shifts

$$\theta_t \omega(\cdot) = \omega(\cdot + t) - \omega(t), \quad t \in \mathbb{R}, \qquad \omega \in \Omega = C_0(\mathbb{R}, \mathbb{R}^d).$$

In this case the measure μ_W is invariant,[1] i.e.,

$$\mu_W(\theta_t^{-1}(A)) = \mu_W(A)$$

for all $A \in \mathcal{F}$. Moreover, this invariant measure is actually *ergodic* with respect to the flow θ.[8]

3. Dynamical systems driven by colored noises

Colored noise, or noise with non-zero correlation ('memory') in time, are common in the physical, biological and engineering sciences.[19] A good candidate for modeling colored noise is the fractional Brownian motion.

3.1. *Fractional Brownian motion*

A fractional Brownian motion (fBM) $B^H(t)$, $t \in \mathbb{R}$, with $H \in (0,1)$ the Hurst parameter, is still a Gaussian process. But it is characterized by the stationarity of its increments and a memory property. The increments of the fractional Brownian motion are not independent, except in the standard Brownian motion case ($H = \frac{1}{2}$). Thus it is not a Markov process except when $H = \frac{1}{2}$. Specifically, $B^H(0) = 0$ a.s., mean $\mathbb{E}B^H(t) = 0$, covariance $\mathbb{E}[B^H(t)B^H(s)] = \frac{1}{2}(|t|^{2H} + |s|^{2H} - |t-s|^{2H})$, and variance $Var\left[B^H(t) - B^H(s)\right] = |t-s|^{2H}$. It also exhibits power scaling and path regularity properties with Hölder parameter H, which are very distinct from Brownian motion. The standard Brownian motion is a special fBM with $H = 1/2$. Figure 3.2 is a sample path of the fractional Brownian motion with $H = 0.25$.

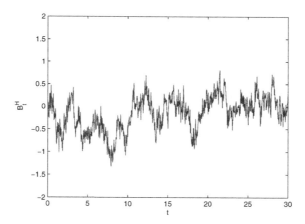

Fig. 3.2.   *A sample path of fractional Brownian motion $B^H(t)$, with $H = 0.25$*

Note that fBM $B^H(t)$ is non-Markov and non-semimartingale. Thus the usual stochastic integration[45] is not applicable, and other integration concepts are needed.[39,40]

3.2. *Canonical sample space*

The stochastic calculus involving fBM is currently being developed; see e.g.[40,55] and references therein. This will lead to more advances in the study of SDEs driven by colored fBM noise:

$$dX_t = b(X_t)dt + \sigma(X_t)dB^H(t). \tag{3.1}$$

Since the fBM $B^H(t)$ is not Markov, the solution process X_t is not Markov either. Thus the usual techniques from Markov processes will not be applicable to the study of SDEs driven by fBms. However, the random dynamical systems approach, as described in §1 above, looks promising.[18,38] The theory of RDS, developed by Arnold and coworkers,[1] describes the qualitative behavior of systems of stochastic differential equations in terms of stability, Lyapunov exponents, invariant manifolds, and attractors.

As in §1 above, the canonical sample space is $\Omega := C_0(\mathbb{R}, \mathbb{R}^n)$, the set of continuous functions that is zero at zero, but the probability measure μ_{fBM} is generated by B_t^H under the compact-open topology as defined in Section 2. The Borel $\sigma-$field is $\mathcal{F} := \mathcal{B}(C_0(\mathbb{R}, \mathbb{R}^n))$.

We can introduce a flow θ_t on this canonical sample space Ω defined by the shifts

$$\theta_t \omega(\cdot) = \omega(\cdot + t) - \omega(t), \quad t \in \mathbb{R}, \qquad \omega \in \Omega = C_0(\mathbb{R}, \mathbb{R}^d).$$

In this case the measure μ_{fBM} is invariant, i.e.

$$\mu_{fBM}(\theta_t^{-1}(A)) = \mu_{fBM}(A)$$

for all $A \in \mathcal{F}$.

4. Dynamical systems driven by non-Gaussian noises

In the last two sections, we considered dynamical systems driven by Gaussian noises (white or colored), in terms of Brownian motion or fractional Brownian motion. In this section, we discuss differential equation driven by non-Gaussian Lévy noises.

4.1. *Lévy Motions*

Gaussian processes like Brownian motion have been widely used to model fluctuations in engineering and science. For a particle in Brownian motion, its sample paths are continuous in time almost surely (i.e., no jumps), its mean square displacement increases linearly in time (i.e., normal diffusion), and its probability density function decays exponentially in space (i.e., light tail or exponential relaxation).[41] But some complex phenomena involve non-Gaussian fluctuations, with peculiar properties such as anomalous diffusion (mean square displacement is a nonlinear power law of time)[7] and heavy tail (non-exponential relaxation).[58] For instance, it has been argued that diffusion in a case of geophysical turbulence[53] is anomalous. Loosely speaking, the diffusion process consists of a series of "pauses", when the particle is trapped by a coherent structure, and "flights" or "jumps" or other extreme events, when the particle moves in a jet flow. Moreover, anomalous electrical transport properties have been observed in some amorphous materials such as insulators, semiconductors and polymers, where transient current is asymptotically a power law function of time.[20,50] Finally, some paleoclimatic data[14] indicates heavy tail distributions and some DNA data[53] shows long range power law decay for spatial correlation.

Lévy motions are thought to be appropriate models for non-Gaussian processes with jumps.[49] Let us recall that a Lévy motion $L(t)$, or L_t, is a non-Gaussian process with independent and stationary increments, i.e., increments $\Delta L(t, \Delta t) = L(t + \Delta t) - L(t)$ are stationary (therefore ΔL has no statistical dependence on t) and independent for any non overlapping time lags Δt. Moreover, its sample paths are only continuous in probability, namely, $\mathbb{P}(|L(t) - L(t_0)| \geq \delta) \to 0$ as $t \to t_0$ for any positive δ. With a suitable modification,[4] these paths may be taken as càdlàg, i.e., paths are continuous on the right and have limits on the left. This continuity is weaker than the usual continuity in time.

This generalizes the Brownian motion $B(t)$ or B_t, as $B(t)$ satisfies all these three conditions. But *Additionally*, (i) Almost every sample path of the Brownian motion is continuous in time in the usual sense and (ii) Brownian motion's increments are Gaussian distributed.

Dynamical systems driven by non-Gaussian Lévy noises have attracted much attention recently.[4,29,51] Under certain conditions, the SDEs driven by Lévy motion generate stochastic flows,[4,35] and also generate random dynamical systems (or cocycles) in the sense of Arnold.[1] Recently, exit time estimates have been investigated by Imkeller and Pavlyukevich[26,27] , and Yang and Duan[57] for SDEs driven by Lévy motion. This shows some qualitatively different dynamical behaviors between SDEs driven by Gaussian and non-Gaussian noises.

Lévy motions are named in honor of the French probabilist Paul Lévy, who first studied them in 1930s. From a mathematical point of view, there are so many reasons why they are so important,[4] such as:

• There are many important examples, such as Brownian motion, the Poisson process, stable processes, and subordinators.

• They are generalizations of random walks to continuous time.

• They are the simplest class of processes whose paths consist of continuous motion interspersed with jump discontinuities of random size appearing at random times.

Definition[4] A stochastic process $L = (L(t), t \geq 0)$ defined on a probability space $(\Omega, \mathcal{F}, P)$ is a Lévy motion if:

(**L1**) $L(0) = 0$ a.s.;

(**L2**) L has independent and stationary increments;

(**L3**) L is *stochastically continuous*, i.e. for all $\epsilon > 0$ and for all s

$$\lim_{t \to s} P(|X(t) - X(s)| > \epsilon) = 0$$

With a suitable modification,[4] these paths may be taken as càdlàg, i.e., paths are continuous on the right and have limits on the left. This continuity is weaker than the usual continuity in time.

Figure 3.3 is a sample path for a Lévy motion.

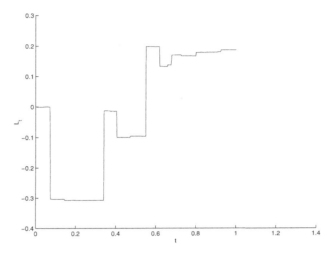

Fig. 3.3. A sample path for a Lévy motion

One way to understand the structure of the Lévy motions is to employ Fourier analysis. It may be shown that each $L(t)$ is *infinitely divisible*. The infinitely divisible random variables are characterized completely through their characteristic functions by a beautiful formula, established by Paul Lévy and A. Ya. Khintchine in the 1930s. The related definitions and theorems are as follows.

Definition[4] The *characteristic function* of a stochastic process $X(t)$ taking values in $\mathbb{R}^d$ is the mapping $\Phi_t : \mathbb{R}^d \to \mathbb{C}$ defined by

$$\Phi_t(u) = \mathbb{E}(e^{iu \cdot X(t)}) := \int_{\mathbb{R}^d} e^{iu \cdot y} p_t(dy)$$

where p_t is the distribution of $X(t)$.

Definition[4] $X(t)$ is *infinitely divisible* if for each $n \in \mathbb{N}$, there exists a probability measure $p_{t,n}$ on $\mathbb{R}^d$ with characteristic function $\Phi_{t,n}$ such that $\Phi_t(u) = (\Phi_{t,n}(u))^n$, for each $u \in \mathbb{R}^d$.

Theorem 4.1 (The Lévy-Khintchine Formula[4]). *If $L = (L(t), t \geq 0)$ is a Lévy motion, then*

$$\Phi_t(u) = e^{t\eta(u)} \qquad \text{for each } t \geq 0, u \in \mathbb{R}^d$$

where

$$\eta(u) = ib \cdot u - \frac{1}{2}u \cdot au + \int_{\mathbb{R}^d - \{0\}} [e^{iu \cdot y} - 1 - iu \cdot y \, \mathbf{I}_{|y|<1}(y)]\nu(dy) \qquad (4.1)$$

for some $b \in \mathbb{R}^d$, a non-negative definite symmetric $d \times d$ matrix a and a Borel measure ν, called Lévy jump measure, on $\mathbb{R}^d - \{0\}$ for which $\int_{\mathbb{R}^d - \{0\}} (|y|^2 \wedge 1)\nu(dy) < \infty$. Here $\mathbf{I}_S$ is the indicator function of the set S.

Conversely, given a mapping of the form (4.1), we can always construct a Lévy motion for which $\Phi_t(u) = e^{t\eta(u)}$ and call it a Lévy motion with the characteristics or generating triple (a, b, ν).

Remark: The so-defined Borel measure ν in (4.1) is called the *Lévy jump measure*, which should not be confused with the *probability measure* induced by the Lévy motion, to be defined below.

Each term in the Lévy-Khintchine formula has a probabilistic meaning. Every Lévy motion is obtained as a sum of independent processes with three types of characteristics $(0, b, 0)$, $(a, 0, 0)$ and $(0, 0, \nu)$. Thus, the Lévy measure accounts for the jumps of L and the knowledge of ν permits to give a probabilistic construction of L; see.[46]

4.2. *Canonical sample space*

Consider a SDE driven by non-Gaussian Levy noise

$$dX_t = b(X_t)dt + \sigma(X_t)dL(t). \qquad (4.2)$$

The canonical space has to be enlarged to include all the *cadlag* functions, i.e. functions that are right-continuous and have left limits, defined on $\mathbb{R}$ and taking values in $\mathbb{R}^d$. This space is denoted as $D(\mathbb{R}, \mathbb{R}^n)$.

We adopt the same point of view as in Section 2 that a stochastic process is also a *random variable*, i.e.

$$L_t(\omega) : \omega \to D(\mathbb{R}, \mathbb{R}^n) \qquad \omega \to L(t), t \in \mathbb{R}.$$

Remark 3.1. Here the compact-open metric defined in (2.4) cannot make $D(\mathbb{R}^+, \mathbb{R}^n)$ separable.[45] For example, if $f_\alpha(t) = 1_{[\alpha,\infty)}(t), f_\beta(t) = 1_{[\beta,\infty)}(t)$, then $\rho(f_\alpha, f_\beta) = 1/2$ for all α, β with $0 \le \alpha < \beta \le 1$, and since there are uncountably many such α, β, the space is not separable.

However, it can be made complete and separable when endowed with the *Skorohod metric*.[6] With this special metric, we call $D(\mathbb{R}, \mathbb{R}^d)$ a Skorohod space. The Skorohod metric on $D(\mathbb{R}, \mathbb{R}^d)$ is defined as

$$d(x, y) := \sum_{m=1}^{\infty} \frac{1}{2^m}(1 \wedge d_m^\circ(x^m, y^m)) \qquad for\ all\ x, y \in D$$

where $x^m(t) := g_m(t)x(t)$, $y^m(t) := g_m(t)y(t)$ with

$$g_m(t) := \begin{cases} 1, & \text{if } |t| \leq m - 1 \\ m - t, & \text{if } m - 1 \leq |t| \leq m, \\ 0, & \text{if } |t| \geq m \end{cases}$$

and

$$d_m^{\circ}(x,y) := \inf_{\lambda \in \Lambda} \left\{ \sup_{-m \leq s < t \leq m} \left| \log \frac{\lambda(t) - \lambda(s)}{t - s} \right| \vee \sup_{-m \leq t \leq m} |x(t) - y(\lambda(t))| \right\},$$

where Λ denotes the set of strictly increasing, continuous functions from $\mathbb{R}$ to itself.

The Borel σ-field under this topology is denoted as $\mathcal{S}$. For studying the *weak convergence* and *tightness* in D, the same approach adopted in C can be applied except that the fact the natural projections are not continuous need to be noticed.[6]

Definition The probability measure, μ_L, in $(D(\mathbb{R}, \mathbb{R}^n), \mathcal{S})$ that makes every element in $D(\mathbb{R}, \mathbb{R}^n)$ a sample Lévy path is called the Lévy probability measure. Note that this measure is not to be confused with the Lévy jump measure ν mentioned above.

We can also introduce a flow $\theta = (\theta_t, t \in \mathbb{R})$ on this canonical sample space Ω by the shifts

$$(\theta_t \omega)(s) := \omega(t + s) - \omega(t). \tag{4.3}$$

The (Lévy) probability measure μ_L is invariant under this flow, i.e.

$$\mu_L(\theta_t^{-1}(A)) = \mu_L(A)$$

for all $A \in \mathcal{F}$; see Applebaum[4] (Page 325) and Liu *et al.*[37] This flow is an ergodic dynamical system[1] with respect to the above probability measure μ_L.

References

1. L. Arnold. *Random Dynamical Systems*. Springer, New York, 1998.
2. L. Arnold, *Stochastic Differential Equations*, John Wiley & Sons, New York, 1974.
3. L. Arnold and M. Scheutzow, Perfect cocycles through stochastic differential equations. *Prob. Theory Rel. Fields* **101** (1995), 65-88.
4. D. Applebaum, *Lévy Processes and Stochastic Calculus*. Cambridge University Press, Cambridge, UK, 2004.
5. R. B. Ash. *Probability and Measure Theory*. Second Edition, Academic Press, New York, 2000.
6. P. Billingsley. *Convergence of Probability Measure*. Wiley Series in probability and Statistics, Chicago, Second Edition, 1999.
7. J. P. Bouchaud and A. Georges, Anomalous diffusion in disordered media: Statistic mechanics, models and physical applications. *Phys. Repts.* **195** (1990), 127-293.
8. P. Boxler, *Stochastische Zentrumsmannigfaltigkeiten*. Ph.D. thesis, Institut Dynamische Systeme, Universitat Bremen, 1988.

9. Z. Brzezniak, and T. Zastawniak. *Basic Stochastic Processes*. Springer, 1998.

10. M. Capinski and E. Kopp. *Measure, Integral and Probability*. Springer-Verlag, New York, 1999.

11. P. L. Chow, *Stochastic Partial Differential Equations*. Chapman & Hall/CRC, New York, 2007.

12. G. Da Prato and J. Zabczyk, *Stochastic Equations in Infinite Dimensions*, Cambridge University Press, 1992.

13. L. Decreusefond and D. Nualart, Flow properties of differential equations driven by fractional Brownian motion. In *Stochastic differential equations: theory and applications*, 249–262, Interdiscip. Math. Sci., 2, World Sci. Publ., Hackensack, NJ, 2007.

14. P. D. Ditlevsen, Observation of α−stable noise induced millennial climate changes from an ice record. *Geophys. Res. Lett.* **26** (1999), 1441-1444.

15. A. Du and J. Duan, Invariant manifold reduction for stochastic dynamical systems. *Dynamical Systems and Applications* **16**(2007), 681-696.

16. R. M. Dudley, *Real Analysis and Probability*. Cambridge University Press, 2002.

17. T. Fujiwara, H. Kunita, Stochastic differential equations of jump type and Leévy flows in diffeomorphisms group,*J. Math. Kyoto Univ.* 25, 71-106, 1985.

18. M. J. Garrido-Atienza, B. Maslowski and B. Schmalfuss, Random attractors for ordinary differential equations driven by a fractional Brownian motion with Hurst parameter greater than 1/2. *Preprint,* 2008.

19. P. Hanggi and P. Jung, Colored Noise in Dynamical Systems. *Advances in Chem. Phys.*, **89**(1995),239-326.

20. M. P. Herrchen, *Stochastic Modeling of Dispersive Diffusion by Non-Gaussian Noise*. Doctoral Thesis, Swiss Federal Inst. of Tech., Zurich, 2001.

21. T. Hida. *Brownian Motion*. Springer, New York, 1974.

22. Z. Huang and J. Yan, *Introduction to Infinite Dimensional Stochastic Analysis*. Science Press/Kluwer Academic Pub., Beijing/New York, 1997.

23. P. Imkeller and C. Lederer, On the cohomology of flows of stochastic and random differential equations. *Probab. Theory Related Fields* **120** (2001), 209–235.

24. N. Ikeda and S. Watanabe. *Stochastic Differential Equations and Diffusion Processes*. North-Holland, New York, 1981.

25. P. Imkeller and A. Monahan (eds.), *Conceptual Stochastic Climate Models*. Special Issue:*Stochastics and Dynamics*, **2**(2002),no.3.

26. P. Imkeller and I. Pavlyukevich, First exit time of SDEs driven by stable Lévy processes. *Stoch. Proc. Appl.* **116** (2006), 611-642.

27. P. Imkeller, I. Pavlyukevich and T. Wetzel, First exit times for Lévy-driven diffusions with exponentially light jumps. arXiv:0711.0982.

28. J. Jacod and A. N. Shiryaev, *Limit Theorems for Stochastic Processes*. Springer, New York, 1987.

29. A. Janicki and A. Weron, *Simulation and Chaotic Behavior of α−Stable Stochastic Processes*, Marcel Dekker, Inc., 1994.

30. O. Kallenberg, *Foundations of Modern Probability*. Second Edition, Applied probability Trust, 2002.

31. I. Karatzs, and Steven E. Shreve. *Brownian Motion and Stochastic Calculus*. Springer, New York, Second Edition, 1991.

32. F. Klebaner. *Introduction to stochastic Calculus with Application*. Imperial College Press, Second Edition, 2005.

33. P. E. Kloeden and E. Platen, *Numerical solution of stochastic differential equations*, Springer-Verlag, 1992; second corrected printing 1995.

34. H. Kunita. *Stochastic flows and stochastic differential equations.* Cambridge University Press, 1990.

35. H. Kunita, Stochastic differential equations based on Lvy processes and stochastic flows of diffeomorphisms. *Real and stochastic analysis* (Eds. M. M. Rao), 305–373, Birkhuser, Boston, MA, 2004.

36. H. Kunita, Stochastic differential equations with jumps and stochastic flows of diffeomorphisms. In *Itô's stochastic calculus and probability theory* (Eds. N. Ikeda and K. It), Springer, Tokyo, 197-211, 1996.

37. X. Liu, J. Duan, J. Liu and P. E. Kloeden, Synchronization of dissipative dynamical systems driven by non-Gaussian Lévy noises. *Preprint*, 2009. arXiv:0901.2446v1 [math.DS]

38. B. Maslowski and B. Schmalfuss, Random dynamical systems and stationary solutions of differential equations driven by the fractional Brownian motion. *Stoch. Anal. Appl.* **22** (2004), 1577- 1607.

39. Y. S. Mishura, *Stochastic Calculus for Fractional Brownian Motion and Related Processes.* Springer, New York, 2008.

40. D. Nualart, Stochastic calculus with respect to the fractional Brownian motion and applications. *Contemporary Mathematics* **336**, 3-39, 2003.

41. B. Oksendal. *Stochastic Differential Equations.* Sixth Ed., Springer-Verlag, New York, 2003.

42. S. Peszat and J. Zabczyk, *Stochastic Partial Differential Equations with Lévy Processes*, Cambridge University Press, Cambridge, UK, 2007.

43. J. von Plato, *Creating Modern Probability.* Cambridge University Press, UK, 1994.

44. C. Prevot and M. Rockner, *A Concise Course on Stochastic Partial Differential Equations*, Lecture Notes in Mathematics, Vol. 1905. Springer, New York, 2007.

45. P. E. Protter, *Stochastic Integration and Differential Equations.* Springer, New York, Second Edition, 2005.

46. D. Revuz, and M. Yor, *Continuous Martingales and Brownian Motion.* Springer, New York, Third Edition, 2005.

47. B. L. Rozovskii, *Stochastic Evolution Equations.* Kluwer Academic Publishers, Boston, 1990.

48. G. Samorodnitsky and M. Grigoriu. *Stable Non-Gaussian Random Processes.* Chapman & Hall, New York, 1994.

49. K.-I. Sato, *Lévy Processes and Infinitely Divisible Distributions*, Cambridge University Press, Cambridge, 1999.

50. H. Scher, M. F. Shlesinger and J. T. Bendler, Time-scale invariance in transport and relaxzation. *Phys. Today* p.26-34, Jan. 1991.

51. D. Schertzer, M. Larcheveque, J. Duan, V. Yanovsky and S. Lovejoy, Fractional Fokker–Planck Equation for Nonlinear Stochastic Differential Equations Driven by Non-Gaussian Levy Stable Noises. *J. Math. Phys.*, **42** (2001), 200-212.

52. M. Scheutzow, On the perfection of crude cocycles. *Random and Computational Dynamics* **4** (1996), 235-255.

53. M. F. Shlesinger, G. M. Zaslavsky and U. Frisch, Lévy Flights and Related Topics in Physics (Lecture Notes in Physics, 450. Springer-Verlag, Berlin, 1995).

54. R. Situ. *Theory of Stochastic Differential Equations with Jumps and Applications* Springer, 2005.

55. C. A. Tudor and F. Viens, Statistical aspects of the fractional stochastic calculus. *Annals of Statistics*, Vol. **35** (3) (2007), 1183-1212.

56. E. Waymire and J. Duan (Eds.), *Probability and Partial Differential Equations in Modern Applied Mathematics.* Springer-Verlag, 2005.

57. Z. Yang, and J. Duan, *An intermediate regime for exit phenomena driven by non-Gaussian Lévy noises. Stochastics and Dynamics* **8** (2008), 583-591.
58. F. Yonezawa, Introduction to focused session on 'anomalous relaxation'. J. Non-Cryst. Solids, **198-200** (1996), 503-506.

Perspectives in Mathematical Sciences
Interdisciplinary Mathematical Sciences, Volume 9, 2009
pp. 71–91

Chapter 4

Epidemic Propagation Dynamics on Complex Networks

Xinchu Fu[*1], Zengrong Liu[2], Michael Small[3] and Chi Kong Tse[3]

[1] *Department of Mathematics, Shanghai University, Shanghai 200444, P. R. China*

[2] *Institute of Systems Biology, Shanghai University, Shanghai 200444, P. R. China*

[3] *Department of Electronic and Information Engineering, Hong Kong Polytechnic University, Hung Hom, Kowloon, Hong Kong*

Dedicated to Professor Youzhong Guo on the occasion of his 75th birthday

This paper provides a partial summary of our recent work on propagation dynamics of complex networks, mainly on constructing and studying network models of disease spreading and related propagation problems. Traditional compartmental models of disease spreading categorize individuals from a population based on their current pathology. These methods provide a population-based description that offers a smooth continuous and exponential response to the presence of an infectious agent. In many cases the available data is inconsistent with the standard models of disease spreading and can be more readily explained using a discrete agent-based model of spreading on complex networks. Moreover, models for diseases spreading are not just limited to SIS or SIR. For instance, for the spreading of AIDS/HIV, the susceptible individuals can be classified into different cases according to their immunity, and similarly, the infected individuals can be sorted into different classes according to their infectivity. In addition, some diseases may develop through several stages, or with mobility property, or with mutually exclusive feature (multi-strain epidemics). So in this paper, in order to better study the dynamical behavior of epidemics, we discuss different epidemic models on complex networks, and provide a mathematical analysis of the epidemic dynamics and spreading behavior, obtaining the epidemic threshold for each case. Some other related diffusion and propagation processes, such as information transmission dynamics, traffic flows, contact processes, etc., are also briefly discussed.

*Corresponding author. E-mail: xcfu@shu.edu.cn
2000 Mathematics Subject Classifications: 34C15, 90B10.
Key words and phrases: Propagation, epidemic, dynamics, complex network, scale-free.
This research was jointly supported by NSFC grants 10672146, 10672093, 10832006 and Shanghai Leading Academic Discipline Project, Project Number: S30104.

1. Introduction

Disease transmission has been extensively studied by the Markov chain and mean-field compartment models. While in the recent decade, great progress has been made by using new results from network science. When disease transmission[12] is modelled over networks,[14,16,18] it is usual to model the infectivity (that is, the rate of transmission between infected and susceptible nodes) by assuming that transmission is equally likely over all links. For an idealized model this is the natural way to consider infectivity. However, when the underlying complex network is scale-free, the situation becomes unrealistic in the extreme tail of the distribution. While it has frequently been observed that real human, social and disease transmission networks exhibit scale-free properties over several orders of magnitude, the tail of the distribution observed from data is always bounded. It is an open question whether these real networks are close to scale-free or only scale-free over a finite domain (note that any real network is of finite size so the degree is bounded).[20] In[25] for example, the observation of a scale-free transmission mechanism for avian influenza is tempered by the fact that the finite available data necessarily limits inference to a bounded distribution. Moreover, when considering transmission of a disease in a finite time period it is natural to suppose that there exists an upper bound on the infectivity of a highly connected individual. It is also quite reasonable to suppose that highly connected (and therefore highly visible) nodes in the network would be the focus of an immunization scheme (even for very limited control measures). Hence, in this paper we consider the case where the infectivity is a non-decreasing, but sub-linear, function of the node degree.

The standard network SIS compartment model (Susceptible-Infected-Susceptible) assumes that each infected node will contact every neighbor once within one time step,[3] that is, the infectivity is equal to the connectivity, or the node degree. In,[27] it is assumed that every individual has the equal infectivity A, in which, at every time step, each infected individual will generate A contacts, where A is a constant. Joo and Lebowitz[10] examined cases where the transmission of infection between nodes depends on their connectivity, and a saturation function $C(k)$ which reduces the infection transmission rate across an edge going from a node with high connectivity k was introduced.

Based on these results, in the present model, we take a more realistic approach. We assume the infectivity is piece-wise linear: when the degree k of a node is relatively small, its infectivity is proportional to k, e.g., αk; when k is big, say, surpasses a constant A/α, then its infectivity is, say, A. We further discuss this model with respect to the effects of various immunization schemes.

Our motivation for this study is the observation that transmission of SARS (Severe Acute Respiratory Syndrome), most notably in Hong Kong during 2003, exhibits characteristics typical of a small world or scale-free network.[21–24] During the SARS outbreak of 2003 several clusters of secondary infections were observed and traced back to a single primary infection. This can either be explained by assuming a highly infectious source or by assuming a highly connected source. The latter case leads naturally to a scale-free model of transmission, and the question of under which conditions a real disease transmitted on an apparently scale-free network will have a finite threshold. It has also recently been observed that the spatial-temporal distribution of avian influenza outbreaks naturally induces a scale-free network connectivity.[25] In this work, the available data exhibits a power law over three orders of magnitude, but nonetheless, the tail of the distribution is bounded because the data is finite.

Of course, the SIS model used here was chosen because it is relatively simple, and also widely applicable. It may also be related to influenza vaccination problems[25] and strategies for dealing with computer viruses[7] among others.

This paper provides a partial summary of our recent series works on propagation dynamics of complex networks, mainly including construction and study of complex network models of disease spreading and related propagation problems.[45,47,53,55,56,58] A more complete list of research papers reflecting our recent works is referred to.[21–25,39–59]

2. The epidemic threshold for SIS model with piecewise linear infectivity

Individuals can be classified into three states, S-susceptible, I-infected and R-recovered (removed). Here we first consider the SIS model.

Let $S_k(t)$ and $I_k(t)$ be the densities of susceptible and infected nodes with degree k at time t, then

$$S_k(t) + I_k(t) = 1,$$

and the mean-field equations for infected nodes with degree k can be written as

$$\frac{dI_k(t)}{dt} = \lambda k(1 - I_k(t))\Theta(k,t) - I_k(t) \tag{2.1}$$

here we take a unit recovery rate, λ is the infection rate, and according to,[4,15,17,18,26] $\Theta(k,t)$ can be written in general as

$$\Theta(k,t) = \sum_{k'} \frac{\varphi(k')P(k'|k)I_{k'}}{k'} \tag{2.2}$$

where $\varphi(k)$ denotes the infectivity of a node with degree k, and $P(k'|k)$ stands for the probability for a node with degree k pointing to a node with degree k'.

An epidemic threshold for (2.1) is the critical value λ_c of the infection rate λ, if λ is below λ_c, the disease will gradually die out, while if λ is above λ_c, the disease will spread on the network.

In,[4,15,17,18] $\varphi(k) = k$, and then the epidemic threshold $\lambda_c = 0$ for sufficiently large networks. If $\varphi(k) = \alpha k$, the threshold λ_c also vanishes. In,[26] $\varphi(k) = A$, where A is a constant, that means every node has the same infectivity, no matter its degree, small or large. In this case, $\lambda_c = \frac{1}{A} > 0$, a positive threshold.

For simplicity, we suppose that the connectivity of nodes is uncorrelated, then $P(k'|k) = k'P(k')/\langle k \rangle$, where $\langle k \rangle = \sum_k kP(k)$. Then (2.2) becomes

$$\Theta = \frac{1}{\langle k \rangle} \sum_{k'} \varphi(k')P(k')I_{k'} \tag{2.3}$$

where for scale-free node distribution $P(k) = C^{-1}k^{-2-\gamma}, 0 < \gamma \leq 1$, where $C \approx \zeta(2 + \gamma)$ is approximately (as some degrees may not appear in a real network) the Riemann's zeta function.[8,25] Note that $\Theta(k,t)$ represents the probability that any given link points to an infected node. For simplified uncorrelated cases, $\Theta(k,t) = \Theta(t)$ doesn't depend on k.

2.1. *Piecewise linear infectivity*

Rather than a piecewise constant infectivity used in,[10] we here take a more realistic piecewise linear infectivity,

$$\varphi(k) = \min(\alpha k, A) \tag{2.4}$$

where α and A are positive constants, $0 < \alpha \leq 1$.

By imposing steady state $\frac{dI_k(t)}{dt} = 0$, from (2.1) we have

$$I_k = \frac{\lambda k \Theta}{1 + \lambda k \Theta} \tag{2.5}$$

Substitute I_k in (2.3) by (2.5), we obtain a self-consistency equation as follows:

$$\Theta = \frac{\lambda \Theta}{\langle k \rangle} \sum_{k'} \frac{k'\varphi(k')P(k')}{1 + \lambda k'\Theta} \equiv f(\Theta) \tag{2.6}$$

Obviously, $\Theta \equiv 0$ is a solution of (2.6), i.e., $f(0) = 0$. Note that

$$f(1) < 1, f'(\Theta) > 0, f''(\Theta) < 0,$$

therefore, a nontrivial solution exists only if

$$\frac{df(\Theta)}{d\Theta}\Big|_{\Theta=0} > 1 \tag{2.7}$$

The value of λ yielding the inequality (2.7) defines the critical epidemic threshold λ_c:

$$\lambda_c = \frac{\langle k \rangle}{\langle k\varphi(k) \rangle} = \frac{\sum\limits_k kP(k)}{\sum\limits_k k\varphi(k)P(k)} \tag{2.8}$$

Approximating the sum in (2.8) on discrete k by continuous integration, and suppose the size of the network is sufficiently large, we can calculate λ_c as

$$\lambda_c = \frac{\int_m^{+\infty} k^{-1-\gamma}dk}{\int_m^{A/\alpha} \alpha k^{-\gamma}dk + \int_{A/\alpha}^{+\infty} Ak^{-1-\gamma}dk} = \begin{cases} \frac{\frac{1-\gamma}{\alpha m}}{(\frac{A}{\alpha m})^{1-\gamma}-\gamma}, & 0 < \gamma < 1, \\ \frac{\frac{1}{\alpha m}}{1+\log\frac{A}{\alpha m}}, & \gamma = 1, \end{cases} \tag{2.9}$$

where m is the minimum connectivity of the network, and $\alpha m < A$.

We remark that when $A \to +\infty$, from the above formula (2.9), $\lambda_c \to 0$, this is consistent with the fact that $\varphi(k)$ approaches to the linear infectivity $\varphi(k) = \alpha k$; and when $\alpha m \geq A$, we can calculate that $\lambda_c = 1/A$, this is consistent with $\varphi(k) = A$ for all k.

From (2.9), we have a positive epidemic threshold λ_c if

$$\alpha m < \frac{1}{\gamma^{\frac{1}{1-\gamma}}} A \ (0 < \gamma < 1) \ \text{ or } \ \alpha m < eA \ (\gamma = 1)$$

If $A \geq \gamma^{\frac{1}{1-\gamma}} m \ (0 < \gamma < 1)$ or $A \geq e^{-1} m \ (\gamma = 1)$, then λ_c is always positive.

2.2. *Piecewise smooth and nonlinear infectivity*

In some cases, the infectivity may take the following piecewise smooth function

$$\varphi_1(k) = \min(\alpha k^\beta, A), \ 0 \leq \beta \leq 1, \alpha > 0.$$

In this case, the epidemic threshold

$$\lambda_c' = \begin{cases} (\frac{A\beta}{\beta - \gamma}(\frac{\alpha m^\beta}{A})^{\frac{\gamma}{\beta}} - \frac{\alpha m^\beta}{\gamma(\beta - \gamma)})^{-1}, \ \beta \neq \gamma \\ (\frac{m\alpha}{\beta} \log \frac{A}{\alpha m^\beta} + \frac{m\alpha}{\beta})^{-1}, \quad \beta = \gamma \end{cases}$$

then we have positive λ_c' if $(\alpha m^\beta)^{\frac{\gamma}{\beta} - 1} > \frac{1}{\gamma(\beta - \gamma)} A^{\frac{\gamma}{\beta}}$ and $\beta > \gamma$ or $\alpha m^\beta < eA \ (\beta = \gamma)$

We can also discuss the epidemic threshold for a smooth nonlinear infectivity, e.g.,

$$\varphi_2(k) = \frac{ak^\beta}{1 + bk^\beta}, \ \ 0 \leq \beta \leq 1, \ a > 0, \ b \geq 0,$$

The details are discussed in.[47] We may also consider the effects of finite scale-free networks on the above discussions.[20,47]

3. Model with different immunities and infectivities

The SIS and SIR models cannot correctly interpret all kinds of diseases. For instance, for the spreading of AIDS/HIV, the susceptible individuals can be classified into different cases according to their immunity, and similarly, the infected individuals can be sorted into different classes according to their infectivity.

To better explore the mechanism of epidemic spreading on complex networks, we suppose that the S and I states can be subdivided into subclasses according to their different immunities, different infectivities and so on. That is, our models can describe $S_i IR$, $SI_i R$ and $SI_{i,1} I_{i,2}, \cdots, I_{i,n} R$, $i = 1, 2, \cdots, n$. In order to make the models more reasonable, we also consider the birth and death of individuals. By using the method as in,[13] we assume that all individuals are distributed on the network, and each node of the network is empty or occupied by at most one individual.

The numbers 0,1,2,3 denote that the node has no individual, a healthy (suscepti-ble) individual, an infected individual and a recovered individual respectively. Each node can change its state with a certain rate. An empty node can give birth to a healthy (susceptible) individual at the rate δ. The susceptible individual can be infected at a rate which is proportional to the number of infected individuals in the neighborhood or die at certain rate α. The infected individual can be cured at certain rate μ or die at certain rate β. If an individual dies, that node will become an empty node again.

3.1. *Multiple susceptible individuals*

We consider the susceptible individuals with several different cases according to their age or immunities. $S_{i,k}, i = 1, \cdots, n$ denote the density of the susceptible individuals with degree k and also belong to the i-th case, I_k and R_k denote the density of the infected individuals and the recovered individuals with degree k, respectively, then

$$
\begin{cases}
\frac{dS_{i,k}}{dt} = \delta_i(1 - \sum\limits_{i=1}^{n} S_{i,k} - I_k - R_k) - \lambda_i S_{i,k} k\Theta - \alpha_i S_{i,k} \\
\frac{dI_k}{dt} = k\Theta \sum\limits_{i=1}^{n} \lambda_i S_{i,k} - (\beta + \mu)I_k \qquad\qquad i = 1, \cdots, n \\
\frac{dR_k}{dt} = \mu I_k - \gamma R_k
\end{cases}
$$

Where $(1 - \sum\limits_{i=1}^{n} S_{i,k} - I_k - R_k)$ is the density of empty nodes which will give birth to nodes with degree k, and $\delta_i, \lambda_i, \alpha_i$ are the birth rates, infectivity rates, and the natural death rates for the i-th case susceptible individuals respectively, β, μ are the natural death rate and the rate from $I \to R$ for infected individuals, and γ is the natural death rate of recovered individuals. Θ takes the form for uncorrelated networks, as discussed above.

Similar to the analysis in Section 2, we have the following inequality

$$
\sum_{i=1}^{n} \frac{\lambda_i \delta_i}{\alpha_i} > \frac{\langle k \rangle (\beta + \mu)(1 + \sum\limits_{i=1}^{n} \frac{\delta_i}{\alpha_i})}{\langle k^2 \rangle}
$$

3.2. *Multiple infected individuals*

We suppose that the infected individuals are classified into several different cases according to their infectivity rates or natural death rates. Let $I_{i,k}, i = 1, \cdots, n$

denote the i-th infected individual with degree k, then

$$\begin{cases} \frac{dS_k}{dt} = \delta(1 - \sum_{i=1}^{n} I_{i,k} - S_k - R_k) - S_k k \sum_{i=1}^{n} \lambda_i \Theta_i - \alpha S_k \\ \frac{dI_{i,k}}{dt} = p_i S_k k \sum_{i=1}^{n} \lambda_i \Theta_i - (\beta_i + \mu_i) I_{i,k} \qquad i = 1, \cdots, n \\ \frac{dR_k}{dt} = \sum_{i=1}^{n} \mu_i I_{i,k} - \gamma R_k \end{cases}$$

Here the new infected individuals will come into the i-th infectivity individuals with probability p_i, so $\sum_{i=1}^{n} p_i = 1$. Other parameters are similar to those in Section 3.1, and

$$\Theta_i = \frac{\Sigma_k k p(k) I_{i,k}}{\langle k \rangle} \quad i = 1, \cdots, n$$

We have

$$\sum_{i=1}^{n} \frac{\lambda_i p_i}{\mu_i + \beta_i} > \frac{\langle k \rangle (\delta + \alpha)}{\langle k^2 \rangle \delta}$$

3.3. *Multiple-staged infected individuals*

As was discussed in,[56] each case of infected individuals can also develop in several stages. So we now discuss the multiple-staged infected individuals models. Let $I_{i,j}, i = 1, \cdots, n, j = 1, \cdots, m$ denote the i-th infected individual which is in the j-th stage.

In order to simplify the computation, we do not consider the natural death rate for $I_{i,j}, i = 1, \cdots, n, j = 1, \cdots, m$, but only suppose that they only go into R state with certain rates. Then the dynamics equations are:

$$\begin{cases} \frac{dS^{(k)}}{dt} = \delta(1 - \sum_{i=1}^{n} \sum_{j=1}^{m} I_{i,j}^{(k)} - S^{(k)} - R^{(k)}) - S^{(k)} k \sum_{i=1}^{n} \sum_{j=1}^{m} \lambda_{i,j} \Theta_{i,j} - \alpha S^{(k)} \\ \frac{dI_{i,1}^{(k)}}{dt} = p_i S^{(k)} k \sum_{i=1}^{n} \sum_{j=1}^{m} \lambda_{i,j} \Theta_{i,j} - \mu_{i,1} I_{i,1}^{(k)} \\ \frac{dI_{i,j}^{(k)}}{dt} = \mu_{i,j-1} I_{i,j-1}^{(k)} - \mu_{i,j} I_{i,j}^{(k)}, \quad i = 1, \cdots, n, \; j = 2, \cdots, m \\ \frac{dR^{(k)}}{dt} = \sum_{i=1}^{n} \mu_{i,m} I_{i,m}^{(k)} - \gamma R^{(k)} \end{cases}$$

Here the individuals' degree k is given as the superscripts to differentiate from the subscripts i, j. The infectivity rates for $I_{i,j}$ on susceptible individuals are $\lambda_{i,j}$, and $\mu_{i,j}$ are the rates of the transformation $I_{i,j} \to I_{i,j+1}, i = 1, \cdots, n, j = 1, \cdots, m-2$, and $\mu_{i,m}$ are the rates of the transformation $I_{i,m} \to R$. Here we suppose that each $I_{i,j}$ can infect susceptible individuals, and new infected individuals will come into the i-th infectivity individuals with probability p_i, so we also have $\sum_{i=1}^{n} p_i = 1$.

$\Theta_{i,j}$ are given by:

$$\Theta_{i,j} = \frac{\Sigma_k k p(k) I_{i,j}^{(k)}}{\langle k \rangle}, \quad i = 1, \cdots, n, j = 1, \cdots, m$$

Then the threshold for the multiple-staged infected model:

$$\sum_{i=1}^{n} \sum_{j=1}^{m} \frac{p_i}{\mu_{i,j}} \lambda_{i,j} > \frac{(\delta + \alpha)\langle k \rangle}{\delta \langle k^2 \rangle}$$

From the above three inequalities, we can obtain the relationship between thresholds of epidemic and the parameters, such as the degree distribution, birth rate, death rate, and so on. In particular, the thresholds for each case are zero when the size of network is sufficiently large, that is, $\langle k^2 \rangle = \Sigma_k k^2 p(k) \to \infty$.

4. SIS model with population mobility

Most previous research on epidemic spreading assumed that a node is an individual, as a result, the deeper structure of networks were neglected, such as the mobility of individuals between different cities was ignored. Most recently, Vittoria Colizza et al studied the behavior of two basic types of reaction-diffusion processes ($B \to A$ and $B + A \to 2B$),[28] they supposed that a node of the network can be occupied by any number of individuals and the individuals can diffuse along the link between nodes. The two basic reaction-diffusion processes can be used to model the spreading of epidemic diseases with SIS model.[28] In the epidemic terminology, a node can be viewed as a city, i.e, all people have the same degree k if they live in the same city (the node with degree k), and the diffusion of particles among different nodes can be considered as the movement of people among different cities. They supposed that the infection may happen inside a city, however, the infection may also happen in different cities by other media, e.g., for the Avian Influenza, in different regions poultry can be infected by migratory birds even though the poultry have no mobility.

We suppose that the infection can also happen in different cities, and study the effect of this kind of epidemic spreading on the epidemic threshold. This can be done by introducing a probability of spreading of the infection to the neighboring nodes without the need of diffusion of infected particles. In fact, as we will show, this mechanism is in part equivalent to the diffusion of the particles.

4.1. *Epidemic spreading without mobility of individuals*

In order to find out the effect of mobility of individuals, we first assume that mobility is zero, then

$$\begin{cases} \frac{dI_k(t)}{dt} = \alpha k S_k \Theta_i + \beta S_k I_k - \mu I_k \\ \frac{dS_k(t)}{dt} = -\alpha k' S_{k'} \Theta_i - \beta S_k I_k + \mu I_k \end{cases} \quad i = 1, 2$$

Here we should note that the total density $S_k + I_k$ is not changed because there is no mobility of individuals among different cities, so we can let $S_k + I_k = 1$. Then we have

$$\frac{\alpha}{\mu - \beta} \frac{\langle k^2 \rangle}{\langle k \rangle} > 1$$

this condition demonstrates that epidemic diseases will always become endemic for a heterogeneous network with sufficiently large size.

For the case Θ_1, we can get

$$\frac{\alpha}{\mu - \beta} > 1$$

this case suggests that the epidemic threshold is irrelevant to the topology of the network, which is similar to the results in.[29,30]

In the following subsections, we take into account the mobility of individuals in different cities, so the individuals' degrees may change, that is, the total density $S_k + I_k$ is not an invariant, but the average density $n = \Sigma_k P(k)(S_k + I_k)$ is.

4.2. *Spreading of epidemic diseases among different cities*

Similar to,[28] we denote the size of the network as V, and N_S and N_I are the numbers of susceptible and infective individuals respectively, so the total number of individuals in the network is $N = N_S + N_I$ and $n = N/V$ is the average density of people. Because the number of individuals on each node is a random non-negative integer, set a_i and b_i as the numbers of S and I stores on node i. In order to take into account the heterogeneous quality of networks we have to explicitly consider the presence of nodes with very different degree k. A convenient representation of the system is therefore provided by the following quantities:

$$S_k = (\Sigma_{i|k_i=k} a_i)/v_k, \qquad\qquad I_k = (\Sigma_{i|k_i=k} b_i)/v_k$$

where v_k is the number of nodes with degree k and the sums run over all nodes i having degree k_i equal to k.

Just as in,[28] we also assume that the mobility of people is unitary time rate 1 along one of the links departing from the node in which they are at a given time. This implies that at each time step an individual occupying a node with degree k will travel to another city with probability $1/k$.

Now the dynamics of epidemic spreading can be described as follows:

$$\begin{cases} \frac{dI_k(t)}{dt} = -I_k(t) + k\Sigma_{k'} P(k'|k)\frac{1}{k'}[(1-\mu)I_{k'}(t) + \alpha k' S_{k'}\Theta_i] \\ \frac{dS_k(t)}{dt} = -S_k(t) + k\Sigma_{k'} P(k'|k)\frac{1}{k'}[S_k(t) + \mu I_{k'}(t) - \alpha k' S_{k'}\Theta_i] \end{cases} \quad i = 1, 2$$

For the case of Θ_1, the threshold for the average density is

$$n_{c_1} = \frac{\mu \langle k \rangle^2}{\alpha \langle k^2 \rangle}$$

For the case of Θ_2, the threshold is

$$n_{c_2} = \frac{\mu \langle k \rangle^3}{\alpha \langle k^2 \rangle^2}$$

We conclude that the epidemic is always endemic for sufficiently large heterogeneous networks, moreover, the prevalence of epidemics with infection rate $\alpha k \Theta_2$ is greater than the infection rate $\alpha k \Theta_1$.

4.3. *Epidemic spreading within and between cities*

Now we assume that the epidemic disease not only occurs within individual cities but also between connected cities. And we also consider two types of epidemic spreadings inside the same cities. In the case of type 1, we consider that each a_i individuals may be infected by all the b_i individuals in the same cities, in this case, the epidemic rate is β when the spreading of the epidemic disease happen in the same cities. In the case of type 2, we consider that each individual has a finite number of contacts with others, in this case the epidemic rate has to be rescaled by the total number of individuals in city i, i.e., β / n_i is the epidemic rate in the same cities, where $n_i = a_i + b_i$ is the total number of individuals in the city i.

4.3.1. *The epidemic rate is β inside the same cities*

In this case, the number of infected individuals generated by the infection taking place in node of the degree class k is $\beta S_k I_k$. Let $T_k = S_k I_k$, we have

$$T = \Sigma_k P(k) T_k = \Sigma_k P(k) S_k I_k$$

Then the dynamics of epidemic spreading can be written as

$$\begin{cases} \frac{dI_k(t)}{dt} = -I_k(t) + k\Sigma_{k'} P(k'|k)\frac{1}{k'}[(1-\mu)I_{k'}(t) + \beta T_k + \alpha k' S_{k'} \Theta_i] \\ \frac{dS_k(t)}{dt} = -S_k(t) + k\Sigma_{k'} P(k'|k)\frac{1}{k'}[S_k(t) + \mu I_{k'}(t) - \beta T_k - \alpha k' S_{k'} \Theta_i] \end{cases} \quad i = 1, 2$$

For the case of Θ_1, the threshold for the prevalence of epidemic is

$$n_{c_3} = \frac{\mu \langle k \rangle^2}{(\alpha + \beta)\langle k^2 \rangle}$$

For the case of Θ_2, the threshold for the prevalence of epidemic disease is

$$n_{c_4} = \frac{\mu \langle k \rangle^3}{(\alpha \langle k^2 \rangle + \beta \langle k \rangle)\langle k^2 \rangle}$$

4.3.2. *The epidemic rate is β / n_i inside the same cities*

In this case, the number of infected individuals generated by the infection taking place in node of the degree class k is $\beta \frac{S_k I_k}{S_k + I_k}$, we also let $T_k = \frac{S_k I_k}{S_k + I_k}$.

We can obtain

$$T = \Sigma_k P(k) T_k = \Sigma_k P(k)\frac{S_k I_k}{S_k + I_k} = \frac{IS}{n}$$

For the Θ_1 case, the prevalence of epidemic disease takes place if $(\beta - \mu)\langle k \rangle^2 + \alpha n \langle k^2 \rangle > 0$, i.e.,

$$n_{c_5} = \begin{cases} 0 & \beta/\mu > 1 \\ \frac{(\mu - \beta)\langle k \rangle^2}{\alpha \langle k^2 \rangle} & \beta/\mu < 1 \end{cases}$$

For the Θ_2 case,

$$n_{c_6} = \begin{cases} 0 & \beta/\mu > 1 \\ \frac{(\mu - \beta)\langle k \rangle^3}{\alpha \langle k^2 \rangle^2} & \beta/\mu < 1 \end{cases}$$

From the above equalities, we can find that the epidemic always occurs whatever the size of networks, when $\beta/\mu > 1$.

5. Multi-strain epidemic models

There are cases of multi-strain epidemics in the real world which we wish to examine, i.e., different strains of the same pathogen transmitting on the same network. For example, the human immunodeficiency virus (HIV) (which can cause AIDS) has many genetic varieties, and can be divided into some strains, such as strain HIV-1 and strain HIV-2.[31] On the one hand, being the same virus, there are many similarities between HIV-1 and HIV-2, such as the modes of HIV-1 and HIV-2 transmission are the same - sexual contact, sharing needles etc. On the other hand, there exist many differences between HIV-1 and HIV-2, for example, HIV-2 seems to weaken the immune system more slowly than HIV-1.

Some multi-strains epidemic dynamics problems on fully mixed species have been generally investigated in.[32–35] During disease spreading processes, a kind of pathogen sometime generates many strains with different spreading features, hence researches on multi-strain epidemic dynamics possess practical significance.

For multi-strain epidemic models, the manner of interaction between two particles with different strain have many types, including co-infection, which means two strains can host in one particle, invasion, which means one strain can host the particle with the other strain, and perfect cross-immunity, which means that two strains are perfectly competing and nodes infected with one strain cannot be infected by the other strain. Previous results have shown that different interaction mechanisms have different effects, such as co-infection, and can induce complex dynamical behaviors, such as chaotic attractors.[36] Super-infection may generate so-called strain replacement phenomenon,[35] but perfect cross-immunity can not produce similar phenomenon even with perfect vaccination (the vaccine provides full protection against all strains[35]). Strain replacement shows that the phenomenon that one strain with smaller basic reproduction number can become prevalent in a long time.

In this section, two-strain epidemic models on complex networks with scale-free connectivity is discussed. We assume that the network has saturated infectivity. The two kinds of strains of the same pathogen can be denoted by strain I and strain J. It may be the case that the two strains have different spreading rates. Let the spreading rate of the strain I be λ_1, and that of the strain J be λ_2. Each individual is represented by a node of the network, and can be in three discrete states, either susceptible or infected by the strain I or J, which allows that infection mechanism to belong to the SIS type,[37] that is, susceptible nodes may be infected owing to contact with an infected node, and infected nodes also may recover into the susceptible state, the recovery rates are β_1 and β_2 for strain I and strain J respectively.

Let $i_k(t)$ and $j_k(t)$ represent the densities at time t of nodes in class with degree k infected by strain I and strain J respectively.

We will focus on two kinds of interaction mechanisms between two strains without co-infection, one is perfect cross-immunity, and the other is super-infection. Then the two-strain models can be written as follows:

Perfect cross-immunity mechanism:

$$
\begin{cases}
\frac{di_k(t)}{dt} = -\beta_1 i_k(t) + \lambda_1 k[1 - i_k(t) - j_k(t)]\Theta_1(t) \\
\frac{dj_k(t)}{dt} = -\beta_2 j_k(t) + \lambda_2 k[1 - i_k(t) - j_k(t)]\Theta_2(t)
\end{cases}
\tag{5.1}
$$

where the probability $0 \leqslant \Theta_1(t) \leqslant 1$ describes a link pointing to an individual infected by the strain I. According to,[45] it satisfies the equality

$$
\Theta_1(t) = \sum_{k'} \frac{P(k'|k)}{k'} \varphi(k') i_{k'}(t)
\tag{5.2}
$$

Similarly, the probability $0 \leqslant \Theta_2(t) \leqslant 1$ describes a link pointing to an individual infected by the strain J, which satisfies the equality

$$
\Theta_2(t) = \sum_{k'} \frac{P(k'|k)}{k'} \varphi(k') j_{k'}(t)
\tag{5.3}
$$

Those nodes with degree k have saturated infectivity, $\varphi(k)$, which satisfies the following three conditions:

(i) $\varphi(k) \leqslant k$; (ii) $\varphi(k)$ is monotonously increasing; (iii) $\lim_{k \to \infty} \varphi(k) = A > 0$.

According to the physical meaning of $\varphi(k)$, the above constraints are reasonable and recover some previous setting about the function. It can be shown that the piecewise linear infectivity introduced in[45] is a special case for the saturated infectivity defined above.

Super-infection mechanism:

For this mechanism, we assume that when nodes infected by the strain I contact nodes infected by strain J they may be reinfected by strain I. The transmission process is asymmetric, that is, strain J cannot reinfect the node infected by strain I. This process is referred to as super-infection. And the asymmetric transmission rate is denoted by δ.

Similar to the perfect cross-immunity mechanism, the two-strain model with super-infection can be described as follows:

$$\begin{cases} \frac{di_k(t)}{dt} = -\beta_1 i_k(t) + \lambda_1 k[1 - i_k(t) - j_k(t)]\Theta_1(t) + \delta k j_k(t)\Theta_1(t) \\ \frac{dj_k(t)}{dt} = -\beta_2 j_k(t) + \lambda_2 k[1 - i_k(t) - j_k(t)]\Theta_2(t) - \delta k j_k(t)\Theta_1(t) \end{cases} \quad (5.4)$$

Now we present detailed analysis on the above models.

5.1. *The two-strain epidemic model with perfect cross-immunity*

Assume the network is uncorrelated about node degree. Therefore

$$P(k'|k) = \frac{k'P(k')}{<k>},$$

so the model (5.1) can be transformed into

$$\begin{cases} \frac{di_k(t)}{dt} = -\beta_1 i_k(t) + \lambda_1 k[1 - i_k(t) - j_k(t)]\Theta_1(t) \\ \frac{dj_k(t)}{dt} = -\beta_2 j_k(t) + \lambda_2 k[1 - i_k(t) - j_k(t)]\Theta_2(t) \end{cases} \quad (5.5)$$

where

$$\Theta_1(t) = \frac{\sum_{k'} \varphi(k')P(k')i_{k'}(t)}{\langle k \rangle} \quad (5.6)$$

and

$$\Theta_2(t) = \frac{\sum_{k'} \varphi(k')P(k')j_{k'}(t)}{\langle k \rangle} \quad (5.7)$$

In the above model, we can see that the system (5.5) has four parameters, which can be reduced to three if using typical time-scale transformation. So this multi-parameters property invokes some different dynamical behaviors, which are different from the case of one strain epidemic. In the following analysis, we find that composed parameter $\sigma_i = \frac{\lambda_i}{\beta_i}, i = 1, 2$ are important, which is referred to as the effective spreading rates for strain I and strain J respectively. It is reasonable to assume that $\sigma_1 \neq \sigma_2$. But the recovery rates β_1, β_2 can not be all eliminated. Hence, the case $\sigma_1 = \sigma_2$ will also be discussed.

5.2. *The case $\sigma_1 \neq \sigma_2$*

5.2.1. *Basic Reproduction Numbers (BRNs)*

In order to obtain the existence of non-trivial equilibria, we define the following two parameters

$$R_1 = \frac{\sigma_1 \langle k\varphi(k) \rangle}{\langle k \rangle}, \qquad R_2 = \frac{\sigma_2 \langle k\varphi(k) \rangle}{\langle k \rangle} \quad (5.8)$$

They are the basic reproduction numbers for the strain I and the strain J respectively. The number R_1 gives the average value of secondary infectious cases produced by the infected individual with strain I during the entire infectious period in a purely susceptible population. The number R_2 has similar meaning. The two BRNs are related by the effective spreading rates. If $\sigma_1 = \sigma_2$, then $R_1 = R_2$. Otherwise, they are different.

Below we list some basic results:

A1. There is always a disease-free equilibrium $E_0 = (1, 0, 0)$;

A2. There is a strain one exclusive equilibrium $E_1 = (s_1^*, i^*, 0)$, if and only if $R_1 > 1$;

A3. There is a strain two exclusive equilibrium $E_2 = (s_2^*, 0, j^*)$, if and only if $R_2 > 1$.

In fact, by imposing steady state $\frac{di_k(t)}{dt} = 0$ and $\frac{dj_k(t)}{dt} = 0$, from (5.5) we have

$$i_k = \frac{\sigma_1 k \Theta_1}{1 + \sigma_1 k \Theta_1 + \sigma_2 k \Theta_2}, \qquad j_k = \frac{\sigma_2 k \Theta_2}{1 + \sigma_1 k \Theta_1 + \sigma_2 k \Theta_2}. \tag{5.9}$$

Substitute (5.9) into (5.6) and (5.7), we can get two self-consistent equations as follows:

$$\Theta_1 = \frac{\sigma_1}{\langle k \rangle} \sum_{k'} \frac{k' \varphi(k') P(k') \Theta_1}{1 + \sigma_1 k' \Theta_1 + \sigma_2 k' \Theta_2} \tag{5.10}$$

and

$$\Theta_2 = \frac{\sigma_2}{\langle k \rangle} \sum_{k'} \frac{k' \varphi(k') P(k') \Theta_2}{1 + \sigma_1 k' \Theta_1 + \sigma_2 k' \Theta_2} \tag{5.11}$$

Obviously, $(\Theta_1, \Theta_2) = (0, 0)$ is a trivial solution of equations (5.10) and (5.11). Since $\sigma_1 \neq \sigma_2$, it can be shown that the equations have no positive solutions. So we need only to consider the other two cases, $\Theta_1 = 0$ and $\Theta_2 = 0$. When $\Theta_2 = 0$, we can only focus on nontrivial solutions of (5.10). Note that (5.10) can be reduced to

$$1 = \frac{\sigma_1}{\langle k \rangle} \sum_{k'} \frac{k' \varphi(k') P(k')}{1 + \sigma_1 k' \Theta_1} \equiv f(\Theta_1). \tag{5.12}$$

Because

$$f(1) < 1$$

and

$$f'(\Theta_1) < 0,$$

a nontrivial solution of (5.12) exists if and only if

$$f(0) > 1 \tag{5.13}$$

So we have $R_1 = f(0)$, i.e., $R_1 = \frac{\sigma_1 <k\varphi(k)>}{<k>} = \frac{\lambda_1 <k\varphi(k)>}{\beta_1 <k>}$. Similarly, by letting $\Theta_1 = 0$, we can get the above result about R_2.

5.2.2. *Asymptotic stability of equilibria*

Now we examine the asymptotic property of the equilibria E_0, E_1 and E_2.
    We rewrite (5.5) as the following matrix form

$$\frac{dU(t)}{dt} = A * U - N(U) \tag{5.14}$$

According to block property of matrices, the zero solution E_0 is locally asymptotically stable, if and only if $R_1 \leq 1$ and $R_2 \leq 1$.
    Perturbing the steady state i_k^* so that $i_k = \varepsilon_k + i_k^*$ and omitting higher powers of ε_k gives the linearization matrix, which determines the stable state.
    It can be shown that the condition $R_1 < 1$ cannot ensure the local stability of E_1. To make E_1 stable, $R_2 < R_1$ must hold.
    According to symmetry, we have that E_2 is locally stable if and only if $R_1 < R_2$.
    When $R_1 < 1$ or $R_2 < 1$ holds, there are no more than two equilibria, system (5.5) is globally stable.

5.2.3. *Invasion Reproduction Numbers(IRNs)*

The IRNs can be written as below:

$$R_3 = \frac{\lambda_2 \beta_2}{\lambda_1 \beta_1} = \frac{R_2}{R_1}, \qquad R_4 = \frac{\lambda_1 \beta_1}{\lambda_2 \beta_2} = \frac{R_1}{R_2}. \tag{5.15}$$

According to biological interpretation,[32] R_3, the IRN of strain one can be considered as the number of secondary cases that one infected individual will produce in a population where strain two is at equilibrium and measures the invasion capability of strain one. Similarly, R_4 is the measure for strain two.
    If $R_3 > 1$, the strain I spread eventually, but the other strain J cannot spread. If $R_4 > 1$ (e.g., $R_3 < 1$), the strain I cannot spread eventually. As for $R_3 = R_4 = 1$, it can be shown numerically that it is a very special case (we omit the details here).
    While $R_3 \ll 1$ (that is $R_4 \gg 1$), the proportion eventually infected by the strain I is not affected by the strain J. In other words, however large the spreading rate of the strain J is, the eventually infected proportion is the same, and it is determined by R_1, that is, the spreading rate λ_1, the recovery rate β_1 and network structure. Of course, when $R_3 \gg 1$ (that is $R_4 \ll 1$), a corresponding result can be also obtained. So we can study only one strain with a much bigger spreading rate, thus the two strains model can be reduced to the one strain model, which is referred as *strain dominance*.
    But when the invasion reproduction numbers are close to 1, we prefer to use the two strains model (5.5) to obtain accurate results about its dynamical behavior for prediction epidemic spreading and optimal containment strategies.

5.2.4. *Uniform immunization strategy*

According to,[38] uniform immunization (or random immunization) strategy is an effective immunization strategy. A selective uniform immunization is proposed to immunize individuals who may be infected by the epidemic with greater spreading rate. If $R_1 > R_2$, one can approximatively use $\lambda_1(1-\varepsilon)$ to substitute λ_1. ε denotes the immunized proportion or immunized rate. The immunized system is as follows:

$$\begin{cases} \frac{di_k(t)}{dt} = -\beta_1 i_k(t) + \lambda_1(1-\varepsilon)k[1 - i_k(t) - j_k(t)]\Theta_1(t) \\ \frac{dj_k(t)}{dt} = -\beta_2 j_k(t) + \lambda_2 k[1 - i_k(t) - j_k(t)]\Theta_2(t) \end{cases} \tag{5.16}$$

Note the above system is just the same as (5.5) if we regard $\lambda_1(1-\varepsilon)$ as a new λ_1. So the BRN for strain J is still R_2, while the BRN for strain I can be written as

$$\hat{R}_{1c} = R_1(1-\varepsilon) < R_1.$$

Therefore, one know that if the immunization rate ε satisfies

$$R_1(1-\varepsilon) < R_2,$$

the strain I evolves from spreading to eliminating eventually. Further, if $1 < R_2 < R_1$, the strain J varies from eliminating to spreading eventually. This alternate spreading phenomenon is actually strain replacement, which is useful to control epidemic outbreak by immunization.[32]

5.3. The case $\sigma_1 = \sigma_2$

For the case $\sigma_1 \neq \sigma_2$ we give a relatively complete analysis, however we still cannot confirm the existence of positive equilibria or the coexistence of two strains. Now we turn to consider the other case, that is, $\sigma_1 = \sigma_2 = \sigma$. In this case, the self-consistent equations still hold. We can find the equilibrium solutions, which can be divided into two cases:

B1. When $R \leq 1$, there is disease-free equilibrium $E_0' = (1, 0, 0)$;

B2. When $R > 1$, there are infinitely many equilibria, the parameter points (Θ_1, Θ_2) of the equilibria form a line segment which connects two endpoints: $(0, \bar{\Theta})$ and $(\bar{\Theta}, 0)$, where $\bar{\Theta}$ satisfies

$$1 = \frac{\sigma}{\langle k \rangle} \sum_{k'} \frac{k'\varphi(k')P(k')}{1 + \sigma k'\bar{\Theta}}. \tag{5.17}$$

At the level of the equilibrium solution, Θ_1, Θ_2 and i_k, j_k are in one-to-one correspondence. So we can transform the above continuous problem to discrete mapping problem.

The self-consistent mapping:

$$F: x \to \frac{\sigma x}{\langle k \rangle} \sum_{k'} \frac{k'\varphi(k')P(k')}{1 + \sigma k'x}$$

The mapping F has the following properties:

(a) There is always the zero solution, that is, $F(0) = 0$ holds. Further, when $R \leq 1$, $x = 0$ is locally stable.

(b) When $R > 1$, there is always a non-zero solution, $x = \bar{x}$, such that $F(\bar{x}) = \bar{x}$ holds. And $x = \bar{x}$ is also locally stable.

We have the following stability results for equilibria:

C1. When $R \leq 1$, the disease-free equilibrium $E_0' = (1, 0, 0)$ is locally asymptotically stable.

C2. When $R > 1$, all non-zero equilibria are locally asymptotically stable; but the zero solution is unstable.

5.4. *The two-strain epidemic model with super-infection*

5.4.1. *The model*

Super-infection is the concurrent or subsequent multiple infection of a host with the same parasite, which may be with identical or different strains.[34] Now we only consider the latter case, that is, super-infection only occurs between different strains. Similar to the case with perfect cross-immunity, we focus on the uncorrelated networks with super-infection mechanism

$$\begin{cases} \frac{di_k(t)}{dt} = -\beta_1 i_k(t) + \lambda_1 k[1 - i_k(t) - j_k(t)]\Theta_1(t) + \delta k j_k(t)\Theta_1(t) \\ \frac{dj_k(t)}{dt} = -\beta_2 j_k(t) + \lambda_2 k[1 - i_k(t) - j_k(t)]\Theta_2(t) - \delta k j_k(t)\Theta_1(t) \end{cases} \quad (5.18)$$

$(\Theta_1, \Theta_2) = (0, 0)$ is a trivial equilibrium. So there is always disease-free equilibrium K_0 (the zero solution) for the system (5.18). Omitting the hight order terms, we confirm that K_0 is locally asymptotically stable when $R_1 \leq 1, R_2 \leq 1$.

We now discuss the nontrivial equilibria. Assuming $\Theta_1 \neq 0, \Theta_2 = 0$, then when $R_1 > 1$, there exists the strain one exclusive equilibrium K_1.

Secondly, assuming $\Theta_2 \neq 0, \Theta_1 = 0$, then there exists the strain two exclusive equilibrium K_2, if and only if $R_2 > 1$.

From the above analysis, we conclude that the two basic reproduction numbers are the same as the case with perfect cross-immunity.

Finally, assuming that $\Theta_2 \neq 0, \Theta_1 \neq 0$, in order to find the positive solutions, we need to discuss the equations determining two implicit functions, which are not easy to be solved. We can obtain that when $R_2 \leq R_1$, there is no positive solutions for the system under consideration. So in what follow, we assume $H : R_2 > R_1$.

To simplify our discussion, we assume that $\beta_1 = \beta_2 = 1$.

Case 1: $0 < p_r \leq 1$ and $0 < p_b \leq 1$;

Since $R_2 > R_1$ and $\beta_1 = \beta_2 = 1$, we get $\lambda_2 > \lambda_1$. So $g'(\Theta_1) < 0$. We have the IRN for positive epidemic state:

$$R_5 = g(0) + g(p_b) - g(0)g(p_b).$$

When $R_5 > 1$, there exists one positive epidemic state for the system.

Case 2: $0 < p_r \leq 1$ and $p_b > 1$;

Similar to the Case 1, we have the IRN:

$$R_6 = g(0) + g(1) - g(0)g(1).$$

Case 3: $p_r > 1$ and $0 < p_b \leq 1$;

The IRN is

$$R_7 = g(p_t) + g(p_b) - g(p_t)g(p_b).$$

Case 4: $p_r > 1$ and $p_b > 1$.

The corresponding IRN can be written as

$$R_8 = g(p_t) + g(1) - g(p_t)g(1).$$

Finally, we remark here that for the case $\beta_1 \neq \beta_2$, theoretical analysis would be quite difficult. We leave this for future research.

At the moment the type A H1N1 influenza has become a pandemic, and the threat of future outbreaks of other emerging diseases or of a human-transmissible version of the H5N1 avian influenza still remain. The problem of how best to respond to disease transmission on a network currently remains unaddressed. And many problems need to be further studied by using the method presented in this paper. Most new approaches are needed for more realistic situations, e.g., on a directed network we may need to distinguish degree distribution between in-degrees and out-degrees, as the infectivity and immunization scheme choice will depend on these quantities. More precisely, in a directed network, the infectivity will depend on out-degree distribution, while the choice of immunization scheme will depend on in-degree distribution.

References

1. R. Albert and A-L. Barabási, Statistical mechanics of complex netowrks. *Reviews of Modern Physics.* **74** (2002) 47-97.
2. F. Brauer. The Kermack-McKendrick epidemic model revisited. *Math. Biosci.* **198** (2005) 119-131.
3. M. Barthelémy, A. Barrat, R. Pastor-Satorras, and A. Vespignani, Velocity and hierarchical spread of epidemic outbreaks in scale-free networks. *Phys. Rev. Lett.,* **92** (2004) 178701.
4. M. Boguñá, R. Pastor-Satorras, and A. Vespignani, Absence of epidemic threshold in scale-free networks with degree correlations. *Phys. Rev. Lett.* **90** (2003) 028701.
5. D.S. Callaway, M.E.J. Newman, S.H. Strogatz, and D.J. Watts, Network robustness and fragility: percolation on random graphs. *Phys. Rev. Lett.* **85** (2000) 5468-5471.
6. R. Cohen, K. Erez, D. ben-Avraham, and S. Havlin, Resilience of the Internet to random breakdowns. *Phys. Rev. Lett.* **85** (2000) 4626-4628.
7. R. Cohen, S. Havlin, and D. ben-Avraham, Efficient immunization strategies for computer networks and populations. *Phys. Rev. Lett.,* **91** (2003) 24791.
8. M. L. Goldstein, S. A. Morris and G. G. Yen, Problems with fitting to the power-law distribution. *Eur. Phys. J. B.* **41** (2004) 255-258.

9. A.N. Hill, I.M. Longini Jr. The critical vaccination fraction for heterogeneous epidemic models. *Math. Biosci.* **181** (2003) 85-106.

10. J. Joo, J.L. Lebowitz, Behavior of susceptible-infected-susceptible epidemics on heterogeneous networks with saturation. Preprint, 2006.

11. N. Madar, T. Kalisky, R. Cohen, D. ben-Avraham, and S. Havlin, Immunization and epidemic dynamics in complex networks. *Eur. Phys. J.*, **B 38** (2004) 269C276.

12. W.O. Kermack and A.G. McKendrick, Proc. Roy. Soc., **A 115** (1927) 700-721.

13. J.Z. Liu, et al, The spread of disease with birth and death on networks. J. Stat. Mech., (2004) P08008.

14. R.M. May and A.L. Lloyd, Phys. Rev. E, **64** (2001) 066112.

15. R.M. May and A.L. Lloyd, Infection dynamics on scale-free networks. *Phys. Rev,* **E 65** (2002) 035108.

16. M.E.J. Newman, Phys. Rev. E, **66** (2002) 016128.

17. R. Pastor-Satorras and A. Vespignani, Epidemic dynamics and endemic states in complex networks. *Phys. Rev.*, **E 63** (2001) 066117.

18. R. Pastor-Satorras and A. Vespignani, Epidemic spreading in scale-free networks. *Phys. Rev. Lett.*, **86** (2001) 3200-3203.

19. R. Pastor-Satorras and A. Vespignani, Immunization of complex networks. *Phys. Rev.*, **E 65** (2002) 036104.

20. R. Pastor-Satorras and A. Vespignani, Epidemic dynamics in finite size scale-free networks. *Phys. Rev.*, **E 65** (2002) 035108.

21. M. Small, P. Shi and C.K. Tse. Plausible models for propagation of the SARS virus. *IEICE Trans. on Fundamentals of Electr., Commun. and Comput. Sci.*, **E87-A** (2004) 2379-2386.

22. M. Small and C.K. Tse. Small world and scale free model of transmission of SARS. *Int. J. of Bifurcat. and Chaos*, **15** (2005) 1745-1755.

23. M. Small and C.K. Tse, Clustering model for transmission of the SARS virus: application to epidemic control and risk assessment. *Physica A*, **351** (2005) 499-511.

24. M. Small, C.K. Tse and D. Walker. Super-spreaders and the rate of transmission of the SARS virus. *Physica D*, **215** (2006) 146-158.

25. M. Small, D.M. Walker and C.K. Tse, Scale free distribution of avian influenza outbreaks. Phys. Rev. Lett., **99** (2007) 188702.

26. R. Yang, J. Ren, W.-J. Bai, T. Zhou, M.-F. Zhang, B.-H. Wang, Epidemic spreading and immunization with identical infectivity. (2006) arXiv: physics/0611095.

27. T. Zhou, J.-G. Liu, W.-J. Bai, G.R. Chen, and B.-H. Wang, Behaviors of susceptible-infected epidemics on scale-free networks with identical infectivity. *Phys. Rev. E* **74** (2006) 056109.

28. V. Colizza, R. Pastor-Satorras, A. Vespignani, Reaction-diffusion processes and metapopulation models in heterogeneous networks. *Nature Physics*, **3** (2007) 276.

29. T. Zhou, J.-G. Liu, et al, Behaviors of susceptible-infected epidemics on scale-free networks with identical infectivity. Phys. Rev. E, **74** (2006) 056109.

30. R. Yang, B.-H.Wang, et al, Epidemic spreading on heterogeneous networks with identical infectivity. Phys. Lett. A, **364** (2007) 189.

31. M. Balter, New HIV strain could pose health threat. *Science*, **281** (5382) (1998) 1425-1426.

32. D.H. Thomasey, M.Martcheva, Serotype replacement of vertically transmitted disease through perfect vaccination. *J. Bio. Sys.*, **16** (2) (2008) 255-277.

33. M. NuNo, Z.Feng, M. Martcheva, and C. Castillo-Chavez, Dynamics of two-strain influenza with isolation and partial cross-immunity. *SIAM J. Appl. Math.*, **65** (3) (2005) 964-982.

34. L.M.Cai, X.Z.Li, J.Y. Yu, A two-strain epidemic model with super infection and vaccination. *Math. Appl.*, **20** (2007) 328-335.

35. M. Martcheva, T.R. Horst, Progression age enhanced backward bifurcation in an epidemic model with infection and perfect vaccination. *Math. Biosci.*, **195** (2005) 23-46.

36. M.Kamo, A. Sasaki, the effect of cross-immunity and seasonal forcing in a multi-strain epidemic model. *Physica D*, **165** (2002) 228-241.

37. F. Brauer, The Kermack-McKendrick epidemic model revisited. *Math. Biosci.*, **198** (2005) 119-131.

38. N. Madar, T. Kalisky, R. Cohen, D. ben-Avrham and S. Havlin, Immunization and epidemic dynamics in complex networks. *Eur. Phys. J.* **B 38** (2004) 269-276.

39. J.-Z. Wang, Z.-R. Liu, J.-H. Xu, Epidemic spreading on uncorrelated heterogeneous networks with non-uniform transmission. *Physica A*, **382** (2007) 715-721.

40. Jiazeng Wang and Zengrong Liu, A Markovia model of the susceptible-infected-removed spreading in networks. Preprint.

41. Jiazeng Wang and Zengrong Liu, Mean-field level analysis of epidemics in directed networks. Preprint.

42. Hai-Feng Zhang, Xin-Chu Fu, The SIR model's epidemic dynamics on complex networks with the effect of immunization. *J. Shanghai Univ. Nat. Sci.*, **13** (2) (2007) 189-192. (in Chinese)

43. M. Small, D.M. Walker, C.K. Tse and X.C. Fu, Scale free distribution of avian influenza outbreaks. 2007 China Symposium on Circuits and Systems, (Guangzhou, China, 15-17 June, 2007). Special Session (in Chinese).

44. Haifeng Zhang, Michael Small and Xinchu Fu, Global behavior of epidemic transmission on heterogeneous networks via two distinct routes. *Nonlinear Biomedical Physics*, 2008, 2:2, DOI: 10.1186/1753-4631-2-2)

45. Xinchu Fu, Michael Small, David M. Walker, Haifeng Zhang, Epidemic dynamics on scale-free networks with piecewise linear infectivity and immunization. *Physical Review E*, **77** (2008) 036113.

46. M. Small, C.K. Tse and X.C. Fu, Transmission of infectious agents on networks. In: Análisis No Lineal De Series Temporales, ed. Grupo de Investigación Interdisciplinar en Sistemas Dinámicos, 2008. In press.

47. Haifeng Zhang and Xinchu Fu, Spreading of epidemics on scale-free networks with nonlinear infectivity. *Nonlinear Analysis TMA*, **70** (2009) 3273-3278.

48. Hai-Feng Zhang, Ke-Zan Li, Xin-Chu Fu and Bing-Hong Wang, An effcient control strategy of epidemic spreading on scale-free networks. *Chinese Physics Letters*, **26** (2009) 068901.

49. Kezan Li and Xinchu Fu, The hybrid models combining complex dynamical networks and epidemic networks. *Draft*, 2009.

50. Ying-Ying Shao and Xin-Chu Fu, The SEIR model with immunization delays. *Draft*, 2009.

51. Qingchu Wu, Xinchu Fu, Estimating method of epidemic threshold value on SW networks. *Preprint*, 2009.

52. Qingchu Wu, Xinchu Fu, et al, Responsive immunizations and epidemic outbreaks in correlated complex networks. *Preprint*, 2009.

53. Qingchu Wu, Xinchu Fu, Zengrong Liu and Michael Small, Two-strain epidemic models on scale-free networks with saturated infectivity. *Preprint*, 2009.

54. Qingchu Wu and Xinchu Fu, A mixed epidemic model on scale-free networks with saturated infectivity. *Preprint*, 2009.

55. Hai-Feng Zhang, Michael Small, Xin-Chu Fu and Bing-Hong Wang, Dynamical be-

havior of SIS model on heterogeneous networks with population mobility. *Chinese Physics B*, **18** (2009) 3633-3640.

56. Haifeng Zhang, Michael Small and Xinchu Fu, Staged progression model for epidemic spread on homogeneous and heterogeneous networks. To appear in *J. of Syst. Sci. and Complexity*. (Accepted in May 2009)

57. Kezan Li, Michael Small, Haifeng Zhang and Xinchu Fu, Epidemic outbreaks on networks with effective contacts. *Nonlinear Anal. Real World Appl.* (In press. DOI: 10.1016/j.nonrwa.2009.01.046)

58. Haifeng Zhang, Michael Small and Xinchu Fu, Different epidemic models on scale-free networks. *Commun. in Theoret. Phys.*, **52** (2009) 180-184.

59. Ying-Ying Shao, Meng Liu, and Xin-Chu Fu, Epidemic dynamics of SIR model with piecewise linear infectivity and immunizations on scale-free networks. To appear in *J. of Dynamics and Control*, (Accepted in Sept. 2008) (in Chinese)

Perspectives in Mathematical Sciences
Interdisciplinary Mathematical Sciences, Volume 9, 2009
pp. 93–113

Chapter 5

Inverse Problems for Equations of Parabolic Type

Zhibin Han, Yongzhong Huang* and Ming Jian

School of Mathematics and Statistics,
Huazhong University of Science and Technology, Wuhan 430074, P.R.China,
hzb7040@sina.com, huang5464@hotmail.com, jm5546@21cn.com

This paper is concerned with a survey and some remarks on the inverse problems for equations of parabolic type. Mainly, the regularization and continuous dependence for the backward ill-posed Cauchy problems are considered by using Lattes-Lions quasi-reversibility method and its variety and by using operator-theoretic methods, in both Hilbert and Banach spaces. Moreover, Ill-posedness of several inverse problems associated with the heat equation is explained, and recovering source term is introduced.

1. Introduction

Based on the background with a lot of practical problems in science and technology, the inverse problems have received much attention since 1960's. The theory of inverse problems for differential equations is bing extensively developed with the framework of mathematical physics: As we know, most of inverse problems are ill-posed. Many books or monographs published, such as,[22,26,37,39,44,47,50,53,58,60,62,68] where[60] is a standard and very accessible reference on ill-posed problems in general. Other good sources are.[22,44,68] There are some surveys, such as.[9,28,53,59][26,37,50] and[62] involve with various equations and methods.

Limited by our knowledge and interests, in this paper we mainly consider the abstract backward Cauchy problems as follows

$$u'(t) + Au(t) = 0 \ (0 < t < T), \quad u(T) = x \qquad (1.1)$$

in both Hilbert and Banach spaces. It also involves with some inhomogeneous case. We do not attempt to cover all aspects, which is impossible.

This paper is organized as follows. In section 2 we introduce three basically inverse problems about the heat equation $u_t = u_{xx}$: backward problem, inverse

---

*Corresponding author.

This project was supported by the NSF of China (Grant No. 10672062)

heat problem, and identification for the inhomogeneous term. They are all ill-posed. Section 3 is our main part. It contains various quasi-reversibility (Q.R.) methods in both Hilbert and Banach spaces, including Lattes-Lions original Q.R.-methods, Gajewski-Zaccharias Q.R.-methods, Miller stabilized Q.R.-method, and Showalter Q.B.V-method, Boussetila-Rebbani modified Q.R.-method. Besides, we also give a simple introduction for some outstanding works to Ames and Hughes *et al* on structural stability. In section 4, we give a simple introduction on recovering source term from.[62]

As we know, the method of abstract differential equations provides proper guidelines for solving various problems with partial differential equations involved. Under the approved interpretation a partial differential equation is treated as an ordinary differential equation in a Banach space.

Here we give two possible examples to end this section. They can be found in many books. Let Ω be a bounded domain in the space R^n, whose boundary is sufficiently smooth. And set $\Omega_T = \Omega \times [0, T]$, $S_T = \partial\Omega \times [0, T]$.

Example 1.1. The initial boundary value problem for the heat conduction equation is as follows

$$\begin{cases} \dfrac{\partial u(t, x)}{\partial t} = \Delta u(t, x) + f(x, t), & (x, t) \in \Omega_T, \\ u(x, 0) = u_0(x), & x \in \Omega, \\ u(x, t) = 0, & (x, t) \in S_T. \end{cases} \quad (1.2)$$

When adopting $X = L^2(\Omega)$, we introduce in X a linear (unbounded) operator $A = \Delta$ with the domain $D(A) = H^2 \cap H_0^1$, where $H^2(\Omega)$ and $H_0^1(\Omega)$ are the classical Sobolev spaces. The functions $u(x, t)$ and $f(x, t)$ are viewed as abstract functions $u(t)$ and $f(t)$ of the variable t with values in X, while $u_0(x)$ is an element $u_0 \in X$. So, the direct problem (1.2) is treated as the Cauchy problem in X for the ordinary differential equation

$$\begin{cases} u'(t) = Au(t) + f(t), & 0 \leq t \leq T, \\ u(0) = u_0. \end{cases}$$

Example 1.2. The initial boundary value problem for the wave equation is as follows. The initial boundary value problem for the heat conduction equation is as follows

$$\begin{cases} \dfrac{\partial^2 w(t, x)}{\partial t^2} = \Delta w(t, x) + f(x, t), & (x, t) \in \Omega_T, \\ w(x, 0) = w_0(x), \quad \dfrac{\partial w}{\partial t}(x, 0) = w_1(x), & x \in \Omega, \\ w(x, t) = 0, & (x, t) \in S_T. \end{cases} \quad (1.3)$$

Take $X = L^2(\Omega)$ and $Y := X \times X$. Along similar lines in Example 1.1, we denote that

$$u(t) = \begin{pmatrix} w(t) \\ w'(t) \end{pmatrix}, \quad F(t) = \begin{pmatrix} 0 \\ f(t) \end{pmatrix}, \quad u_0 = \begin{pmatrix} w_0 \\ w_1 \end{pmatrix}, \quad \mathcal{A}u = \begin{pmatrix} 0 & I \\ A & 0 \end{pmatrix}.$$

So, the direct problem (1.1) is treated as the Cauchy problem in Y for the ordinary differential equation

$$\begin{cases} u'(t) = \mathcal{A}u(t) + F(t), & 0 \le t \le T, \\ u(0) = u_0. \end{cases}$$

2. Some simple inverse problems with heat equation

We consider the inverse problems about the heat equation $u_t = u_{xx}$. Several important types of inverse problems arise.

2.1. *Backward problem*

The problem of determining the initial temperature from later measurements, mathematically speaking, the *backwards heat equation*.

$$\begin{cases} \dfrac{\partial u(x,t)}{\partial t} = \dfrac{\partial^2 u(x,t)}{\partial x^2}, & (x,t) \in (0,1) \times (0,T), \\ u(x,T) = \varphi(x), & x \in [0,1], \\ u(0,t) = u(1,t) = 0, & t \in (0,T). \end{cases} \tag{2.1}$$

It is well known that the problem (2.1) is ill-posed: the solution (if it exists) does not depend continuously on the initial $\varphi(x)$. In fact, the solution of (2.1) can be written as follows

$$u(x,t) = \sum_{n=1}^{\infty} [2 \int_0^1 \varphi(x)\sin(n\pi x)dx] \exp(n^2\pi^2(T-t))\sin(n\pi x)$$

for a "good" final function $\varphi(x)$ by separation of variables. Frequently, subject to the usual L^2-norm, we have

$$\|u(\cdot,t)\|^2 = \sum_{n=1}^{\infty} |a_n \exp(n^2\pi^2(T-t))|^2$$

where $a_n := 2\int_0^1 \varphi(x)\sin(n\pi x)dx$. Noting $t < T$, we can see that the requirement $\|u(x,\cdot)\| < \infty$ implies rapid decay of a_n^2. This induces that the solution u is unstable if u exists (obviously, the existence of u needs some conditions on the function φ).

Many regularization methods have been developed for (2.1) (e.g.[12,15,16,23,29,40,59,67]), these methods include Tikhonov method,[68] quasi-reversibility method,[47] logarithmic convexity method,[1,14] and numerical and programming method,[12,30,40] *etc.* Payne[59] gave a simple and clear introduction more than these methods mentioned above. Further, some modified methods above and new techniques are applied to deal with (2.1), such as mollification method,[29] 'optimal error estimate',[67] iterative algorithm,[42] operator-splitting method,[43] variational scheme,[76] new filter (spectral regularization) method,[75] heatlets,[33] quasi Tikhonov regularization,[16] *etc.* Moreover,[15] provides some interesting information.

2.2. *Inverse heat problem*

The inverse heat problem or sideways heat problem is of determining the surface temperature (or heat flux) on an inaccessible part of the boundary from heat flux and temperature measurements on other parts of the boundary.[8] Typical practical applications include determination of the temperature and the heat flux at the highly heated outer surface of a reentry vehicle in the atmosphere from measurements taken inside the body, calorimeter-type instrumentation, combustion chambers, etc.

It is now to determine the temperature $u(x,t)$ for $x \in [0,1)$ from temperature measurements $\varphi = u(1,\cdot)$ and heat-flux measurements $\psi = u_x(1,\cdot)$, where $u(x,t)$ satisfies

$$\begin{cases} \dfrac{\partial u(x,t)}{\partial t} = \dfrac{\partial^2 u(x,t)}{\partial x^2}, & x \geq 0,\ t \geq 0, \\ u(x,0) = 0, & x \geq 0, \\ u(1,t) = \varphi(t), & t \geq 0, \\ \lim_{x \to \infty} u(x,\cdot) \quad bounded. \end{cases} \tag{2.2}$$

By the technique of Fourier transform, the (formal) solution of (2.2) can be written as follows[19,63]

$$u(x,t) = \int_{-\infty}^{\infty} \exp(\sqrt{2\pi i \omega}(1-x))\hat{g}(\omega)d\omega,$$

where $\sqrt{z}$ denotes the principal square root of z, the Fourier transform

$$\hat{g}(\omega) := \int_{-\infty}^{\infty} g(t)\exp(-2\pi i \omega t)dt, \quad -\infty < \omega < +\infty$$

and we define $g(t) \equiv 0$ for $t < 0$. Further, if we define the norm

$$\|h\|^2 := \int_{-\infty}^{\infty} |h(t)|^2 dt = \int_{-\infty}^{\infty} |\hat{h}(\omega)|^2 d\omega,$$

we then get

$$\|u(x,\cdot)\|^2 = \int_{-\infty}^{\infty} |\exp(\sqrt{2\pi i \omega}(1-x))\hat{g}(\omega)|^2 d\omega.$$

It is now seen that since the real part of $\sqrt{2\pi i \omega}$ is non-negative and tends to infinity as $|\omega|$ tends to infinity, the requirement $\|u(x,\cdot)\| < \infty$ implies rapid decay of $\hat{g}(\omega)$ at high frequencies. In other words, small high frequency perturbations in g are blown up catastrophically.[19] Hence the problem (2.2) is ill-posed and some regularization methods must be applied. These methods include Tikhonov regularization,[13] filtering method,[21] wavelets,[63] 'optimal approximation',[66] and in[20] where a survey of numerical solutions is presented. More development, see[74] and its references.

2.3. *Inhomogeneous heat equation: identification for the inhomogeneous term*

It is well known that the identification for the inhomogeneous term of heat equation involves the heat source. At first, we consider an identification of finding a pair of function (u, f) satisfying the following initial-boundary values problem[69]

$$\begin{cases} \dfrac{\partial u(x,t)}{\partial t} = \dfrac{\partial^2 u(x,t)}{\partial x^2} - \varphi(t)f(x), & (x,t) \in (0,1) \times (0,1), \\ u(1,t) = 0, \quad u_x(0,t) = u_x(1,t) = 0, \\ u(x,0) = 0, \quad u(x,1) = g(x), \end{cases} \tag{2.3}$$

where φ and g are two given functions. From,[69] (2.3) is equivalent to finding a function f satisfying an integral equation

$$g(x) = -\int_0^1 \int_0^1 N(x,1;\xi,\tau)\varphi(\tau)f(\xi)d\xi d\tau, \tag{2.4}$$

where $N(x,t;\xi,\tau)$ is given by

$$N(x,t;\xi,\tau) = \frac{1}{2\sqrt{\pi(t-\tau)}}\left(\exp\left(-\frac{(x-\xi)^2}{4(t-\tau)}\right) + \exp\left(-\frac{(x+\xi)^2}{4(t-\tau)}\right)\right).$$

The problem (2.4) is ill-posed from.[27] Based on the spectral cutoff method and Fourier transform, a regularized solution $f_\varepsilon(x)$ was given in,[69] where

$$f_\varepsilon(x) = -\frac{1}{\pi}\int_{-\lambda(\varepsilon)}^{\lambda(\varepsilon)} e^{\lambda^2} \int_0^1 g(s)\cos(\lambda s)\left(\int_0^1 e^{\lambda^2 t}\varphi(t)dt\right)^{-1} e^{i\lambda x}d\lambda,$$

and $\lambda(\varepsilon) = \sqrt[7]{\pi}\varepsilon^{\frac{2}{7}(\gamma-1)}$, $0 < \gamma < 1$.

Moreover, the discrepancy in[69] between the regularized solution $f_\varepsilon(x)$ and the exact solution $f_0(x)$ is, depending on the degree of smoothness of the exact solution $f_0(x)$, of the order $(-\ln \varepsilon)^{-1}$ or $\sqrt[8]{\varepsilon}$, $0 < \varepsilon < 1$, if the discrepancy between $\varphi(t)$ (respectively $g(x)$) and its exact solution $\varphi_0(t)$ (respectively $g_0(x)$) is of the order ε for the norm $\|\cdot\|_{L^2(0,1)}$.

More information on this topic, see[70] and its references herein. It is worth noting the monograph[62] contains plenty of identification for the inhomogeneous term, both the abstract case and application to PDE.

3. **Abstract backward parabolic problems**

As mentioned in 1, the method of abstract differential equations provides proper guidelines for solving various problems with partial differential equations involved.

In this section, we consider the following final value problem

$$u'(t) + Au(t) = 0 \ (0 \le t < T), \quad u(T) = x, \tag{3.1}$$

where the unbounded operator A satisfies that $-A$ is the generator of a uniformly bounded analytic semigroup $S(t)$ on a Banach space X, and x is one prescribed final value in X. It is well known that the problem (3.1) is ill-posed. In fact, the solution does not necessarily exist for $x \in X$. However, as is conventional in the theory of ill-posed problems, we assume that a solution $u(t)$ of the problem (3.1) is a priori known to a certain final value x. Then $S(T-t)u(t) = x$ from.[35] This means (3.1) is not stable because $S(t)^{-1}$ is unbounded operator for each $t > 0$.[49] Thus the problem (3.1) can lead to a general ill-posed problem which was introduced by Tikhonov (see[68]).

Many regularization methods have been developed for (3.1), these methods include quasisolution method,[46,68] quasi-reversibility method[47] and its modified methods,[11,17,25,34–36,39,56,64] logarithmic convexity method,[1,3,4,6,57,59] iterative procedures,[7,45] and the C-regularized semigroups technique,[6,31,53,54] *etc.*

However, the final value itself is given with an error $\delta > 0$, which means that x_δ is given instead of x and $\|x - x_\delta\| \le \delta$. The error δ is called the final-data error. (In the most applied problems, we are given only approximate final data and the error is defined by the accuracy of measuring instruments.) In this setting of the problem, it is required to construct a regularizing operator of the problem, i.e., an operator providing an approximate solution for a given x_δ such that the approximate solutions converge to the exact one as the final-data error tends to zero.

Definition 3.1.[35,53]   An operator $R_\alpha(t) : X \to X$ depending on the parameters $t \in [0, T]$ and $\alpha > 0$ is called the regularizing operator of the Cauchy problem (3.1) if the following conditions are fulfilled:

(1) for arbitrary $\alpha > 0$ and $t \in [0, T]$, the operator $R_\alpha(t)$ is bounded in X and $R_\alpha x \in C([0, T]; X)$ for each $x \in X$;

(2) there is a dependence $\alpha = \alpha(\delta) \ (\alpha(\delta) \to 0$ as $\delta \to 0)$ such that

$$\|R_{\alpha(\delta)}(t)x_\delta - u(t)\| \to 0 \ (\delta \to 0), \quad t \in [0, T].$$

The parameter α is called the regularizing parameter for problem (3.1). We also call $\{R_\alpha(t)\}_{t \ge 0}$ a family of regularizing operators.

All the mentioned methods allow one to construct regularizing operators

$$R_\alpha x_\delta := u_\alpha$$

for the problem (3.1) such that $\alpha = \alpha(\delta) \to 0$ and

$$\|R_{\alpha(\delta)} x_\delta - u\| \to 0 \ \text{as} \ \delta \to 0.$$

In this section we mainly introduce quasi-reversibility method for (3.1) and continuous dependence for the solution of (3.1), in both Hilbert and Banach space.

3.1. *Description of quasi-reversibility method (Q.R.-method)*

One method for approaching such problems (3.1) is *quasi-reversibility*, introduced first by Lattes and Lions[47] in 1960s. The idea is to replace (3.1) with an approximate problem which is well posed, then use the solutions of this new problem to construct approximate solutions to (3.1). After that, some modified quasi-reversibility methods appear, such as Gajewski and Zachirias quasi-reversibility,[25,36,64] Miller[56] *stabilized quasi-reversibility*, Showalter[17,18,65] *quasi-boundary-value method* and Boussetila and Rebbani[11,34] *modified quasi-reversibility, et al.*

More precisely, the description of the (modified) quasi-reversibility method is as follows.

Step 1   Let v_α be the solution of the following perturbed problem

$$v'_\alpha(t) + g_\alpha(A)v_\alpha(t) = 0 \ (0 \leq t < T), \quad v_\alpha(T) = x. \tag{3.2}$$

by choosing an operator $g_\alpha(A)$ which guarantees the problem (3.2) well-posed.

Step 2   Use the initial value

$$v_\alpha(0) = \varphi_\alpha$$

in the problem

$$u'_\alpha(t) + Au_\alpha(t) = 0 \ (0 < t \leq T), \quad u_\alpha(0) = \varphi_\alpha. \tag{3.3}$$

Step 3   Show that

$$\|u_\alpha(T) - x\| \to 0 \ (\alpha \to 0^+).$$

Step 4   Finally, construct the regularizing operator

$$R_\alpha(t)x := u_\alpha(t) = S(t)S_\alpha(T)x$$

and give the estimate of $\|R_\alpha(t)\|$, where $S(t)$ and $S_\alpha(t)$ are semigroups generated by $-A$ and $g_\alpha(A)$, respectively.

For $\alpha > 0$, $g_\alpha(A) = A - \alpha A^2$ in;[47] $g_\alpha(A) = A(I + \alpha A)^{-1}$ in;[25] $g_\alpha(A) = -\frac{1}{pT}\ln(\alpha + e^{-pTA})$, $p \geq 1$ in.[11] As for,[17] it is a special case while $p = 1$, see[11] or Remark 3.5 in.[34]

3.2. *Hilbert space case for (3.1)*

In the following, assume that H is a separable Hilbert space and A is self-adjoint operator on H such that $-A$ generates a compact contraction semi-group on H, and that 0 is in the resolvent set of A. Let $S(t)$ be the compact contraction semi-group generated by $-A$. Since A^{-1} is compact, there is an orthonormal eigenbasis φ_n for H and eigenvalues $\frac{1}{\lambda_n}$ of A^{-1} such that $A^{-1}\varphi_n = \frac{1}{\lambda_n}\varphi_n$. Then the eigenvalues of $-A$ are $-\lambda_n$ and those for $S(t)$ are $e^{-t\lambda_n}$ (and possibly zero).[61]

It is useful to know exactly when (3.1) has a solution. The following lemma answers this question.

Lemma 3.2.[17] *If* $f = \sum_{i=1}^{\infty} b_i \phi_i$, *then (3.1) has a solution if and only if* $\sum_{i=1}^{\infty} b_i^2 e^{2T\lambda_i}$ *converges.*

3.2.1. *The original Q.R.-method*

$g_\alpha(A) = A - \alpha A^2$, $u_\alpha(t) = S(t)S_\alpha(T)x$. So, $R_\alpha(t) = S(t)S_\alpha(T)$. Because

$$R_\alpha(t)x = \Sigma_{n=1}^{\infty} \exp((T-t)\lambda_n - \alpha\lambda_n^2) < x, \varphi_n >$$

and

$$\max_{0 < \lambda < \infty} \{(T-t)\lambda - \alpha\lambda^2\} = \frac{(T-t)^2}{4\alpha},$$

we have

$$\|R_\alpha(t)\| \le \exp\left(\frac{(T-t)^2}{4\alpha}\right). \tag{3.4}$$

In addition, Ames and Hughes have the following regularization result with Hölder-continuous dependence from Theorem 2 in:[5] If $\|u(0)\| \le k$ for some constant k, then there exist constants C and M, independent of $\alpha > 0$, such that for $0 < t \le T$,

$$\|u(t) - u_\alpha(t)\| \le C\alpha^{1-\frac{t}{T}} M^{t/T}.$$

It is interesting that the regularized solution $u_\alpha(t)$ can be represented by using heatlets from.[33] The heatlets may be computed and stored, and the regularized solution $u_\alpha(t)$ requires evaluation of $e^{T(A-\alpha A^2)}\Psi_{j,n}^h(\cdot, T-t)$, rather than $e^{T(A-\alpha A^2)}e^{(t-T)A}x$.

3.2.2. *Gajewski and Zachirias quasi-reversibility*

$g_\alpha(A) = A(I + \alpha A)^{-1}$ in,[25,64] and

$$R_\alpha(t)x = S(t)S_\alpha^{gz}(T)x = \Sigma_{n=1}^{\infty} \exp(-\lambda_n t + \lambda_n(1 + \alpha\lambda_n)^{-1}(T-t) < x, \varphi_n > .$$

Similarly,

$$\|R_\alpha(t)\| \le \exp\left(\frac{T - 2\sqrt{t(T-t)}}{\alpha}\right). \tag{3.5}$$

3.2.3. *Quasi-boundary value method*

In,[17] approximate (3.1) with the *quasi-boundary value problem*

$$\begin{cases} u'(t) + Au(t) = 0,\, 0 < t < T \\ \alpha u(0) + u(T) = f. \end{cases} \qquad (QBVP)$$

Define $u_\alpha^{co}(t) = S(t)(\alpha I + S(T))^{-1}x$, for x in H, $\alpha > 0$ and t in $[0, T]$. Then $u_\alpha^{co}(t)$ is the unique solution of (QBVP) and it depends continuously on x. Further, for all x in H, $\alpha > 0$, and t in $[0, T]$ one has that (see[17])

$$\|u_\alpha^{co}(t)\| \le \alpha^{\frac{t-T}{T}} \|x\|.$$

Hence,

$$\|R_\alpha^{co}(t)\| = \|S(t)(\alpha I + S(T))^{-1}\| \le \alpha^{\frac{t-T}{T}}. \qquad (3.6)$$

In,[18] approximate the problem (3.1) by replacing the final condition by

$$u(T) - \alpha u'(0) = x,$$

and

$$\|R_\alpha(t)x\| = \|u_\alpha(t)\| \le \frac{T}{\alpha(1 + \ln(T/\alpha))} \|x\|.$$

3.2.4. *Modified quasi-reversibility method*

In,[11] $g_\alpha(A) = -\frac{1}{pT} \ln(\alpha + e^{-pTA})$, $\alpha > 0$, $p \ge 1$, and $u_\alpha^{br}(t) = S(t)(\alpha I + S(pT))^{-1/p}x$, for x in H, and t in $[0, T]$. Moreover,

$$\|R_\alpha^{br}(t)\| = \|S(t)(\alpha I + S(pT))^{-1/p}\| \le \alpha^{\frac{t-T}{pT}}. \qquad (3.7)$$

Obviously, when $p = 1$, the modified quasi-reversibility method in[11] amounts to the quasi-boundary value method in.[17]

From (3.4)-(3.7), we observe that the error factor $e(\alpha) := R_\alpha(0)$ introduced by small changes in the final value x are of order

$$e^{\frac{T^2}{4\alpha}}, \quad e^{\frac{T}{\alpha}}, \quad \frac{1}{\alpha}, \quad (\frac{1}{\alpha})^{1/p},$$

respectively.

Remark 3.3. It seems to show that the method in[11] has a nice regularizing effect and gives a better approximation with comparison to the methods developed in.[17,25,47] However, the regularized solution $u_\alpha^{br}(t)$ involves the fractional power of bounded operator if $p > 1$. On the other hand, we can obtain better convergent order for $\|u_\alpha(t) - u(t)\|$ if we impose smooth conditions on the final value x (c.f.[10,11,17,18] *etc.*).

3.3. *Banach space case for (3.1)*

In this subsection we assume that $-A$ generates a uniformly bounded analytic semigroup on a Banach space X. In general Banach space, it needs a suit functional calculus of unbounded operator in order to obtain approximate solution for the problem (1.3). Hence difficulties may arise. Krein and Prozorovskaja[46] in 1960 discussed the relation between the analytic semigroup and the ill-posed Cauchy problem (3.1). The conclusion is that if $-A$ generates an analytic semigroup, then the problem (3.1) is well-posed in the class of bounded solutions $u \in C^1([0,T], X)$. Here we notice that such solutions imply $u(T) \in D(A)$. However, our purpose is to treat the regularization for (3.1), where the final data need not even belong to $D(A)$ and the solution $u(t)$ is not restricted to be bounded. Indeed, as seen above, (3.1) is ill-posed in general. Miller[57] in 1975 obtained the logarithmic convexity inequality for the solution of (3.1) by Carleman inequality for analytic functions, provided that u is small at T and bounded at 0. Further development, see[6,31,34–36,38,52,54] and the monographs.[39,53,55]

In this subsection we extend last subsection to a general Banach space X based on,[34–36,52] and in the next subsection consider the continuous dependence based on.[6,31]

3.3.1. *Lattes-Lions quasi-reversibility method*

As mentioned above, the *quasi-reversibility method*, first introduced by Lattes and Lions,[47] leads to the regularization of the problem (3.1) by using the solution of the well-posed Cauchy problem

$$u'_\alpha(t) + (A - \alpha A^b)u_\alpha(t) = 0 \ (0 \le t < T), \quad u_\alpha(T) = x \tag{3.8}$$

to construct the regularized solution of (3.1), where $\alpha > 0$ and A^b $(b > 1)$ is defined as the fractional power.

Mel'nikova *et al*(c.f.[39,52,53,55]) applied the method to (3.1), in which $-A$ is the generator of an analytic semigroup of angle θ with $\pi/4 < \theta < \pi/2$. Precisely, the result is contained in the following theorem.

Theorem 3.4.[39,52,53,55] *Let A be a densely defined linear operator whose spectrum belongs to the region*

$$\Lambda 1 = \{\lambda \in \mathbf{C} : |\arg \lambda| < \beta < \frac{\pi}{4}\}.$$

Let the estimate of the resolvent of A

$$\|R(\lambda)\| \le C(1 + |\lambda|)^{-1}$$

hold for arbitrary $\lambda \notin \Lambda 1$ and a certain $C > 0$. Then the operator

$$R_\alpha(t)x_\delta = u_{\alpha,\delta}(t) = U_{A_\alpha}(t)x_\delta = -\frac{1}{2\pi i} \int_{\partial \Lambda 1} e^{(\lambda - \alpha \lambda^2)t} R(\lambda)x_\delta d\lambda$$

constructed by the quasi-reversibility method as a solution of the Cauchy problem

$$u'_{\alpha,\delta}(t) + (A - \alpha A^2)u_{\alpha,\delta}(t) = 0, \, 0 \le t < T, \quad u_{\alpha,\delta}(T) = x_\delta$$

is a regularizing operator for the ill-posed problem (3.1). Here $A_\alpha := A - \alpha A^2$ and $\{U_{A_\alpha(t)}, t \ge 0\}$ is a C_0-semigroup with the generator A_α.

However, when $0 < \theta \le \pi/4$ it is false to choose $b = 2$ in (3.8). In fact, since the spectrum of $-A^2$ may contain a sector of angle $> \pi/2$, it does not necessarily generate a semigroup. To this end, we in[35] choose $1 < b < \pi/(\pi - 2\theta)$, where we notice $\pi/(\pi - 2\theta) \le 2$. This means that we have to face the fractional power operators $-A^b$ and $A - \alpha A^b$.

Remark 3.5. It is worthwhile to relax the restriction $\pi/4 < \theta < \pi/2$ to $0 < \theta < \pi/2$. On the one hand, if $-A$ generates an analytic semigroup of angle θ with $\pi/4 < \theta < \pi/2$, then it also generates an analytic semigroup of angle θ with $0 < \theta < \pi/4$. But the inverse is not true. On the other hand, the result in[35] is more convenient in the application to differential operators. Indeed, it is not always easy to obtain the exact value of angle θ, for example, in the case that $-A$ is a general strongly elliptic differential operator with Dirichlet boundary condition.

In,[35] the main result is

Theorem 3.6.[35] *Suppose $-A$ is the generator of an analytic semigroup. Then there exists a family of regularizing operators for (3.1).*

Indeed, the family of regularizing operators is defined by

$$R_{\alpha,t} = -\frac{1}{2\pi i} \int_{\Gamma(\gamma)} e^{(\mu - \alpha\mu^b)t} R(\mu, A) d\mu, \qquad 0 < t \le T.$$

and $R_{\alpha,0} = I$. Where the path $\Gamma(\gamma)$, $\frac{\pi}{2} - \theta < \gamma < \pi$, connects the points $\infty e^{-i\gamma}$ and $\infty e^{i\gamma}$ in $\rho(A)$, while avoiding the negative real axis and the origin.

The proof of Theorem 3.6 depends on following results.

Theorem 3.7.[35] *Let $-A$ be the generator of a bounded analytic semigroup of angle θ $(0 < \theta < \pi/2)$, and let $0 \in \rho(A)$. Then $-A^b$ is the generator of a bounded analytic semigroup of angle $\frac{\pi}{2} - (\frac{\pi}{2} - \theta)b$, where $b \in (1, \frac{\pi}{\pi - 2\theta})$.*

Theorem 3.8.[35] *Suppose that the operator A satisfies the conditions of Theorem 3.7. Let $A_\alpha = A - \alpha A^b$, where $\alpha > 0$ and $b \in (1, \frac{\pi}{\pi - 2\theta})$. Then for any $\beta \in (0, \frac{\pi}{2} - (\frac{\pi}{2} - \theta)b)$, A_α is the generator of an analytic semigroup $\{V_\alpha(t)\}$ of angle β. Moreover, $\|V_\alpha(t)\| \le M \exp(C\alpha^{1/(1-b)}t)$ for $t \ge 0$, where M and C are positive constants independent of α.*

Theorem 3.7 was proved directly by constructing related semigroup. Theorem 3.8 was proved by a moment inequality and perturbation.

3.3.2. *Gajewski and Zachirias quasi-reversibility method*

Assume that $-A$ is the generator of an analytic semigroup of angle θ, $\pi/4 \leq \theta < \pi/2$, we in[36] obtain the regularization of (3.1) by using the solution of the well-posed Cauchy problem

$$u'_\alpha(t) + \alpha A u'_\alpha(t) + A u_\alpha(t) = 0 \ (0 < t < T), \quad u_\alpha(T) = x \qquad (3.9)$$

to construct the regularized solution of (3.1), where $\alpha > 0$.

Set $A_\alpha = A(I + \alpha A)^{-1}$, then A_α is bounded. We thus can define a C_0-group by

$$S_\alpha(t) = \exp(-tA_\alpha), \qquad -\infty < t < \infty,$$

where we use the power series to define the exponential function.

The key result is

Theorem 3.9.[36]   *For each $\alpha > 0$ let $E_\alpha(t) = S(t)S_\alpha(-t)$ $(t \geq 0)$. Then $E_\alpha(t)$ is an analytic semigroup on X and $A_\alpha - A$ is its generator. In addition, there exists a positive constant M independent of α such that $\|E_\alpha(t)\| \leq M$ $(t \geq 0)$.*

Define $R_\alpha(t) = S_\alpha(-t)$ for $\alpha > 0$ and $0 \leq t \leq T$, where $S_\alpha(t)$ is the group generated by $-A_\alpha$. Then $\{R_\alpha(t); \alpha > 0, t \in [0, T]\} \subset B(X)$.

The main result is

Theorem 3.10.[36]   *The operator family $\{R_\alpha(t)\}$ defined above is a family of regularizing operators for (3.1).*

In addition, we in $[,$[36] Remark 3.3] point out that restriction $\pi/4 \leq \theta < \pi/2$ can not be relaxed to $0 < \theta < \pi/2$ by this method.

3.3.3. *Modified quasi-reversibility method*

As mentioned above, Boussetila and Rebbani[11] propose a modified quasi-reversibility method based on the perturbation

$$g_\alpha(A) = -\frac{1}{pT} \ln(\alpha + e^{-pTA}), \qquad \alpha > 0, \ p \geq 1$$

in a Hilbert space H. The advantage of this perturbation is that the amplification factor of the error resulting from the approximated problem is better by comparison with other results(cf.[11]).

In,[11] A is a positive $(A \geq \gamma > 0)$ self-adjoint, unbounded linear operator on H. In,[34] we extend it to a more general case. Precisely, we consider this method in a Banach space X and assume that $-A$ generates a uniformly bounded analytic semigroup. To end this, we need the functional calculus for A introduced by deLaubenfels (see[48]). We now give some results in.[34]

Proposition 3.11.[34] (1) $\|g_\alpha(A)\| \leq -\frac{3}{pT}\ln\alpha$ *for* $0 < \alpha < \frac{\sqrt{5}-1}{2}$;
(2) $\forall x \in D(A)$, $\lim_{\alpha\to 0}\|g_\alpha(A)x - Ax\| = 0$;
(3) $\forall x \in X$, $S_\alpha(t)x = e^{-tg_\alpha(A)}x \to S(t)x$ *as* $\alpha \to 0^+$, *uniformly in* t *on* $[0,T]$.

By the Proposition, $S_\alpha(t)$ $(t \in \mathbf{R})$ is a group generated by $-g_\alpha(A)$. Consequently, the solution of (3.2) can be represented as follows:

$$v_\alpha(t) = S_\alpha(t - T)x = (\alpha + S(pT))^{-\frac{T-t}{pT}}x,$$

and it depends continuously on x. In fact,

$$\|S_\alpha(-t)\| = \|(\alpha + S(pT))^{-\frac{t}{pT}}\| \leq \alpha^{-\frac{3t}{pT}} \leq \alpha^{-\frac{3}{p}}.$$

Using $v_\alpha(0) = S_\alpha(-T)x$ as the initial value, it follows that $u_\alpha(t) = S(t)S_\alpha(-T)x$ is the unique solution of (3.3).

Theorem 3.12.[34] *Let* ε *be a fixed positive numbers, then* $\|u_\alpha(t)\| \leq \alpha^{\frac{t-T}{pT}-\varepsilon}\|x\|$, $0 < t \leq T$, *for small* α.

Proposition 3.13.[34] *Let* $x \in R(S(t_0))$ *for some* $t_0 \in (0,T)$. *Then* $\lim_{\alpha\to 0^+} u_\alpha(T) = x$.

By the functional calculus, $S(t)$ and $S_\alpha(r)$ commute with each other. Let $u(t)$ be a solution of (3.1) with the final value $u(T) = x$, then $S(T-t)u(t) = x$. Hence

$$u_\alpha(t) = S(t)S_\alpha(-T)S(T-t)u(t) = S(T)S_\alpha(-T)u(t).$$

This induces a good result.

Theorem 3.14.[34] *Let* $u(t)$ *be a solution of (3.1) with the final value* $u(T) = x$, *then* $\lim_{\alpha\to 0} u_\alpha(t) = u(t)$ *for each* $t \in (0,T]$.

Remark 3.15. Define $R_\alpha(t) = S(t)S_\alpha(-T)$, $t \geq 0$, $\alpha > 0$. By Theorem 3.12 and Theorem 3.14, it is easy to show that $R_\alpha(t)$ is a family of regularizing operators for (3.1). For details, see.[11,35,36]

3.4. *Structural stability for (3.1)*

Structural stability is the continuous dependence for the difference between solutions of certain ill-posed and approximate well-posed problems. That is to give the estimates of

$$\|u(t) - v(t)\|, \quad \|u(t) - u_\alpha(t)\|$$

with some order in α, see[41] or[6] and its references.

Ames and Hughes in[6] proved Hölder-continuous dependence in both Hilbert and Banach spaces by using operator-theoretic methods, including regularized semigroups and functional calculus, and logarithmic convexity technique. We introduce it simply in Hilbert spaces H as follows.

Consider the Cauchy problem

$$u'(t) = Au(t), \quad u(0) = x, \quad 0 \leq t < T. \tag{3.10}$$

Its "approximate problem" and well-posed final-value problem are

$$v'(t) = f(A)v(t), \quad v(0) = x, \quad 0 \leq t < T. \tag{3.11}$$

and

$$w'(t) = Aw(t), \quad w(T) = v(t) = e^{Tf(A)}x, \tag{3.12}$$

respectively. For the most part, we work under the assumption that the solution $u(t)$ exists and is bounded in the norm of H. This is essentially necessary in order to obtain any meaningful error estimates for approximating the exact solution. If we make no assumptions about the solution of (3.10), then we can only bound the error between the regularized solution and the approximation.

In the spirit of Millers work,[56] one seeks functions f for which solutions to the corresponding well-posed Cauchy problem (3.11), approximate known solutions of the original ill-posed problem (3.10). Note that (3.10) is same to (3.1) by setting $t = T - s$.

Definition 3.16.[6] Let $f : [0, \infty) \to \mathbf{R}$ be a Borel function, and assume that there exists $\omega \in \mathbf{R}$ such that $f(\lambda) \leq \omega$ for all $\lambda \in [0, \infty)$. Then f is said to satisfy Condition (A) if there exist positive constants β, δ, with $0 < \beta < 1$, for which $D(A^{1+\delta}) \subset D(f(A))$, and

$$\|(-A + f(A))\psi\| \leq \|A^{1+\delta}\psi\|,$$

for all $\psi \in D(A^{1+\delta})$. Here, $A^{1+\delta}$ and $f(A)$ are defined by means of the functional calculus for self-adjoint operators.

Now, assume $u(t)$ is a known solution of (3.10), with initial data $x \in H$.

The following theorem shows that the Hölder-continuous dependence on modeling and the perturbation parameter, respectively.

Theorem 3.17.[6] *Let A be a positive self-adjoint operator acting on H, let f satisfy Condition (A), and assume that there exists a constant γ, independent of β and ω, such that $(g(A)\psi, \psi) \leq \gamma$, for all $\psi \in H$. If $u(t)$, $v(t)$ and $w(t)$ are solutions of (3.10), (3.11) and (3.12), respectively, and $\|u(T)\| \leq \tilde{M}$, then there exist constants C and M, independent of β, such that for $0 \leq t < T$,*

$$\|u(t) - v(t)\|, \|u(t) - w(t)\| \leq C\beta^{1-t/T}M^{t/T}.$$

Remark 3.18. (1) This work builds on the earlier results in §3.2.1 and §3.2.2. by several authors, including Lattes and Lions,[47] Miller,[56,57] and Showalter,[64] who use quasi-reversibility methods to approximate the original ill-posed problem. Ames *et al*[3] construct numerical approximations using the quasi-boundary value regularization of Clark and Oppenheimer[17] and present several sample calculations ([32]).

(2) Theorem 4 and Theorem 5 in[6] are generalized results of Theorem 3.17 to Banach space X. It is remarkable that these results are very interesting. However, there has a limited condition: $-A$ generates an analytic semigroup of $\theta \in (\pi/4, \pi/2]$ on X. Indeed, it is not always easy to obtain the exact value of angle θ, see Remark 3.5.

(3)[31] extended the results in[6] to inhomogeneous case in Banach space. Very recently,[32] proved continuous dependence on modeling for the nonlinear Cauchy problem

$$u'(t) = Au(t) + h(t, u(t)), \quad 0 \le t < T, \quad u(0) = x,$$

where A is a positive self-adjoint operator on a Hilbert space H, $x \in H$, and $h : [0, T) \times H \to H$. Assume that h is Lipschitz continuous in both variables, i.e.

$$\|h(t_1, u) - h(t_2, v)\| \le Const.(|t_1 - t_2| + \|u - v\|).$$

The results are obtained by extending the solutions into the complex plane following[6] and, following the approach in,[1] introducing a related holomorphic function whose growth properties yield the desired Hölder continuous dependence.

(4) The approximation is done in[4] by casting (3.10) as a first-kind Fredholm integral equation, using the kernel for the corresponding forward operator. It amounts to a direct application of Tikhonov regularization (much of the theory developed in Groetsch[27] can be applied to the regularization), but they obtained their results using a log-convexity the approximating operator is shown to be logarithm convex. Further, the numerical solution of backward parabolic problems is given by solving the corresponding discrete least-squares system with using an eigenvalue/eigenvector approach, or by an iterative technique.

It is remarkable that although the notion of the kernel is fundamental to the reformulation of the problem, it does not need a closed-form representation of the kernel in order to compute approximations. On the other hand, one would obtain the same solutions by solving the direct PDE problem

$$\begin{cases} u'_\alpha + Au_\alpha = \alpha Bu_\alpha, & t > 0, \\ u_\alpha(T) = (1 + \alpha)^{-1}\varphi, \end{cases} \tag{3.13}$$

where $B = -2(K * K + \alpha I)^{-1}A$ and $K = S(T)$. For the details, see.[4]

(5) Although there are many works on the linear homogeneous case of the backward problem, the literature on the linear inhomogeneous and the nonlinear case of the problem are quite scarce. We here give some references:.[2,24,31,32,51,69,71–73]

(6) Melnikova *et al* have many works on the regularization of abstract ill-posed Cauchy problems, see.[53,55]

4. The linear inverse problem: recovering a source term

We extend 2.3 to a more general case, with both partial differential equation and abstract first order equation. This section is from.[62] Our purpose is give a simple introduction to them, see[62] for the details. It is worth noting the monograph[62] which covers the basic of equations: elliptic, parabolic and hyperbolic. Special emphasis is given to the Navier-Stokes equations as well as to the well-known kinetic equations: Boltzman equation and neutron transport equation.

At first, let Ω be bounded domain in R^n with boundary $\partial\Omega \in C^2$. And set $\Omega_T = \Omega \times [0,T]$, $S_T := \partial\Omega \times [0,T]$.

Consider the following inverse problem in Ω_T of finding a pair of the functions $u(x,t)$ and $p(x)$

$$
\begin{cases}
\dfrac{\partial u}{\partial t} = \displaystyle\sum_{i,j=1}^{n} \dfrac{\partial}{\partial x_i}\left(a_{ij}(x)\dfrac{\partial u}{\partial x_j}\right) + b(x)u + f(x,t), & (x,t) \in \Omega_T, \\[2mm]
u(x,0) = u_0(x), \quad u(x,T) = u_1(x), & x \in \Omega, \\[2mm]
f(x,t) = \Phi(x,t)p(x) + F(x,t), & (x,t) \in \Omega_T, \\[2mm]
u(x,t) = 0, & (x,t) \in S_T.
\end{cases}
\tag{4.1}
$$

Assumptions are as follows

(1) $a_{ij} \in C^1(\Omega)$, $a_{ij} = a_{ji}$, $\displaystyle\sum_{i,j=1}^{n} a_{ij}(x)\xi_i\xi_j \geq c\sum_{i=1}^{n}\xi_i^2$, $c > 0$;

(2) $b \in C(\Omega)$, $b \leq 0$;

(3) Φ, $\Phi_t \in C(\overline{\Omega}_T)$, $\Phi > 0$, $\Phi_t > 0$;

(4) $F \in C^1([0,T]; L^2(\Omega)) + C([0,T]; H^2(\Omega) \cap H_0^1(\Omega))$;

(5) u_0, $u_1 \in H^2(\Omega) \cap H_0^1(\Omega)$.

Define the operator A as follows:

$$
Au = \sum_{i,j=1}^{n} \frac{\partial}{\partial x_i}\left(a_{ij}(x)\frac{\partial u}{\partial x_j}\right) + b(x)u,
$$

$$
D(A) = H^2(\Omega) \cap H_0^1(\Omega).
$$

Then the operator A is self-adjoint and negative and A^{-1} is compact. Frequently, the operator A is a generator of a compact C_0-semigroup.

Proposition 4.1(,[62]Cor.9.4.1). *If conditions (1)-(5) hold, then a solution u, p of the problem (4.1) exists and is unique in the class of functions*

$$
u \in C^1([0,T]; L^2(\Omega)) \cap C([0,T]; H^2(\Omega)),
$$
$$
p \in L^(\Omega).
$$

Next we return to the abstract case. Let X and Y be Banach spaces. Assume that the operator A generates a C_0-semigroup $\{S(t)\}_{t\geq 0}$. We consider the inverse problem of finding a pair of the functions $u \in C([0,T]; X)$ and $p \in C([0,T]; Y)$ from the problem

$$\begin{cases} u'(t) = Au(t) + f(t), & 0 \le t \le T, \\ u(0) = u_0, \quad f(t) = \Phi(t)p(t) + F(t), & 0 \le t \le T, \\ Bu(t) = \psi(t), & 0 \le t \le T. \end{cases} \tag{4.2}$$

Theorem 4.2 ([62] Th.6.2.1 and Th.6.2.2). *Assume that B, $\overline{BA} \in L(X,Y)$ and $\Phi \in C([0,T]; L(Y,X))$, $F \in C([0,T]; X)$, $u_0 \in X$ and $\psi \in C^1([0,T]; Y)$. If for any $t \in [0,T]$ the operator $B\Phi(t)$ is invertible,*

$$(B\Phi)^{-1} \in ([0,T]; L(Y))$$

and $Bu_0 = \psi(0)$, then a solution u, p of the inverse problem (4.2) exists and is unique in the class of functions

$$u \in C([0,T]; X), \qquad p \in C([0,T]; Y).$$

Moreover, there exists a positive constant $M = M(A, B, \Phi, T)$ satisfies the estimates

$$\|u\|_{C([0,T];X)} \le M\Big(\|u_0\|_X + \|\psi\|_{C^1([0,T];Y)} + \|F\|_{C([0,T];X)}\Big),$$

$$\|p\|_{C([0,T];Y)} \le M\Big(\|u_0\|_X + \|\psi\|_{C^1([0,T];Y)} + \|F\|_{C([0,T];X)}\Big).$$

The sketch of the proof. By the theory of semigroup, one has

$$u(t) = S(t)u_0 + \int_0^t S(t-s)\Phi(s)ds + \int_0^t S(t-s)F(s)ds.$$

Therefore, the over-determination condition $Bu(t) = \psi(t)$ is equivalent to the following equation

$$B\Big(S(t)u_0 + \int_0^t S(t-s)\Phi(s)ds + \int_0^t S(t-s)F(s)ds\Big) = \psi(t).$$

This implies that

$$p(t) = p_0(t) + \int_0^t K(t,s)p(s)ds, \tag{4.3}$$

where $K(t,s) = -(B\Phi(t))^{-1}\overline{BA}S(t-s)\Phi(s)$ and

$$p_0(t) = (B\Phi(t))^{-1}\Big(\psi'(t) - \overline{BA}S(t)u_0 - \overline{BA}\int_0^t S(t-s)F(s)ds - BF(t)\Big).$$

Noting that the function p_0 is continuous on $[0,T]$ and $K(t,s)$ is strongly continuous for $0 \le s \le t \le T$, we obtain the existence and uniqueness of the solution to the integral equation (4.3) in the class of continuous functions.

Remark 4.3. (1) The higher order smoothness of the solution u, p for the problem (4.2) may be obtained by improving the smoothness of the functions Φ, F and ψ (see Corollary 6.2.2 in[62]).

(2) Similarly, nonlinear inverse problems were studied in.[62] Say, $f(t)$ is replaced by $f(t, u(t), p(t))$.

References

1. S. Agmon and L. Nirenberg, Properties of solutions of ordinary differential equations in Banach spaces, *Comm. Pure Appl. Math.*, **16**, 121-139, (1963).
2. D. D. Ang, Stabilized approximate solution to the inverse time problem for a parabolic evolution equation, *J. Math. Anal. Appl.*, **111**, 148-155, (1985).
3. K. A. Ames, G. W. Clark, J. F. Epperson, and S. F. Oppenheimer, A comparison of regularizations for an ill-posed problem, *Math. Computation*, **67**, 1451-1471, (1998).
4. K. A. Ames and J. F. Epperson, A kernel-based method for the approximate solution of backward parabolic problems, *SIAM J. Numer. Anal.*, **34**, 1357-1390, (1997).
5. K. A. Ames and R. J. Hughes, Continuous dependence results for ill-posed problems, *Semigroups of Operators: Theory and Applications, Second International Conference*, Rio de Janeiro, Brazil, September 10-14, 2001, (Optimization Software, Inc. Publications, New York and Los Angeles, 2002), pp. 1-8.
6. K. A. Ames and R. J. Hughes, Structural stability for ill-posed problems in Banach space, *Semigroup Forum*, **70**, 127-145, (2005).
7. J. Baumeister, and A. Leitao, On iterative methods for solving ill-posed problems modeled by partial differential equations, *J. Inv. Ill-Posed Problems*, **9**(1), 1-17, (2001).
8. J.V. Beck, B. Blackwell and S.R. Chair, *Inverse heat conduction: ill-posed problems*, (John Wiley, New York, 1985).
9. F. Ben Belgacem, Why is the Cauchy problem severely ill-posed? *Inverse Problems*, **23**, 823-836, (2007).
10. N. Boussetila and F. Rebbani, Optimal regularization method for ill-posed Cauchy problems, *Electronic J. Diff. Eqns.*, **2006**, 1-15, (2006).
11. N. Boussetila and F. Rebbani, The modified quasi-reversibility method for a class of ill-posed Cauchy problems, *Georgian Math. J.*, **14**(4), 627-642, (2007).
12. A. S. Carraso, J. Sanderson and J. Hyman, Digital removal of random media image degradations by solving the diffusion equation backwards in time, *SIAM J. Numer. Anal.*, **15**, 344-367, (1978).
13. A.S. Carraso, Determining surface temperature from interior observations, *SIAM J. Appl. Math.*, **42**, 558-574, (1982).
14. A. S. Carraso, Logarithmic convexity and the slow evolution constraint in ill-posed initial value problems, *SIAM J. Math. Anal.*, **30**(3), 479-496, (1999).
15. L.D. Chiwiaciwsky and H.F. de Campos velho, Different approaches for the solution of a backward heat conduction problem, *Inverse Problems in Eng.*, **11**(6), 471-494, (2003).
16. J. Cheng and J. Liu, A quasi Tikhonov regularization for a two-dimensional backward heat problem by a fundamental solution, *Inverse Problems*, **24**(6), 065012(18pp), (2008).
17. G. Clark and S. Oppenheimer, Quasi-reversibility type methods for non-well posed problems, *Elect. J. Diff. Eqs.*, **1994**, 1-9, (1994).
18. M. Denche and K. Bessila, A modified quasi-boundary value method for ill-posed problems, *J. Math. Anal. Appl.*, **301**, 419-426, (2005).

19. L. Eldén, Approximation for a Cauchy problem for the heat equation, *Inverse Problems*, **3**, 263-273, (1987).

20. L. Eldén, Numerical solution of the sideways heat equation, In eds. H. Engle and W. Rundell, Eds., *Inverse Problems in Diffusion Processes*, SIAM, Philadelphia, PA, (1995).

21. L. Eldén and T.I. Seidman, An optimal filtering method for the sideways heat equation, *Inverse Problems*, **6**, 681-696, (1990).

22. H. W. Engel and W. Rundel, Eds., *Inverse problems in diffusion processes*, (1995). SIAM, Philadelphia.

23. R. E. Ewing, The approximation of certain parabolic equations backward in time by Sobolev equations, *SIAM J. Math. Anal.*, **6**(2), 283-294, (1975).

24. X. L. Feng, Z. Qian and C. L. Fu, Numerical approximation of solution of nonhomogeneous backward heat conduction problem in bounded region, *Math. Comp. Simul.*, **79**, 177-188, (2008).

25. H. Gajewski and K. Zacharias, Zur Regularisierung einer Klasse nichtkorrekter Probleme bel Evolutionsgleichungen, *J. Math. Anal. Appl.*, **38**, 784-789, (1972).

26. V. B. Glasko, *Inverse Problems of Mathematical Physics*, (1988). American Institute of Physics, New York.

27. C. Groetsch, *The theory of Tikhonov regularization for Fredholm equations of the first kind*, (1994). Pitman, Boston, MA.

28. M. Hanke and P. C. Hansen, Regularization methods for large-scale problems, *Surveys Math. Indust.*, **3**, 252-315, (1993).

29. D .N. Hào, A Mollification method for ill-posed problems, *Numer. Math.*, **68**(4), 469-506, (1994).

30. A. Hassanov, J. L. Mueller, A numerical method for backward parabolic problems with nonselfadjoint elliptic operator, *Applied Numerical Mathematics*, **37**, 55-78, (2001).

31. B. Hetrick and R. Hughes, Continuous dependence results for inhomogeneous ill-posed problems in Banach space, *J. Math. Anal. Appl.*, **331**, 342-357, (2007).

32. B. Hetrick and R. Hughes, Continuous dependence on modeling for nonlinear ill-posed problems, *J. Math. Anal. Appl.*, **349**, 420-435, (2009).

33. B. Hetrick and R. Hughes and E. Mcnabb, Regularization of the backward heat equation via heatlets, *Electronic J. Diff. Eqns.*, (130), 1-8, (2008).

34. Y. Huang, Modified quasi-reversibility method for final value problems in Banach spaces, *J. Math. Anal. Appl.*, **340**(2), 757-769, (2008).

35. Y. Huang and Q. Zheng, Regularization for ill-posed Cauchy problems associated with generators of analytic semigroups, *J. Diff. Eqs.*, **203**(1), 37-54, (2004).

36. Y. Huang and Q. Zheng, Regularization for a class of ill-posed Cauchy problems, *Proc. Amer. Math. Soc.*, **133**, 3005-3012, (2005).

37. V. Isakov, *Inverse Problems for Partially Differential Equations.* (Springer, Berlin,1998).

38. V. K. Ivanov, Ill-posed problems and divergent processes, *Uspekhi Mat. Nauk.*, **40**, 165-166, (1985).

39. V. K. Ivanov, I. V. Mel'nikova and A.I. Filinkov, *Differential Operator Equations and Ill-posed Problems.* (Nauka, Moscow, 1995). (In Russian)

40. F. John, Numerical solution of the heat equation for preceding time, *Ann. Math. Pura Appl.*, **40**,129-142, (1955).

41. F. John, Continuous dependence on data for solutions of partial differential equations with a prescribed bound, *Comm. Pure Appl. Math.*, **13**, 551-585, (1960).

42. M. Jourhmane and N.S. Mera, An iterative algorithm for the backward heat conduction problem based on variable relaxition factors, *Inverse Probl. Sci. Eng.*, **10**(4),

293-308, (2002).

43. S. M. Kirkup and M. Wadsworth, Solution of inverse diffusion problems by operator splitting methods, *Applied Mathematical Modelling*, **26**(10), 1003-1018, (2002).

44. A. Kirsch, *An introduction to the mathematical theory of inverse problems.* (Springer-Verlag, New York, 1996).

45. V. A. Kozlov and V. G. Mazya, On the iterative method for solving ill-posed boundary value problem that preserve differential equations, *Leningrad Math. J.*, **1**, 1207-1228, (1990).

46. S. G. Krein and O. I. Prozorovskaja, Analytic semi-groups and incorrect problems for evolutionary equations, *Dokl. Akad. Nauk SSSR*, **133**, 277-280, (1960).

47. R. Lattes and J.-L. Lions, *The method of quasi-reversibility applications to partial differential equations.* (Amer. Elsevier Publ. Co., New York, 1969).

48. R. deLaubenfels, Functional Calculus for generators of uniformly bounded holomorphic semigroups, *Semigroup Forum*, **38**, 91-103, (1989).

49. R. deLaubenfels, *Existence Families, Functional Calculi and Evolution Equation.* (Springer-Verlag, Berlin, 1994).

50. M. M. Lavrent'ev, V. G. Romanov and S. P. Shishatskiĭ, *Ill-posed Problem of Mathematical Physics and Analysis*, (1986). Amer. Math. Soc., Providence, R. I.

51. N. Long and A. Dinh, Approximation of a parabolic non-linear evolution equation backwards in time, *Inverse Problems*, **10**, 905-914, (1994).

52. I. V. Mel'nikova, General theory of ill-posed Cauchy problem, *J. Inverse and Ill-posed Problems*, **3**(2), 149-171, (1995).

53. I. V. Mel'nikova and U. A. Anufrieva, Peculiarities and regularization of ill-posed Cauchy problems with differential operators, *J. Math. Sci.*, **148**(4), 481-632, (2008).

54. I. V. Mel'nikova and S. V. Bochkareva, C-semigroups and regularization of an ill-posed Cauchy problems, *Russian Acad. Sci. Dokl. Math.*, **47**, 228-232, (1993).

55. I. V. Mel'nikova and A. Filinkov, *Abstract Cauchy Problems: Three Approaches.* (Chapman and Hall, London, 2001).

56. K. Miller, Stabilized quasireversibility and other nearly best-possible methods for non-well-posed problems, "Symposium on Non-Well-Posed Problems and Logarithmic Convexity, Lecture Notes in Mathematics," 316, pp.161-176, (Springer-Verlag, Berlin-Heidelberg-New York, 1973).

57. K. Miller, Logarithmic convexity results for holomorphic semigroups, *Pacific J. Math.*, **58**, 549-551, (1975).

58. V. A. Morozov, *Methods for solving incorrectly posed problems*, (Springer-Verlag, New York, 1984).

59. L. E. Payne, Some general remarks on improperly posed problems for partial differential equations, "Symposium on Non-Well-Posed Problems and Logarithmic Convexity, Lecture Notes in Mathematics," 316, pp.1-30, (Springer-Verlag, Berlin-Heidelberg-New York, 1973).

60. L. E. Payne, *Improperly Posed Problems in Partial Differential Equations*, 1-17, CBMS Regional Conference Series in Applied Mathematics, 22, (1975). Society for Industrial and Applied Mathematics, Philadelphia.

61. A. Pazy, *Semigroups of linear operators and applications to partial differential equations.* (Springer-Verlag, New York, 1983).

62. A. I. Prilepko, D. G. Orlovsky and I.A. Vasin, *Methods for solving inverse problems in mathematical physics*, Monographs and textbooks in pure and applied mathematics 222, (Marcel Dekker, Inc., New York, 2000).

63. T. Reginska, Sideway heat equation and wavelets, *J. Comput. Appl. Math.*, **63**, 209-214, (1995).

64. R. E. Showalter, The final problem for evolution equations, *J. Math. Anal. Appl.*, **47**, 563-572, (1974).

65. R. E. Showalter, *Cauchy problem for hyper-parabolic partial differential equation*, "Trends in the theory and the practice of non-linear analysis", (Elsevier, 1983).

66. U. Tautenhahn, Optimality stable approximations for the sideways heat equation, *J. Inverse and Ill-posed Problems*, **5**, 287-307, (1997).

67. U. Tautenhahn and T. Schräter, On optimal regularization methods for the backward heat equation., *Zeitschrift Analysis und ihre Anwendungen*, **15**(2), 475-493, (1996).

68. A. N. Tikhonov and V. Y. Arsenin, *Solution of Ill-posed Problems*. (Winston and Sons, Washington D.C., 1977).

69. D. Trong, N. Long and P. Alain, Nonhomogeneous heat equation: identification and regularization for the inhomogeneous term, *J. Math. Anal. Appl.*, **312**, 93-104, (2005).

70. D. Trong, N. Long and P. Alain, Determine the special term of a two-dimensional heat source, http://arxiv.org/abs/0807.1806v1.

71. D. Trong, P. Quan, T. Khanh and N. Tuan, A nonlinear case of the 1-D backward heat problem: Regularization and error estimate, *Zeitschrift Analysis und ihre Anwendungen*, **26**(2), 231-245, (2007).

72. D. Trong and N. Tuan, Regularization and error estimates for nonhomogeneous backward heat problems, *Electronic J. Diff. Eqns.*, (4), 1-10, (2006).

73. D. Trong and N. Tuan, Stabilized quasi-reversibility method for a class of nonlinear ill-posed problems, *Electronic J. Diff. Eqns.*, (84), 1-12, (2008).

74. X.T. Xiong, C.L. Fu and J. Cheng, Spectral regularization methods for solving a sideways parabolic equation within the framework of regularization theory, *Math. Comp. Simul.*, **79**, 1668-1678, (2009).

75. X.T. Xiong, C.L. Fu and Z. Qian, On three spectral regularization methods for a backward heat conduction problem, *J. Korean Math. Soc.*, **44**(6), 1281-1290, (2007).

76. B. Yildiz, H. Yetiskin, and A. Sever, A stability estimate on the regularized solution of the backward heat equation, *Appl. Math. Comput.*, **135**, 561-567, (2003).

Perspectives in Mathematical Sciences
Interdisciplinary Mathematical Sciences, Volume 9, 2009
pp. 115–133

Chapter 6

The Existence and Asymptotic Properties of Nontrivial Solutions of Nonlinear (2 − q)-Laplacian Type Problems with Linking Geometric Structure

Gongbao Li *,** and Zhaofen Shen

School of Mathematics and Statistics,
Central China Normal University,
Wuhan 430079, P. R. China

In this paper, we study the existence of nontrivial solutions and the asymptotic properties of solutions to the following nonlinear elliptic problems with singular perturbation in a bounded domain $\Omega \subset R^N$:

$$\begin{cases} -\triangle u - a(x)u - \varepsilon \triangle_q u = f(x, u), \ x \in \Omega \\ u|_{\partial\Omega} = 0 \end{cases} \tag{1.1}_\varepsilon$$

where $\varepsilon \geq 0$, $\triangle_q u = div(|\nabla u|^{q-2}\nabla u)$, $1 < q < 2 < N$, and that $a \in L^{\frac{N}{2}}(\Omega)$, $f \in C^0(\overline{\Omega} \times R^1, R^1)$, $f(x,t)$ is super-linear at $t = 0$ and subcritical at $t = \infty$ and $(1.1)_\varepsilon$ possesses the so-called linking geometric structure. We prove that there exists an $\varepsilon_0 > 0$ such that for any $0 \leq \varepsilon \leq \varepsilon_0$, $(1.1)_\varepsilon$ possesses at least one nontrivial solution u_ε and for any sequence ε_n with $\varepsilon_n \to 0^+$, there is a subsequence $\{u_{\varepsilon_{n_k}}\}_{k=1}^{+\infty} \subseteq \{u_{\varepsilon_n}\}_{n=1}^{+\infty}$ such that $u_{\varepsilon_{n_k}} \to \tilde{u}$ in $H_0^1(\Omega)$ for some nontrivial solution $\tilde{u}$ of $(1.1)_0$. Our result generalizes a classical result in [10] about the existence of nontrivial solutions of $(1.1)_0$ to the $(2 - q)$-Laplacian type problem $(1.1)_\varepsilon$ and provides a new result for the asymptotic property of the solutions u_ε as $\varepsilon \to 0$.

Keywords: $(2 - q)$-Laplacian; nonlinear elliptic problem; linking geometric structure; nontrivial solutions; asymptotic properties.

1. Introduction and main results

In this paper, we study the existence of nontrivial solutions and the asymptotic properties of solutions as $\varepsilon \to 0^+$ to the following nonlinear $(2 - q)$-Laplacian type elliptic problems with singular perturbation in a bounded domain $\Omega \subset \mathbb{R}^N$:

$$\begin{cases} -\triangle u - a(x)u - \varepsilon \triangle_q u = f(x, u), \ x \in \Omega \\ u|_{\partial\Omega} = 0 \end{cases} \tag{1.1}_\varepsilon$$

* Partially supported by NSFC Grant No: 10571069 and NSFC Grant No:10631030 and The Lab of Mathematical Sciences, Central China Normal University, Hubei Province, China
** Corresponding author. Email address: ligb@mail.ccnu.edu.cn

where $\varepsilon \geq 0$, $\Delta_q u = div(|\nabla u|^{q-2}\nabla u)$, $1 < q < 2 < N$, $a(x) \in L^{\frac{N}{2}}(\Omega)$, $f \in C^0(\bar{\Omega} \times R^1, R^1)$ and $(1.1)_\varepsilon$ possesses the so-called linking geometric structure.

To state some conditions imposed on $f(x,t)$ satisfies, we first recall some knowledge about the eigenvalues of elliptic operators.

According to the theory of spectrum of compact operator (see e.g. Chapter 4 of [4]or Lemma 2.12 below), let

$$-\infty < \lambda_1 < \lambda_2 \leqslant \lambda_3 \leqslant \cdots$$

be the sequence of all eigenvalues of the following eigenvalue problem

$$\begin{cases} -\triangle u - a(x)u = \lambda u, \, x \in \Omega \\ u|_{\partial\Omega} = 0 \end{cases} \tag{1.2}$$

where each eigenvalue is repeated according to its multiplicity, $\lim_{j\to\infty} \lambda_j = +\infty$ and let $e_1, e_2, ..., e_n, ...$ be the corresponding eigenfunctions in $H_0^1(\Omega)$ normalized in the sense of $L^2(\Omega)$, that is

$$\int_\Omega e_i e_j dx = \delta_{ij} = \begin{cases} 1, i = j \\ 0, i \neq j \end{cases},$$

hence for any i and j we have

$$\int_\Omega [\nabla e_j \cdot \nabla e_i - a(x)e_j e_i]dx = \lambda_j \int_\Omega e_i e_j dx = \lambda_j \delta_{ij}.$$

In this paper, we study the case where $(1.1)_\varepsilon$ possesses the so-called linking geometric structure, so we assume that $\lambda_1 \leq 0$, and there exists an $n \in N$, such that

$$\lambda_1 < \lambda_2 \leqslant \lambda_3 \leqslant \cdots \leq \lambda_n \leqslant 0 < \lambda_{n+1} \leqslant \cdots \quad . \tag{1.3}$$

Now we give some conditions imposed on $f(x,t)$.

(f_1) Suppose that $f \in C^0(\bar{\Omega} \times R^1, R^1)$ and there are constants C_1, C_2 and $1 < p < \frac{N+2}{N-2}$, such that

$$|f(x,t)| \leq C_1 + C_2|t|^p.$$

for all $(x,t) \in \bar{\Omega} \times R^1$

(f_2) (Ambrosetti–Rabinowitz condition) There are $\theta \in (0, \frac{1}{2})$ and $M > 0$ such that

$$0 \leq F(x,t) \leq \theta t f(x,t), \, for \, |t| \geq M, x \in \bar{\Omega} \quad .$$

(f_3) Condition $\lim_{t\to 0} \frac{f(x,t)}{t} = 0$ uniformly with respect to $x \in \bar{\Omega}$.

(f_4) $\frac{\lambda_n}{2}t^2 \leq F(x,t)$, $\forall(x,t) \in \bar{\Omega} \times R^1$ where $F(x,t) = \int_0^t f(x,s)ds$.

We call $u \in H_0^1(\Omega)$ a weak solution to $(1.1)_\varepsilon$, if

$$\int_\Omega [\nabla u \cdot \nabla v - a(x)uv]dx + \varepsilon \int_\Omega |\nabla u|^{q-2}\nabla u \cdot \nabla v dx = \int_\Omega f(x,u)v dx, \quad \forall v \in H_0^1(\Omega).$$

By hypothesis (f_3), we see that $f(x,0) = 0$, so $u \equiv 0$ is a trivial solution of $(1.1)_\varepsilon$. We are interested in getting nontrivial solutions to $(1.1)_\varepsilon$. Let $g(x,t) = a(x)t + f(x,t)$, then $(1.1)_\varepsilon$ can be viewed as a perturbation of the following problem:

$$\begin{cases} -\triangle u = g(x,u), \, x \in \Omega \\ u|_{\partial\Omega} = 0. \end{cases} \tag{1.4}$$

Problem (1.4) as a classical typical problem has been widely studied during the last thirty years. Since it possesses a variational structure, weak solution to (1.4) correspond to critical points of the following variational functional

$$I(u) = \frac{1}{2}\int_\Omega |\nabla u|^2 dx - \int_\Omega G(x,u)dx,$$

defined on the *Sobolev* space $H_0^1(\Omega)$ where $G(x,u) = \int_0^u g(x,t)dt$. During the past decadesthe critical point theory has become a standard method to study the existence of nontrivial solutions of elliptic equations of variational type. In 1973, *Ambrosetti* and *Rabinowitz* proposed the Mountain Pass Theorem in [2] and established a theoretical framework for obtaining a critical point for C^1 functional on Banach spaces with the so-called mountain-pass geometry hence solved the existence of a nontrivial weak solution to (1.4) when $f(x,t)$ is of super-linear at t=0 and subcritical at $t = \infty$ such that it possesses the mountain-pass geometry. In 1978, *Rabinowitz* proposed the so-called Linking Theorem in [3] and [8]) which resulted in the existence of at least one nontrivial solution to (1.4) when it possesses the linking geometry.

On the other hand, the $p − Laplacian$ type problem

$$\begin{cases} -\triangle_p u = g(x,u), \, x \in \Omega \\ u|_{\partial\Omega} = 0 \end{cases} \tag{1.5}$$

have been widely studied and many results about problem (1.4) have been generalized to (1.5) and there are a lot of literatures in this respect (see [9] and the references therein).

The $(p − q)$-Laplacian type nonlinear elliptic problem

$$\begin{cases} -\Delta_p u - \Delta_q u = g(x,u), \, x \in \Omega \\ u\,|_{\partial\Omega} = 0 \end{cases} \tag{1.6}$$

for $1 < p < \infty, 1 < q < \infty$ comes, for example, from a general reaction-diffusion system

$$u_t = div[D(u)\nabla u] + c(x,u), \tag{1.7}$$

where $D(u) = (|\nabla u|^{p-2} + |\nabla u|^{q-2})$. The system has a wide range of applications in Physics and related sciences such as Biophysics, plasma Physics and chemical reaction design. In such applications, the function u describes a concentration, the first term on the right-hand side of (1.7) corresponds to the diffusion with a diffusion coefficient $D(U)$ whereas the second one is the reaction term and related to source and loss processes. Usually the reaction term $c(x, u)$ is a polynomial of u with variable coefficients. Equilibrium solutions to (1.7) are solutions to equation like (1.6). In recent years, the study of (1.6) and its extension on unbounded domain have been found in a variety of literatures. Roughly speaking, the existence of nontrivial solutions of (1.6) with mountain-pass geometry has been obtained (see [5][6][7]. To guarantee that $I(u)$ possesses the mountain-pass geometry around the trivial solution $u = 0$, it is natural to assume that $g(x, t)$ is of superlinear at $t = 0$, that is $\lim_{t \to 0} \frac{g(x,t)}{t} = 0$ uniformly with respect to $x \in \bar{\Omega}$.

When $\varepsilon = 0$, $(1.1)_\varepsilon$ can be regarded as a special case of (1.4) for $g(x, t) = a(x)t + f(x, t)$. By the hypothesis (f_3), we see that $\lim_{t \to 0} \frac{g(x,t)}{t} = a(x)$ on $\bar{\Omega}$, therefore equation $(1.1)_\varepsilon$ generally has no mountain-pass geometric structure when $\varepsilon = 0$ and we could not get solution of it by using the Mountain Pass Theorem. When $f(x, t)$ is not of superlinear at $t = 0$, one usually uses the Linking Theorem to get nontrivial solutions as in [3][10]. Theorem 4.2 in [3] shows that $(1.1)_0$ possesses at least one nontrivial solution if $a(x) \in C^\gamma(\bar{\Omega})$ for some $0 < \gamma < 1$, $a(x) > 0$ for $x \in \bar{\Omega}$ and $f(x, t) \in C^\gamma(\bar{\Omega} \times R^1)$ satisfies (f_1)-$(f_3)(f_4)'$ together with

$(f_4)' : tf(x, t) \geq 0, \forall (x, t) \in \bar{\Omega} \times R^1$

On the other hand, [10, Theorem 2.18] proves the existence of at least one nontrivial solution of $(1.1)_0$ if $a(x) \in L^{\frac{N}{2}}(\Omega)$ and $f \in C^0(\bar{\Omega} \times R^1, R^1)$ satisfies $(f_1)-(f_4)$. [10, Theorem 2.18] is a generalization of Theorem 4.2 in [3]. The purpose of this paper is to generalize the main results in [10] to the $(2 - q)$-Laplacian type problem. To state our main result, we first give some notations.

For $u \in H_0^1(\Omega)$, let $\|u\| \triangleq (\int_\Omega |\nabla u|^2 dx)^{\frac{1}{2}}$ be its norm and let

$$I(u) = \frac{1}{2} \int_\Omega [|\nabla u|^2 - a(x)u^2] dx - \int_\Omega F(x, u) dx,$$

$$J(u) = \frac{1}{q} \int_\Omega |\nabla u|^q dx,$$

$$I_\varepsilon(u) = I(u) + \varepsilon J(u).$$

Under the assumptions (1.3) and (f_4), let

$$Y \triangleq span\{e_1, \cdots, e_n\},$$

$$Z \triangleq \{u \in H_0^1(\Omega)| \int_\Omega uv dx = 0, v \in Y\}.$$

For any $\rho > r > 0$, we note $M^\rho := \{u = y + \lambda z : \|u\| \leq \rho, \lambda \geq 0, y \in Y\}$, $M_0^\rho :=$ $\{u = y + \lambda z : y \in Y, \|u\| = \rho, \lambda \geq 0$ or $\|u\| \leq \rho, \lambda = 0\}$, $N_r := \{u \in Z : \|u\| = r\}$, and

$$c_\varepsilon := \inf_{\gamma \in \Gamma} \max_{u \in M^\rho} I_\varepsilon(\gamma(u)),$$

$$\Gamma := \{\gamma \in C(M^\rho, H_0^1(\Omega)) : \gamma \mid_{M_0^\rho} = id\}.$$

Our main result is as follows.

Theorem 1.1 Let $\Omega \subset R^N$ be a bounded smooth domain, $1 < q < 2 < N, a(x) \in L^{\frac{N}{2}}(\Omega)$, $-\infty < \lambda_1 < \lambda_2 \leqslant \cdots \leqslant \lambda_n \leqslant 0 < \lambda_{n+1} \leqslant \cdots$ be the sequence of eigenvalues of (1.2) and $f \in C^0(\bar{\Omega} \times R^1, R^1)$ satisfies $(f_1) - (f_4)$. Then there exists an $\varepsilon_0 > 0$ such that for any $0 \leq \varepsilon \leq \varepsilon_0$, $(1.1)_\varepsilon$ possesses a nontrivial solution u_ε with the property that $I_\varepsilon(u_\varepsilon) = c_\varepsilon$ and for any sequence $\varepsilon_n \to 0^+$, there exists a subsequence $\{u_{\varepsilon_{n_k}}\}_{k=1}^{+\infty}$ of $\{u_{\varepsilon_n}\}_{n=1}^{+\infty}$ and some nontrivial solution $\tilde{u}$ of $(1.1)_0$ such that $u_{\varepsilon_{n_k}} \to \tilde{u}$ in $H_0^1(\Omega)$ and $I_0(\tilde{u}) = c_0$.

Remark 1.2: Theorem 1.1 is a new result which generalize Theorem 2.18 of [10] to the case where $\varepsilon > 0$ and provides an asymptotic properties of u_ε as $\varepsilon \to 0^+$.

Next we briefly describe the idea of the proof of Theorem 1.1.

Problem $(1.1)_\varepsilon$ can be viewed as a perturbation of $(1.1)_0$. By conditions $(f_1) - (f_4)$, we know that $I, J, I_\varepsilon \in C^1(H_0^1(\Omega), R^1)$. We hope to obtain the nontrivial critical points of $I_\varepsilon(u)$, which are solutions to $(1.1)_\varepsilon$, by means of the critical point theory and variational method. For C^2 functional of the form $K_\varepsilon = H(u) + \varepsilon G(u)$, the existence of critical points can be obtained by means of general perturbation theory (see e.g. Theorem 2.12 and Theorem 2.25 in [1]). However, since $1 < q < 2$, $J(u)$ is not a C^2 functional and under the assumption that $f \in C^0(\bar{\Omega} \times R^1, R^1)$ one see that $I(u), I_\varepsilon(u)$ are not always C^2 functional, so the method in [1] can not be used to prove our results.

We basically follow the main idea of the proof of the existence theorem for $(1.1)_0$ in [10] and use Linking Theorem (see Proposition 2.8 in this paper) to get a nontrivial critical point of the functional $I_\varepsilon(u)$ on $H_0^1(\Omega)$. To guarantee that $I_\varepsilon(u)$ satisfy the conditions of Linking Theorem, we need to require that the parameter ε appeared in $(1.1)_\varepsilon$ be suitably small i.e. $0 \leq \varepsilon \leq \varepsilon_0$ for some $\varepsilon_0 > 0$. In order to prove the asymptotic property of the solutions u_ε as $\varepsilon \to 0^+$, we carefully construct the "linking structure" of I_ε uniformly with respect to ε for $0 \leq \varepsilon \leq \varepsilon_0$ so that the linking structure depends only on $a(x)$ and $f(x,t)$. Then we manage to prove that u_ε is uniformly bounded in $H_0^1(\Omega)$ and the strong convergence of a subsequence of $u\varepsilon$ in $H_0^1(\Omega)$.

We use standard notations. For example, $\|u\|_s = (\int_\Omega |u|^s dx)^{\frac{1}{s}}$, denotes the norm of u in $L^s(\Omega)$ for $1 < s \leq \infty$, "$\sim$" denotes two norms in a linear space are equivalent; $\to$ and $\rightharpoonup$ denote strong and weak convergence respectively in a corresponding function space; $|E|$ denotes the N-dimensional *Lebesgue* measure of

$E \subset R^N$; $C, \tilde{C}$ denote arbitrary positive constants and $o(1)$ denotes the infinitesimal, that is, $\lim_{n \to +\infty} o(1) = 0$.

The paper is organized as follows, in §2 we give some preliminary results and in §3, we prove our main result.

2. Preliminary Results

In this section we give some definitions and preliminary results which will be used in §3 for the proof of our main result.

The Sobolev space $H^1(R^N) := \{u \in L^2(R^N) : \nabla u \in L^2(R^N)\}$ endowed with the inner product

$$(u, v)_1 := \int_{R^N} (\nabla u \cdot \nabla v + uv) dx$$

and corresponding norm

$$\|u\|_1 = \left(\int_{R^N} (|\nabla u|^2 + u^2) dx \right)^{\frac{1}{2}}$$

is a *Hilbert* space. Let $\Omega \subset R^N$ be an open set, $H_0^1(\Omega)$ is defined as the closure of $C_0^\infty(\Omega)$ in $H^1(R^N)$ with respect to the norm. $H_0^1(\Omega)$ is a Hilbert space with inner product

$$(u, v) := \int_{R^N} \nabla u \cdot \nabla v dx$$

and an equivalent norm in $H_0^1(\Omega)$ is given by $\|u\| = (\int_\Omega |\nabla u|^2 dx)^{\frac{1}{2}}$.

Proposition 2.1 (Rellich embedding theorem, [10, Theorem 1.9])   If Ω is a bounded domain in R^N, then the following embeddings are compact:

$$H_0^1(\Omega) \hookrightarrow L^p(\Omega), 1 \le p < 2^*.$$

Where $2^* = \frac{2N}{N-2}$, if $N > 2$; $2^* = +\infty$, if $N \le 2$.

Proposition 2.2 (Lemma 2.16 of [10]])   Assume that $|\Omega| < \infty$, $f \in C(\bar{\Omega} \times R)$, $C > 0$ such that

$$|f(x, u)| \le C(1 + |u|^{p-1}),$$

where $1 < p < \infty$, if $N = 1, 2$ and $1 < p \le 2^*$, if $N \le 3$. Then the functional φ

$$\varphi(u) := \int_\Omega F(x, u) dx$$

defined on $H_0^1(\Omega)$ satisfies that $\varphi \in C^1(H_0^1(\Omega), R)$ and

$$\langle \varphi'(u), v \rangle = \int_\Omega f(x, u) v dx, \forall u, v \in H_0^1(\Omega),$$

where $F(x, t) = \int_0^t f(x, s) ds$.

By **Proposition 2.2**, it is easy to see that the functional $I_\varepsilon \in C^1(H_0^1(\Omega), R)$ for any ε and

$$\langle I_\varepsilon'(u), v \rangle = \int_\Omega [\nabla u \cdot \nabla v - a(x) uv] dx + \varepsilon \int_\Omega |\nabla u|^{q-2} \nabla u \cdot \nabla v dx - \int_\Omega f(x, u) v dx,$$

$$\forall u, v \in H_0^1(\Omega)$$

where $\langle \cdot, \cdot \rangle$ denotes the paring between $H_0^1(\Omega)$ and its dual.

Definition 2.3 Let $(X, \| \cdot \|)$ be a *Banach* space, $(X^*, \| \cdot \|_*)$ be its dual space and $\varphi \in C^1(X, R)$. The functional φ satisfies the (P.S) condition if any sequence $\{x_n\} \subset X$ such that

$$|\varphi(x_n)| \leq C < +\infty, \ \varphi'(x_n) \to 0 \text{ in } X^*,$$

has a convergent subsequence $\{x_{n_k}\}$ in X. For $c \in R^1$, we say that φ satisfies the $(PS)_c$ condition if any sequence $\{x_n\} \in X$ such taht

$$\varphi(x_n) \to c, \ \varphi'(x_n) \to 0 \text{ in } X^*,$$

has a convergent subsequence $\{x_{n_k}\}$ in X.

Proposition 2.4 (Linking Theorem, Theorem 2.12 in [10]) Let $X = Y \oplus Z$ be a *Banach* space with $\dim Y < \infty$. Let $\rho > r > 0$ and let $z \in Z$ be a fixed element such that $\|z\| = r$. Define

$$M := \{u = y + \lambda z; \|u\| \leq \rho, \lambda \geq 0, y \in Y\},$$

$$M_0 := \{u = y + \lambda z; y \in Y, \|u\| = \rho, \lambda \geq 0 \quad \text{or} \quad \|u\| \leq \rho, \lambda = 0\},$$

$$N_r := \{u \in Z; \|u\| = r\}.$$

Let $\varphi \in C^1(X, R)$ be such that

$$b := \inf_{N_r} \varphi > a := \max_{u \in M_0} \varphi.$$

If φ satisfies $(PS)_c$ condition with

$$c := \inf_{\gamma \in \Gamma} \max_{u \in M} \varphi(\gamma(u)),$$

$$\Gamma := \{\gamma \in C(M, X) : \gamma \mid M_0 = id\},$$

then c is a critical value of φ.

Remark 2.5 $c \geq b$ in Proposition 2.8.

By **Proposition 2.1**, it is easy to see that the following result is true.

Proposition 2.6 If $N \geq 3$ and $a(x) \in L^{\frac{N}{2}}(\Omega)$, then the real functional

$$\chi(u) := \int_\Omega a(x) u^2 dx$$

defined on $H_0^1(\Omega)$ is weakly continuous.

Proposition 2.7 (Lemma 2.15 in [10]]) If Ω is a bounded domain in R^N, $N \geq 3$ and $a(x) \in L^{\frac{N}{2}}(\Omega)$, then

$$\lambda_1 = \inf_{u \in H_1^0(\Omega), \|u\|_2 = 1} \int_\Omega [|\nabla u|^2 - a(x)u^2] dx > -\infty.$$

Lemma 2.8 If $a(x) \in L^{\frac{N}{2}}(\Omega)$. Then the sequence of all eigenvalues $\{\lambda_j\}_{j=1}^{+\infty}$ of the problem

$$\begin{cases} -\triangle u - a(x)u = \lambda u, x \in \Omega \\ u|_{\partial\Omega} = 0 \end{cases}$$

satisfies

$$-\infty < \lambda_1 < \lambda_2 \leqslant \lambda_3 \leqslant \cdots,$$

and $\lim_{j \to \infty} \lambda_j = +\infty$.

Proof. By Proposition 2.11, it follows that $\lambda_1 > -\infty$. Therefore, there is a λ_0 large enough such that

$$\int_\Omega [|\nabla u|^2 - a(x)u^2] dx + \int_\Omega \lambda_0 u^2 dx > 0.$$

for any $u \in H_0^1(\Omega)$. So we can define an equivalent inner product on $H_0^1(\Omega)$ by

$$(u, v)_{\lambda_0} = \int_\Omega [\nabla u \cdot \nabla v - a(x)uv] dx + \int_\Omega \lambda_0 uv dx, \forall \ u, v \in H_0^1(\Omega).$$

By *Poincare* inequality and *Riesz* representation theorem, we know that any $u \in L^2(\Omega)$, there exists a unique $w \in H_0^1(\Omega)$ such that

$$\int_\Omega uv dx = (w, v)_{\lambda_0}, \forall v \in H_0^1(\Omega).$$

For $u \in H_0^1(\Omega)$ define $K_{\lambda_0} : L^2(\Omega) \longrightarrow H_0^1(\Omega)$ by $w = K_{\lambda_0}u$, then K_{λ_0} is a bounded linear operator. If $i : H_0^1(\Omega) \longrightarrow L^2(\Omega)$ is the natural embedding operator, then *Rellich* theorem shows that i is a compact operator and for any $u, v \in H_0^1(\Omega)$ we have

$$(K_{\lambda_0} \circ i(u), v)_{\lambda_0} = \int_\Omega uv dx.$$

Since $K_{\lambda_0} \circ i$ is a compact operator from $H_0^1(\Omega)$ to $H_0^1(\Omega)$ and $(K_{\lambda_0} \circ i(u), u)_{\lambda_0} > 0$ for $u \neq 0$, we see that by *Hilbert − Schmidt* theory (see e.g. Section 4 of Chapter 4 in [4]), it follows that the sequence of all eigenvalues $\{\mu_j\}_{j=1}^{+\infty}$ of $K_{\lambda_0} \circ i$ satisfies $\mu_1 > \mu_2 > \mu_3 > \ldots > \mu_n > \ldots > 0, \mu_j \to 0$ (as $j \to +\infty$), and

$$\lambda_j = \frac{1}{\mu_j} - \lambda_0 \quad , (j = 1, 2, 3 \ldots)$$

is the sequence of all eigenvalues of (1.2) and the corresponding eigenfunctions satisfy

$$\int_\Omega e_i e_j dx = \delta_{ij}.$$

$\square$

Lemma 2.9 (Corollary 1.4.17 in [4])) Let $\| \cdot \|_1$ and $\| \cdot \|_2$ be two norms on a normed linear space X. $\| \cdot \|_1$ is equivalent to $\| \cdot \|_2$ if and only if there are two constants $C_1, C_2 > 0$ such that

$$C_1 \|x\|_1 \leq \|x\|_2 \leq C_2 \|x\|_1, \quad \forall x \in X.$$

3. The Proof of Theorem 1.1

In this section we prove our main result Theorem 1.1.
According to Theorem 2.8, let

$$-\infty < \lambda_1 < \lambda_2 \leqslant \lambda_3 < \cdots \lambda_n < \lambda_{n+1} \leq \lambda_{n+2} < \cdots$$

be the sequence of all eigenvalues of the problem:

$$\begin{cases} -\triangle u - a(x)u = \lambda u, & x \in \Omega, \\ \qquad\quad u|_{\partial\Omega} = 0, \end{cases}$$

with $\lim_{j \to \infty} \lambda_j = +\infty$ and let $e_1, e_2, ..., e_n, ...$ be all the corresponding eigenvectors such that

$$\int_\Omega e_i e_j dx = \delta_{ij}.$$

Following the notation in the proof of Lemma 2.8, we denote an equivalent inner product in $H_0^1(\Omega)$ as

$$(u, v)_{\lambda_0} = \int_\Omega [\nabla u \cdot \nabla v - a(x)uv]dx + \int_\Omega \lambda_0 uv dx, \forall u, v \in H_0^1(\Omega),$$

where $\lambda_0 + \lambda_1 > 0$, and

$$\lambda_1 = \inf_{u \in H_0^1(\Omega), \|u\|_2 = 1} \int_\Omega [|\nabla u|^2 - a(x)u^2]dx.$$

In addition, we know that from the definition of Y, Z and Lemma 2.8 that $\dim Y < +\infty$, $H_0^1(\Omega) = Y \oplus Z$.

Lemma 3.1 (Lemma 2.15 in [10]])

$$\delta := \inf_{u \in Z, \|\nabla u\|_2 = 1} \int_\Omega [|\nabla u|^2 - a(x)u^2]dx > 0. \tag{3.1}$$

Proof.   For every $u \in Z$, we have $\int_\Omega u e_i dx = 0$ $(1 \leq i \leq n)$. Let

$$u = \sum_{i=n+1}^{+\infty} c_i e_i,$$

where $c_i = \int_\Omega u e_i dx \ (i = n+1, n+2...)$ hence

$$
\begin{aligned}
(u,u)_{\lambda_0} &= (\sum_{i=n+1}^{+\infty} c_i e_i, \sum_{j=n+1}^{+\infty} c_j e_j)_{\lambda_0} \\
&= \sum_{i=n+1}^{+\infty} \sum_{j=n+1}^{+\infty} c_i c_j (e_i, e_j)_{\lambda_0} \\
&= \sum_{i=n+1}^{+\infty} \sum_{j=n+1}^{+\infty} c_i c_j [\int_\Omega (\nabla e_i \cdot \nabla e_j - a(x) e_i e_j) dx + \lambda_0 \int_\Omega e_i e_j dx] \\
&= \sum_{i=n+1}^{+\infty} \sum_{j=n+1}^{+\infty} c_i c_j (\lambda_i + \lambda_0) \int_\Omega e_i e_j dx \\
&= \sum_{i=n+1}^{+\infty} c_i^2 (\lambda_i + \lambda_0) \\
&\geq (\lambda_{n+1} + \lambda_0) \sum_{i=n+1}^{+\infty} c_i^2 \\
&= (\lambda_{n+1} + \lambda_0) \int_\Omega u^2 dx,
\end{aligned}
$$

So for every $u \in Z$, we have

$$\int_\Omega [|\nabla u|^2 - a(x) u^2] dx \geq \lambda_{n+1} \int_\Omega u^2 dx. \tag{3.2}$$

Take the minimizing sequences $\{u_n\}_{n=1}^{+\infty} \subset Z$ such that

$$\|u_n\| = \|\nabla u_n\|_2 = 1, 1 - \int_\Omega a(x) u_n^2 dx \to \delta,$$

Without loss of generality, let

$$u_n \rightharpoonup u \quad on \quad H_0^1(\Omega).$$

By *Rellich* theorem, we may assume that

$$u_n \to u \quad on \quad L^2(\Omega).$$

So we get

$$\delta = 1 - \int_\Omega a(x) u^2 dx \geq \int_\Omega [|\nabla u|^2 - a(x) u^2] dx \geq \lambda_{n+1} \int_\Omega u^2 dx.$$

If $u = 0$, then $\delta = 1$. If $u \neq 0$, then $\delta \geq \lambda_{n+1} \int_\Omega u^2 dx > 0$. $\qquad \square$

Lemma 3.2 (Lemma 2.18] in [10]) For every $u \in Y$, we have

$$\int_\Omega [|\nabla u|^2 - a(x) u^2] dx \leq \lambda_n \int_\Omega u^2 dx.$$

Proof.   For $u \in Y$, we can write

$$u = \sum_{i=1}^{n} c_i e_i,$$

where $c_i = \int_\Omega u e_i dx$ $(i = 1, 2 \dots n)$. Then

$$
\begin{aligned}
(u, u)_{\lambda_0} &= \left(\sum_{i=1}^{n} c_i e_i, \sum_{j=1}^{n} c_j e_j \right)_{\lambda_0} \\
&= \sum_{i=1}^{n} \sum_{j=1}^{n} c_i c_j (e_i, e_j)_{\lambda_0} \\
&= \sum_{i=1}^{n} \sum_{j=1}^{n} c_i c_j \left[\int_\Omega (\nabla e_i \cdot \nabla e_j - a(x) e_i e_j) dx + \lambda_0 \int_\Omega e_i e_j dx \right] \\
&= \sum_{i=1}^{n} \sum_{j=1}^{n} c_i c_j (\lambda_i + \lambda_0) \int_\Omega e_i e_j dx \\
&= \sum_{i=1}^{n} c_i^2 (\lambda_i + \lambda_0) \\
&\le (\lambda_n + \lambda_0) \sum_{i=1}^{n} c_i^2 \\
&= (\lambda_n + \lambda_0) \int_\Omega u^2 dx.
\end{aligned}
$$

Hence it follows that

$$\int_\Omega [|\nabla u|^2 - a(x) u^2] dx \le \lambda_n \int_\Omega u^2 dx. \tag{3.3}$$

for $u \in Y$. $\square$

Lemma 3.3 Suppose that $a(x), f(x, t)$ are given as in Theorem 1.1. Then the real functional

$$I_\varepsilon(u) := \frac{1}{2} \int_\Omega [|\nabla u|^2 - a(x) u^2] dx + \frac{\varepsilon}{q} \int_\Omega |\nabla u|^q dx - \int_\Omega F(x, u) dx$$

defined on $H_0^1(\Omega)$ belongs to $C^1(H_0^1(\Omega), R)$ for any $\varepsilon \ge 0$. Moreover $I_\varepsilon(u)$ satisfies the $(PS)_c$ condition for any $c \in R$.

Proof.   By $1 < q < 2$ and *Hölder* inequality, there is a constant $C = C(|\Omega|, q, 2)$ such that

$$\int_\Omega |\nabla u|^q dx \le C \|u\|^q , \ \forall u \in H_0^1(\Omega).$$

According to hypothesis (f_1) and Proposition 2.2, it is easy to see that $I_\varepsilon(u) \in C^1(H_0^1(\Omega), R^1)$ and

$$\langle I_\varepsilon'(u), v \rangle = \int_\Omega \nabla u \cdot \nabla v dx + \varepsilon \int_\Omega |\nabla u|^{q-2} \nabla u \cdot \nabla v dx$$

$$- \int_\Omega a(x) uv dx - \int_\Omega f(x, u) v dx, \quad \forall u, v \in H_0^1(\Omega). \tag{3.4}$$

By hypothesis $(f_2)(Ambrosetti - Rabinowitz$ condition) and the continuity of f, there are $C_1 > 0$ and $\alpha = \frac{1}{\theta} > 2$, such that

$$F(x, u) \geq C_1(|u|^\alpha - 1). \tag{3.5}$$

Fix $\beta \in (\frac{1}{\alpha}, \frac{1}{2})$.

If $\{u_n\}_{n=1}^{+\infty} \subset H_0^1(\Omega)$ is a $(PS)_c$ sequence of $I_\varepsilon(u)$ such that

$$I_\varepsilon(u_n) \to c, I_\varepsilon'(u_n) \to 0.$$

By Proposition 2.9, Proposition 3.1 (3.4) and that $H_0^1(\Omega) = Y \oplus Z$, let $u_n = y_n + z_n$ with $y_n \in Y, z_n \in Z$, then there exist $C_2, C_3 > 0$ with n large enough such that

$$c + 1 + \|u_n\| \geq I_\varepsilon(u_n) - \beta \langle I_\varepsilon'(u_n), u_n \rangle$$

$$= (\tfrac{1}{2} - \beta) \int_\Omega [|\nabla u_n|^2 - a(x) u_n^2] dx + (\tfrac{1}{q} - \beta) \varepsilon \int_\Omega |\nabla u_n|^q dx$$

$$+ \int_\Omega [\beta f(x, u_n) u_n - F(x, u_n)] dx \tag{3.6}$$

$$\geq (\tfrac{1}{2} - \beta)(\delta \|z_n\|^2 + \lambda_1 \|y_n\|_2^2) + (\alpha\beta - 1) \int_\Omega F(x, u_n) dx - C_2$$

$$\geq (\tfrac{1}{2} - \beta)(\delta \|z_n\|^2 + \lambda_1 \|y_n\|_2^2) + C_1(\alpha\beta - 1) \|u_n\|_\alpha^\alpha - C_3.$$

Hence there are $\tilde{C}_1, \tilde{C}_2, \tilde{C}_3 > 0$ such that

$$\tilde{d} + \|u_n\| + \tilde{C}_2 \|y_n\|_2^2 \geq \tilde{C}_1 \|z_n\|^2 + \tilde{C}_3 \|u_n\|_\alpha^\alpha.$$

Since $\int_\Omega y_n z_n dx = 0$, we have

$$\|u_n\|_2^2 = \|y_n\|_2^2 + \|z_n\|_2^2.$$

By $\alpha > 2, |\Omega| < +\infty$, *Hölder* inequality and *Young* inequality, we get

$$\|y_n\|_2^2 \leq \|u_n\|_2^2 \leq C_4 \|u_n\|_\alpha^2 \leq \varepsilon \|u_n\|_\alpha^\alpha + C_\varepsilon,$$

$$\|y_n\|_2 \leq \|u_n\|_2 \leq C_5 \|u_n\|_\alpha \leq \varepsilon \|u_n\|_\alpha^\alpha + C_\varepsilon.$$

Since Y is a finite dimensional space, any two norms in Y are equivalent, thus

$$\|y_n\|_2 \sim \|y_n\|_\alpha \sim \|y_n\|.$$

By Proposition 2.9, there exists $C_6 > 0$ such that

$$\|y_n\| \le C_6 \|y_n\|_2 \le \varepsilon \|u_n\|_\alpha^\alpha + C_\varepsilon,$$

hence

$$\tilde{C}_1 \|z_n\|^2 + \tilde{C}_3 \|u_n\|_\alpha^\alpha \le \tilde{d} + \|u_n\| + \tilde{C}_2 \|y_n\|_2^2$$

$$\le \tilde{d} + \|y_n\| + \|z_n\| + \tilde{C}_2 \|y_n\|_2^2$$

$$\le \tilde{d} + \|z_n\| + C_6 \|y_n\|_2 + \tilde{C}_2 \|y_n\|_2^2$$

$$\le \tilde{d} + \varepsilon \|u_n\|^2 + \varepsilon \|u_n\|_\alpha^\alpha + C_\varepsilon + \varepsilon \|u_n\|_\alpha^\alpha + C_\varepsilon$$

$$\le \tilde{d} + \varepsilon \|z_n\|^2 + 2\varepsilon \|u_n\|_\alpha^\alpha + 2C_\varepsilon.$$

That is

$$(\tilde{C}_1 - \varepsilon)\|z_n\|^2 + (\tilde{C}_3 - 2\varepsilon)\|u_n\|_\alpha^\alpha \le \tilde{d} + 2C_\varepsilon. \tag{3.7}$$

From (3.7), it follows that there exists a $C_7 > 0$ such that

$$\|z_n\| \le C_7, \|u_n\|_\alpha^\alpha \le C_7.$$

Hence there exists $\quad 0 < C < \infty \quad$ such that

$$\|u_n\| \le C.$$

Thus we may assume that there exists $u_0 \in H_0^1(\Omega)$ such that

$$u_n \rightharpoonup u_0 \quad in \quad H_0^1(\Omega).$$

By $|\Omega| < +\infty$ and Proposition 2.1 (*Rellich* embedding theorem), we may assume that

$$\begin{cases} u_n \to u_0 & \text{on } L^p(\Omega)(2 \le p < 2^*), \\ u_n \to u_0 & a.e \text{ on } \Omega. \end{cases} \tag{3.8}$$

By (3.8) and *Lebesgue's* Dominated Convergence Theorem, we have that

$$\begin{cases} \int_\Omega f(x, u_n)u_n dx \to \int_\Omega f(x, u_n)u_0 dx, \\ \int_\Omega f(x, u_n)u_0 dx \to \int_\Omega f(x, u_0)u_0 dx. \end{cases} \tag{3.9}$$

On the one hand

$$\|u_n - u_0\|^2 + \varepsilon \int_\Omega (|\nabla u_n|^{q-2}\nabla u_n - |\nabla u_0|^{q-2}\nabla u_0)(\nabla u_n - \nabla u_0)dx$$

$$= \langle I_\varepsilon'(u_n) - I_\varepsilon'(u_0), u_n - u_0 \rangle + \int_\Omega a(x)(u_n - u_0)^2 dx$$

$$+ \int_\Omega [f(x, u_n) - f(x, u_0)](u_n - u_0)dx. \tag{3.10}$$

By $I'_\varepsilon(u_n) \to 0$ and $u_n \rightharpoonup u_0$ *in* $H^1_0(\Omega)$, we know that

$$\langle I'_\varepsilon(u_n) - I'_\varepsilon(u_0), u_n - u_0 \rangle \to 0.$$

By (3.9), we obtain

$$|\int_\Omega [f(x, u_n) - f(x, u_0)](u_n - u_0)dx| \to 0.$$

By Proposition 2.1 we obtain

$$|\int_\Omega a(x)(u_n - u_0)^2 dx| \to 0.$$

It follows from (3.10) that

$$u_n \to u_0 \quad on \quad H^1_0(\Omega).$$

So we have proved that $I_\varepsilon(u)$ satisfies $(PS)_c$ condition for any $c \in R$. $\square$

The Proof of Theorem 1.1.

We will complete the proof by applying the Linking Theorem (Proposition 2.4).

Proof.   Let $\rho > r > 0$, $z = \frac{e_{n+1}}{\|e_{n+1}\|}r \in Z$ and define

$$M^\rho := \{u = y + \lambda z : \|u\| \leq \rho, \lambda \geq 0, y \in Y\},$$

$$M^\rho_0 := \{u = y + \lambda z : y \in Y, \|u\| = \rho, \lambda \geq 0 \ \ or \ \ \|u\| \leq \rho, \lambda = 0\},$$

$$N_r := \{u \in Z : \|u\| = r\},$$

$$c_\varepsilon := \inf_{\gamma \in \Gamma} \max_{u \in M^\rho} I_\varepsilon(\gamma(u)),$$

$$\Gamma := \{\gamma \in C(M^\rho, H^1_0(\Omega)) : \gamma\mid_{M^\rho_0} = id\}.$$

We hope to find $0 < r < 1 < \rho$ and $\varepsilon_0 > 0$ such that

$$\inf_{N_r} I_\varepsilon > \max_{M^\rho_0} I_\varepsilon.$$

when $0 \leq \varepsilon \leq \varepsilon_0$. By hypothesis $(f_1)(f_3)$, for $\forall \sigma > 0$, there exists a $C_\sigma > 0$ such that

$$|F(x, u)| \leq \sigma|u|^2 + C_\sigma|u|^{p+1}. \tag{3.11}$$

For every $u \in N_r$, we have that $u \in Z$ and $\|u\| = r$. We deduce from Lemma 3.1 and *Sobolev* embedding theorem that

$$\begin{aligned}
I_\varepsilon(u) &= \frac{1}{2}\int_\Omega \left[|\nabla u|^2 - a(x)u^2\right]dx + \frac{\varepsilon}{q}\int_\Omega |\nabla u|^q dx - \int_\Omega F(x, u)dx \\
&\geq \frac{\delta}{2}\|u\|^2 + \frac{\varepsilon}{q}\int_\Omega |\nabla u|^q dx - \sigma \int_\Omega |u|^2 dx - C_\sigma \|u\|^{p+1}_{p+1} \\
&\geq \frac{\delta}{2}\|u\|^2 - C\sigma\|u\|^2 - \tilde{C}_\sigma\|u\|^{p+1} \\
&\geq \frac{\delta}{2}r^2 - o(r^2).
\end{aligned}$$

If $u \in M_0^\rho$ with $u = y + \lambda z, \|u\| \leq \rho, \lambda = 0$, then $u = y \in Y$. By Lemma 3.2 and (f_4), we known that

$$
\begin{aligned}
I_\varepsilon(u) &= \frac{1}{2} \int_\Omega \left[|\nabla u|^2 - a(x)u^2 \right] dx + \frac{\varepsilon}{q} \int_\Omega |\nabla u|^q dx - \int_\Omega F(x,u) dx \\
&\leq \frac{\lambda_n}{2} \int_\Omega |u|^2 dx + \frac{\varepsilon}{q} \int_\Omega |\nabla u|^q dx - \int_\Omega F(x,u) dx \\
&\leq \frac{\varepsilon}{q} \int_\Omega |\nabla u|^q dx \leq \frac{\varepsilon}{q} |\Omega|^{\frac{2-q}{2}} \|u\|^q \leq \frac{\varepsilon}{q} |\Omega|^{\frac{2-q}{2}} \rho^q.
\end{aligned} \tag{3.12}
$$

By hypothesis (f_2), there exist $C_1, C_2 \geq 0$ such that

$$
F(x,u) \geq C_1 |u|^{\frac{1}{\theta}} - C_2. \tag{3.13}
$$

If $u \in M_0^\rho$ with $u = y + \lambda z, \lambda \geq 0$, then we have

$$
\begin{aligned}
I_\varepsilon(u) &= \frac{1}{2} \int_\Omega \left[|\nabla u|^2 - a(x)u^2 \right] dx + \frac{\varepsilon}{q} \int_\Omega |\nabla u|^q dx - \int_\Omega F(x,u) dx \\
&\leq \frac{1}{2} \|u\|^2 + C\|u\|^2 + \frac{\varepsilon}{q} C\|u\|^q - C \int_\Omega |u|^{\frac{1}{\theta}} dx - \tilde{C}.
\end{aligned} \tag{3.14}
$$

Since $M_0^\rho \subset Y \oplus Rz$ with $\dim(Y \oplus Rz) < \infty$ and any two norms in the finite dimensional space $Y \oplus Rz$ are equivalent, we see that there exists a $\bar{C} > 0$ such that

$$
\int_\Omega |u|^{\frac{1}{\theta}} dx \geq \bar{C} \|u\|^{\frac{1}{\theta}}. \tag{3.15}
$$

for every $u \in Y \oplus Rz$.

By (3.14) and (3.15), for every $u \in M_0^\rho$ with $\|u\| = \rho$, we have that

$$
I_\varepsilon(u) \leq \frac{1}{2} \|u\|^2 + C\|u\|^2 + C\|u\|^q - C\|u\|^{\frac{1}{\theta}} - C \to -\infty,
$$

as $\rho \to +\infty$.

Now we fix $\rho > 1 > r$ such that

$$
\max_{M_0^\rho} I_\varepsilon(u) \leq \frac{\varepsilon}{q} |\Omega|^{\frac{2-q}{2}} \rho^q.
$$

Since

$$
\inf_{N_r} I_\varepsilon(u) \geq \frac{\delta}{2} r^2 - o(r^2),
$$

we can take $0 < r < 1$ sufficiently small such that

$$
\frac{\delta}{2} r^2 - o(r^2) > \frac{\delta}{4} r^2
$$

hence there exists an $\varepsilon_0 = \varepsilon_0(|\Omega|, q, \rho, r) > 0$ such that

$$
\inf_{N_r} I_\varepsilon(u) > \frac{\delta}{4} r^2 > \frac{\varepsilon}{q} |\Omega|^{\frac{2-q}{2}} \rho^q \geq \max_{M_0^\rho} I_\varepsilon(u).
$$

when $0 \leq \varepsilon \leq \varepsilon_0$.

By Lemma 3.3 we know that $I_\varepsilon(u)$ satisfies $(PS)_c$ condition on $H_0^1(\Omega)$ for any $c \in R^1$, so from the Linking Theorem (Proposition 2.4) it follows that

$$c_\varepsilon := \inf_{\gamma \in \Gamma} \max_{u \in M^\rho} I_\varepsilon(\gamma(u))$$

with

$$\Gamma := \{\gamma \in C(M^\rho, H_0^1(\Omega)) : \gamma \mid_{M_0^\rho} = id\},$$

is a critical value of I_ε and

$$c_\varepsilon \geq \inf_{N_r} I_\varepsilon(u) > 0.$$

for $0 < \varepsilon \leq \varepsilon_0$.

Thus we have proved that for any $0 < \varepsilon \leq \varepsilon_0$, there exists a nontrivial solution $u_\varepsilon \in H_0^1(\Omega) \setminus \{0\}$ to $(1.1)_\varepsilon$ with $I_\varepsilon(u_\varepsilon) = c_\varepsilon$.

Finally, we prove that for any sequence $\varepsilon_n \to 0^+$, there exists a subsequence $\{u_{\varepsilon_{n_k}}\}_{k=1}^{+\infty}$ of $\{u_{\varepsilon_n}\}_{n=1}^{+\infty}$ and some nontrivial solution $\tilde{u}$ of $(1.1)_0$ such that $u_{\varepsilon_{n_k}} \to \tilde{u}$ in $H_0^1(\Omega)$ with $I_0(\tilde{u}) = c_0$. To this end, we notice that $M_0^\rho, M^\rho, N_r, \Gamma$ depends only on $a(x), f(x,t), |\Omega|, q$ which makes us possible to prove $\|u_\varepsilon\| \leq C < +\infty$ for some constant C as what we did for any $(PS)_c$ sequence in the proof of Lemma 3.3.

As M^ρ, Γ do not depend on ε for $0 < \varepsilon \leq \varepsilon_0$, we see that

$$c_0 = \inf_{\gamma \in \Gamma} \max_{u \in M^\rho} I_0(\gamma(u)) \leq \inf_{\gamma \in \Gamma} \max_{u \in M^\rho} I_\varepsilon(\gamma(u))$$

$$= c_\varepsilon \leq \inf_{\gamma \in \Gamma} \max_{u \in M^\rho} I_{\varepsilon_0}(\gamma(u)) = c_{\varepsilon_0}.$$

Since $\langle I_\varepsilon'(u_\varepsilon), u_\varepsilon \rangle = 0$ so

$$\int_\Omega [|\nabla u_\varepsilon|^2 - a(x)u_\varepsilon^2]dx + \varepsilon \int_\Omega |\nabla u_\varepsilon|^q dx = \int_\Omega f(x, u_\varepsilon)u_\varepsilon dx.$$

As $u_\varepsilon \in H_0^1(\Omega) = Y \oplus Z$, there exist are $y_\varepsilon \in Y, z_\varepsilon \in Z$ such that $u_\varepsilon = y_\varepsilon + z_\varepsilon$. By Proposition 2.7, Lemma 3.1 and (3.13), there are $C_{M_1}, C_{M_2} > 0$ and some $\beta \in (\frac{1}{\alpha}, \frac{1}{2})$. such that

$$c_{\varepsilon_0} + 1 + \|u_\varepsilon\| \geq c_\varepsilon = I_\varepsilon(u_\varepsilon) = I_\varepsilon(u_\varepsilon) - \beta\langle I_\varepsilon'(u_\varepsilon), u_\varepsilon\rangle$$

$$= (\frac{1}{2} - \beta) \int_\Omega [|\nabla u_\varepsilon|^2 - a(x)u_\varepsilon^2]dx + \varepsilon(\frac{1}{q} - \beta) \int_\Omega |\nabla u_\varepsilon|^q dx$$

$$+ \int_\Omega [\beta f(x, u_\varepsilon)u_\varepsilon - F(x, u_\varepsilon)]dx$$

$$\geq (\frac{1}{2} - \beta)(\delta\|z_\varepsilon\|^2 + \lambda_1\|y_\varepsilon\|_2^2) + \varepsilon(\frac{1}{q} - \beta) \int_\Omega |\nabla u_\varepsilon|^q dx$$

$$+ (\frac{\beta}{\theta} - 1) \int_\Omega F(x, u_\varepsilon)dx - C_{M_1}$$

$$\geq (\frac{1}{2} - \beta)(\delta\|z_\varepsilon\|^2 + \lambda_1\|y_\varepsilon\|_2^2) + C_1(\frac{\beta}{\theta} - 1)\|u_\varepsilon\|_{\frac{1}{\theta}}^{\frac{1}{\theta}} - C_{M_2}.$$

Similar to what we did in the proof of Lemma 3.3, we can show that there exists a $C > 0$ such that $\|u_\varepsilon\| \leq C < +\infty$.

Thus for any sequence $\varepsilon_n \to 0^+$, by Proposition 2.1 (the *Rellich* embedding theorem), we may assume that there is a $\tilde{u} \in H_0^1(\Omega)$ and a subsequence of $\{u_{\varepsilon_n}\}_{n=1}^{+\infty}$, still denoted by $\{u_{\varepsilon_n}\}_{n=1}^{+\infty}$, such that

$$\begin{cases} u_{\varepsilon_n} \rightharpoonup \tilde{u} & \text{on} \quad H_0^1(\Omega), \\ u_{\varepsilon_n} \to \tilde{u} & \text{on} \quad L^p(\Omega)(2 \leq p < 2^*). \end{cases}$$

From $I_\varepsilon'(u_\varepsilon) = 0$, it is easy to show that

$$\int_\Omega [\nabla \tilde{u} \nabla v - a(x)\tilde{u}v - f(x, \tilde{u})v]dx = 0, \qquad \forall v \in H_0^1(\Omega).$$

Thus $\tilde{u}$ is a weak solution of $(1.1)_0$.

If $\tilde{u} = 0$, then

$$\int_\Omega f(x, u_{\varepsilon_n})u_{\varepsilon_n}dx \to 0.$$

$$\int_\Omega a(x)u_{\varepsilon_n}^2 dx \to 0.$$

Hence

$$\int_\Omega |\nabla u_{\varepsilon_n}|^2 dx + \varepsilon_n \int_\Omega |\nabla u_{\varepsilon_n}|^q dx = \int_\Omega [f(x, u_{\varepsilon_n})u_{\varepsilon_n} + a(x)u_{\varepsilon_n}^2]dx \to 0,$$

which shows that $u_{\varepsilon_n} \to 0$ in $H_0^1(\Omega)$ and

$$c_{\varepsilon_n} = I_{\varepsilon_n}(u_{\varepsilon_n}) \to 0.$$

But

$$c_{\varepsilon_n} \geq c_0 > 0,$$

which is a contradiction. Thus $\tilde{u} \neq 0$ is a nontrivial solution to $(1.1)_0$.

Finally we prove that $I_0(\tilde{u}) = c_0$.

From $c_{\varepsilon_n} \geq c_0 > 0$,it follows that for any $\gamma \in \Gamma$,

$$c_{\varepsilon_n} \leq \max_{u \in M^\rho} I_\varepsilon(\gamma(u)) \leq \max_{u \in M^\rho}(I(\gamma(u)) + \varepsilon_n J(\gamma(u)))$$
$$\leq \max_{u \in M^\rho} I(\gamma(u)) + \varepsilon_n \max_{u \in M^\rho} J(\gamma(u))),$$

which gives

$$c_0 \leq \liminf_{n \to +\infty} c_{\varepsilon_n} \leq \limsup_{n \to +\infty} c_{\varepsilon_n}$$
$$\leq \limsup_{n \to +\infty} \max_{u \in M^\rho}(I(\gamma(u)) + \limsup_{n \to +\infty}(\varepsilon_n \max_{u \in M^\rho} J(\gamma(u))))$$
$$= \max_{u \in M^\rho} I_0(\gamma(u)).$$

As $\gamma \in \Gamma$ is arbitrary, we see that

$$c_0 \leq \liminf_{n \to +\infty} c_{\varepsilon_n} \leq \limsup_{n \to +\infty} c_{\varepsilon_n} \leq \inf_{\gamma \in \Gamma} \max_{u \in M^\rho} I_0(\gamma(u)) = c_0.$$

Thus

$$\lim_{n \to +\infty} c_{\varepsilon_n} = c_0.$$

In fact, we have shown that

$$\lim_{\varepsilon \to 0} c_\varepsilon = c_0 > 0.$$

Since $u_{\varepsilon_n} \rightharpoonup \tilde{u}$ in $H_0^1(\Omega)$ and $I_0'(\tilde{u}) = 0$ we see that

$$\int_\Omega |\nabla u_{\varepsilon_n}|^2 dx + \varepsilon_n \int_\Omega |\nabla u_{\varepsilon_n}|^q dx = \int_\Omega [f(x, u_{\varepsilon_n})u_{\varepsilon_n} + a(x)u_{\varepsilon_n}^2] dx$$
$$\to \int_\Omega [f(x, \tilde{u})\tilde{u} + a(x)\tilde{u}^2] dx = \int_\Omega |\nabla \tilde{u}|^2 dx.$$

As

$$\int_\Omega |\nabla u_{\varepsilon_n}|^q dx \leq C \|u_{\varepsilon_n}\|^q \leq \tilde{C},$$

so

$$\varepsilon_n \int_\Omega |\nabla u_{\varepsilon_n}|^q dx \to 0,$$

hence

$$\int_\Omega |\nabla u_{\varepsilon_n}|^2 dx \to \int_\Omega |\nabla \tilde{u}|^2 dx.$$

which together with the fact that $u_{\varepsilon_n} \rightharpoonup \tilde{u}$ in $H_0^1(\Omega)$ shows that

$$u_{\varepsilon_n} \to \tilde{u} \quad in \quad H_0^1(\Omega)$$

and

$$c_{\varepsilon_n} = I_{\varepsilon_n}(u_{\varepsilon_n}) \to I_0(\tilde{u}).$$

Hence $I_0(\tilde{u}) = c_0$. This completes the proof.   □

Remark 3.4 If $\lambda_1 > 0$, then it suffices to use the Mountain Pass Theorem instead of the Linking Theorem to prove the existence of a nontrivial solutions to $(1.1)_\varepsilon$.

References

1. A. Ambrosetti, A. Malchiodi, Perturbation methods and semilinear elliptic problems on $\mathbb{R}^N$, Birkhäuser Verlag, Boston • Basel • Berlin, 2006.
2. A. Ambrosetti, P. Rabinowitz, Dual variational methods in critical points theory and applications, J. Funct. Anal. 14(1973), 349-381.
3. K. C. Chang, Critical point theory and its applications, Shanghai Sci. Tech. Press, 1986 (in Chinese).
4. K. C. Chang, Q. Y. Lin, Functional Analysis, Beijing University Press, Beijing, 1987 (in Chinese).
5. L. Cherfils, Y. Ìlyasov, On the stationary solution of generalized reaction diffusion equation with p- and q-Laplacian, Comm. Pure. Appl. Anal, 4(1) (2005), 9-22.
6. Chenjun He, Gongbao Li, The existence of a nontrivial solution to the p and q-Laplacian problem with nonlinearity asymptotic to u^{p-1} at infinity in R^N, Nonlinear Analysis, TMA, 68(2008), 1100-1119.
7. Gongbao Li, Xiaoyan Liang, The existence of nontrivial solutions to nonlinear elliptic equations of p and q-Laplacian type on R^N, Nonlinear Analysis, 71(2009), 2316-2334.
8. P. Rabinowitz, Some minimax theorems and application to nonlinear partial elliptic differential equation in Nonlinear Analysis: A collection of papers in honor of Erich Rothe, Academic Press, New York, 1978, 161-177.
9. Y. T. Shen, S. S. Yan, Variational methods of quasilinear elliptic equations, South China University of Technology Press, 1995 (in Chinese).
10. M. Willem, Minimax Theorems, Birkhäuser, Boston • Basel • Berlin, 1996.

Perspectives in Mathematical Sciences
Interdisciplinary Mathematical Sciences, Volume 9, 2009
pp. 135–144

Chapter 7

Chaotic Dynamics for the Two-component Bose-Einstein Condensate System

Jibin Li[1,2]

[1]*Department of Mathematics, Zhejiang Normal University,*
Jinhua, Zhejiang 321004,
P. R. China

[2]*Kunming University of Science and Technology,*
Kunming, Yunnan 650093, P. R. China

E-mail address: lijb@zjnu.cn or jibinli@gmail.com

Dedicated to Professor Youzhong Guo on the occasion of his 75th birthday

In this paper chaotic behavior of the two-component Bose-Einstein condensate system is considered. Under given parameter conditions, the existence of chaotic motions and the bifurcations of subharmonic solutions are rigorously proven. To verify this new strange attractor, two numerical examples are demonstrated.

1. Introduction

[Raghavan, S. et al, 1999] discussed the coherent atomic oscillations between two weakly coupled Bose-Einstein condensates. [Marino, I. et al, 1999] continuously considered Josephson-like tunneling phenomena. The model equations were given by

$$\dot{z} = -\sqrt{1-z^2}\sin\phi, \quad \dot{\phi} = \Lambda z + \Delta E + \frac{z}{\sqrt{1-z^2}}\cos\phi, \tag{1}$$

where z is fractional population imbalance, ϕ is the relative phase difference, Λ and ΔE are constant parameters (see the above two references).

2000 Mathematics Subject Classifications: 34C25-28, 58F05, 58F14, 58F30.
Key words and phrases: Chaotic behavior, subharmonic bifurcation, Melnikov method, Bose-Einstein condensate system.
This research was supported by the National Natural Science Foundation of China (10671179) and the (10831003).

[Xie, Q. T. et al, 2003] considered the chaotic behavior for the system

$$\dot{z} = -2K(t)\sqrt{1-z^2}\sin\phi, \quad \dot{\phi} = \Lambda z + \Delta E(t) + \frac{2K(t)z}{\sqrt{1-z^2}}\cos\phi, \tag{2}$$

where

$$\Delta(t) = \Delta E_0 + \Delta E_1 \sin\omega t, \quad K(t) = K_0 + K_1 \sin\omega t, \tag{3}$$

ΔE_0, K_0 are constants and $|K_1| \ll 1$, $|\Delta E_1| \ll 1$.

Recently, [Boli Xia et al 2006] investigated the stability and chaotic behavior of a periodically driven Bose-Einstein Condensate (BEC) with two hyperfine states as follows:

$$\frac{d\eta}{dt} = -2K\sqrt{1-\eta^2}\sin\phi, \quad \frac{d\phi}{dt} = \gamma_0 + \gamma_1\sin(\omega t) + U\eta + 2K\eta(1-\eta^2)^{-\frac{1}{2}}\cos\phi, \tag{4}$$

where $\gamma = \gamma_0 + \gamma_1 \sin(\omega t)$ is relative energy fluctuation between the two hyperfine states. The authors gave some numerical results for the chaotic regimes of (4).

To our knowledge, the complete and rigorous analysis of bifurcation behavior in the parametric space of (2) and (4) has not be considered. In this paper we shall consider this problem for the system

$$\frac{d\eta}{dt} = -\alpha\sqrt{1-\eta^2}\sin\phi, \quad \frac{d\phi}{dt} = \gamma + \beta\eta + \alpha\eta(1-\eta^2)^{-\frac{1}{2}}\cos\phi, \tag{5}$$

where, for simplicity, we use parameters β and α to substitute the mean-field interaction parameter U and the population transfer $2K$ in [Boli Xia et al 2006], respectively. Noting system (2), we are also interesting to the perturbed system

$$\frac{d\eta}{dt} = -\alpha_0\sqrt{1-\eta^2}\sin\phi - \epsilon\alpha_1\sqrt{1-\eta^2}\sin\phi\sin\omega t,$$

$$\frac{d\phi}{dt} = \beta_0\eta + \alpha_0\eta(1-\eta^2)^{-\frac{1}{2}}\cos\phi$$
$$+ \epsilon[\gamma_0 + \beta_1\eta + (\gamma_1 + \alpha_1\eta(1-\eta^2)^{-\frac{1}{2}}\cos\phi)\sin(\omega t)], \tag{6}$$

where we assume in (4) that γ_0 and γ_1 are small and $\alpha = \alpha_0 + \epsilon\alpha_1 \sin\omega t$, $\beta = \beta_0 + \epsilon\beta_1, \epsilon \ll 1$.

We notice that system (5) has two straight line solutions $\eta = \pm 1$ in the phase plane (ϕ, η) on them $\frac{d\phi}{dt}$ has no definition. In addition, (5) is periodic in ϕ. Hence, the state (ϕ, η) can be viewed on a bounded phase cylinder $S^1 \times (-1, 1)$, where $S^1 = [-\pi, \pi]$ with $-\pi, \pi$ identified.

Notice that system (5) has the Hamiltonian quality

$$H(\phi, \eta) = \gamma\eta + \frac{\beta}{2}\eta^2 - \alpha\sqrt{1-\eta^2}\cos\phi = h. \tag{7}$$

This paper is organized as follows. In section 2 we study the bifurcations of phase portraits of (5) in the three-parametric space (α, β, γ) and give all parametric representations of the phase orbits for system (5) in the case $\gamma = 0$. In section 3, we calculate the Melnikov integrals to establish bifurcation conditions and give some numerical results.

2. The bifurcations of phase portraits of system (5)

In this section, we discuss the bifurcations of phase portraits of (5) when the parameters α, β and γ are varied. We always assume that $\beta > 0$. Otherwise, by using the transformation $\phi \to -\phi$ in (5), we can get this case. In addition, the case $\gamma < 0$ or $\alpha < 0$ can be discussed, by using the transformation $\phi \to \phi + \pi$ in (5). So that, we can always assume that $\alpha > 0$, $\beta > 0$, $\gamma \geq 0$.

I. The case $\gamma > 0$.
We see from (5) that an equilibrium point (ϕ_e, η_e) of (5) lies in the straight lines $\phi = 0$ and $\phi = \pi$, respectively, and satisfies $-1 < \eta_e < 1$, where η_e is an intersection point of two curves $u = \beta\eta + \gamma$ and $u = \mp\dfrac{\alpha\eta}{\sqrt{1-\eta^2}}$, respectively. It is easy to show that there exists only one equilibrium point $O(0, \eta_{e1})$ when $\phi = 0$. For $\alpha < \beta$, we denote that $\eta_c = \sqrt{1 - (\frac{\alpha}{\beta})^{\frac{2}{3}}}$, $\gamma_c = \eta_c \left(\dfrac{\alpha}{\sqrt{1-\eta_c^2}} - \beta \right)$. Then, for a fixed pair (α, β), we have the following conclusions.

1) When $\gamma < \gamma_c$, there exist three equilibrium points $P_\pi^a(\pi, \eta_{e2})$, $P_\pi^b(\pi, \eta_{e3})$, $P_\pi^c(\pi, \eta_{e4})$ with $\eta_{e2} < \eta_{e3} < 0 < \eta_{e4}$.

2) When $\gamma = \gamma_c$ there exist two equilibrium points $P_\pi^{ab}(\pi, \eta_c)$, $P_\pi^c(\pi, \eta_{e4})$ with $\eta_{e2} = \eta_{e3} \equiv \eta_c < 0 < \eta_{e4}$.

3) When $\gamma > \gamma_c$ there exists only one equilibrium points $P_\pi^c(\pi, \eta_{e4})$ with $0 < \eta_{e4} < 1$.

Let $M(\phi_e, \eta_e)$ be the coefficient matrix of the linearized system of (5) at an equilibrium point (ϕ_e, η_e) and $J(\phi_e, \eta_e)$ be its Jacobian determinant. Then, we have $\text{Trace}(M(v_e, \eta_e) = 0$ and

$$J(0, \eta_{e1}) = \alpha\sqrt{1 - \eta_{e1}^2} \left(\beta + \frac{\alpha}{\sqrt{(1-\eta_{e1}^2)^3}} \right),$$

$$J(\pi, \eta_{ej}) = -\alpha\sqrt{1 - \eta_{ej}^2} \left(\beta - \frac{\alpha}{\sqrt{(1-\eta_{ej}^2)^3}} \right), \quad j = 2, 3, 4.$$

Noting $J(\pi, \eta_c) = 0$, it follows that for $\alpha < \beta$,, we have $J(\pi, \eta_{e2}) > 0$, $J(\pi, \eta_{e3}) < 0$, $J(\pi, \eta_{e4}) > 0$.

By the theory of planar dynamical systems, we know that for an equilibrium point of a planar Hamiltonian system, if $J < 0$ then the equilibrium point is a saddle point; if $J > 0$, then it is a center point; if $J = 0$ and the Poincare index of the equilibrium point is 0 then it is a cusp.

Write that

$$h_O = H(0, \eta_{e1}) = -\left[\alpha\sqrt{1 - \eta_{e1}^2} + \eta_{e1}^2\left(\frac{\alpha}{\sqrt{1 - \eta_{e1}^2}} + \frac{\beta}{2}\right)\right] < 0,$$

$$h_j^\pi = H(\pi, \eta_{ej}) = \alpha\sqrt{1 - \eta_{ej}^2} + \eta_{ej}^2\left(\frac{\alpha}{\sqrt{1 - \eta_{ej}^2}} - \frac{\beta}{2}\right), \ j = 2, 3, 4.$$

$$h_5 = H(\phi, -1) = -\gamma + \frac{\beta}{2}, \quad h_6 = H(\phi, 1) = \gamma + \frac{\beta}{2}.$$

When $\beta > 2\alpha$, as γ increasing from $0 < \gamma < \gamma_c$ to $\gamma > \gamma_c$, we have the bifurcation of phase portraits of (5) shown in Fig. 1.

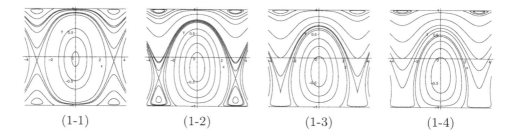

$\qquad\quad$ (1-1) $\qquad\qquad\quad$ (1-2) $\qquad\qquad\quad$ (1-3) $\qquad\qquad\quad$ (1-4)

Fig. 1. The bifurcations of phase portraits of system (5) for $\beta > 2\alpha$

(1-1) $h_O < h_2^\pi < h_5 < h_1^\pi < h_6 < h_3^\pi$, $\gamma < \gamma_c$. (1-2) $h_O < h_5 < h_2^\pi < h_1^\pi < h_6 < h_3^\pi$, $\gamma < \gamma_c$.
(1-3) $h_O < h_5 < h_2^\pi = h_1^\pi < h_6 < h_3^\pi$, $\gamma = \gamma_c$. (1-4) $h_O < h_5 < h_6 < h_3^\pi$, $\gamma > \gamma_c$.

We see from Fig. 1 (1-1) that in the phase cylinder, there exist five families of periodic orbits of (5) as follows:

(i) The family of oscillating periodic orbits $\{\Gamma_1^h\}$ enclosing the equilibrium point $O(0, \eta_{e1})$ which is given by $H(\phi, \eta) = h$, $h \in (h_O, h_2^\pi)$.

(ii) Two families of rotating periodic orbits $\{\Gamma_{r\pm}^h\}$, which are given by $H(\phi, \eta) = h$, $h \in (h_3^\pi, h_5)$ and $h \in (h_3^\pi, h_6)$, respectively.

(iii) Two families of oscillating periodic orbits $\{\Gamma_{2\pm}^h\}$ enclosing the equilibrium points $P_\pi^a(\pi, \eta_{e2})$ and $P_\pi^c(\pi, \eta_{e4})$, respectively, which are given by $H(\phi, \eta) = h$, $h \in (h_5, h_2^\pi)$ and $h \in (h_6, h_4^\pi)$, respectively.

Similarly, we know from Fig. 1 (1-2)-(1-4) that there exist different families of periodic orbits of (5).

When $\alpha < \beta < 2\alpha$, as γ increasing from $0 < \gamma < \gamma_c$ to $\gamma > \gamma_c$, we have the bifurcation of phase portraits of (5) shown in Fig. 2.

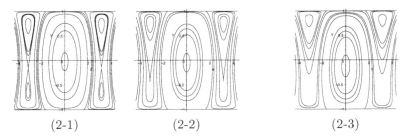

(2-1)   (2-2)   (2-3)

Fig. 2. The bifurcations of phase portraits of system (5) for $\alpha < \beta < 2\alpha$

(2-1) $h_O < h_5 < h_6 < h_2^\pi < h_1^\pi < h_6 < h_3^\pi$, $\gamma < \gamma_c$.
(2-2) $h_O < h_5 < h6 < h_2^\pi = h_1^\pi < h_3^\pi$, $\gamma = \gamma_c$. (2-3) $h_O < h_5 < h_6 < h_3^\pi$, $\gamma > \gamma_c$.

When $0 < \beta < \alpha$, for $\gamma > 0$, we have the phase portrait of (5) shown in Fig. 3.

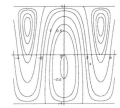

Fig. 3. The phase portrait of system (5) for $\alpha > \beta > 0$ where

$$h_O < h_5 < h_6 < h_3^\pi, \ \gamma > 0.$$

We see from (7) that $\cos \phi = \frac{h - \gamma\eta - \frac{\beta}{2}\eta^2}{-\alpha\sqrt{1-\eta^2}}$. By using the first equation of (5), we have

$$\frac{d\eta}{dt} = -\sqrt{(\alpha^2 - h^2) + 2h\gamma\eta + (h\beta - \alpha_2 - \gamma^2)\eta^2 - \beta\gamma\eta^3 - \frac{1}{4}\beta^2\eta^4}.$$

Hence, we can use this formula to obtain all parametric representations of periodic orbits and homoclinic orbits of (5) in the different parameter regions of the 3-parameter space (α, β, γ). Because we are interested the symmetry case i.e, the case $\gamma = 0$. We omit these parametric representations.

II. The case $\gamma = 0$.

We see from (5) that if $\alpha \geq \beta$, there exist two equilibrium points $O(0,0)$ and $P(\pm\pi, 0)$(which are identified) of $(5)_{\gamma=0}$. If $\alpha < \beta$, there exist four equilibrium points O, P and $P_\pi^\pm(\pm\pi, \pm\eta_1)$, where $\eta_1 = \sqrt{1 - \frac{\alpha^2}{\beta^2}}$. Now, we have

$$J(0,0) = \alpha(\alpha + \beta), \quad J(\pi,0) = -\alpha(\beta - \alpha), \quad J(\pi,\eta_1) = \beta^2 - \alpha^2.$$

$$h_0 = H(0,0) = -\alpha, \quad h_\pi = H(\pi,0) = \alpha, \quad h_1 = H(\pi,\eta_1) = \frac{\alpha^2 + \beta^2}{2\beta}$$

and for $y = \pm 1$, $h = h_2 = \frac{\beta}{2}$.

By using the above fact, we have the bifurcations of phase portraits of $(5)_{\gamma=0}$ shown in Fig. 4 (4-1)-(4-4).

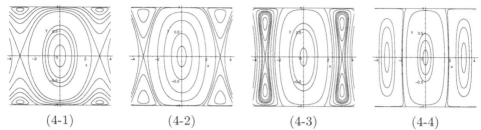

(4-1)   (4-2)   (4-3)   (4-4)

Fig. 4. The bifurcations of phase portraits of system $(5)_{\gamma=0}$ for $\alpha > 0$, $\beta > 0$

(4-1) $\beta > 2\alpha$, $h_0 < h_\pi < \frac{\beta}{2} < h_1$. (4-2) $\beta = 2\alpha$, $h_0 < \frac{\beta}{2} = h_\pi < h_1$.

(4-3) $\alpha < \beta < 2\alpha$, $h_0 < \frac{\beta}{2} < h_\pi < h_1$. (4-4) $\beta \leq \alpha$, $h_0 < \frac{\beta}{2} < h_1$.

We next consider the parametric representations of orbits of the vector fields defined by $(5)_{\gamma=0}$ in different parameter conditions. We see from (7) that $\cos\phi = \frac{h - \frac{\beta}{2}\eta^2}{-\alpha\sqrt{1-\eta^2}}$. By using the first equation of $(5)_{\gamma=0}$, we have

$$\frac{d\eta}{dt} = -\sqrt{(\alpha^2 - h^2) + (h\beta - \alpha_2)\eta^2 - \frac{1}{4}\beta^2\eta^4} = -\frac{\beta}{2}\sqrt{(a^2 - \eta^2)(\eta^2 - b^2)}, \qquad (8)$$

where

$$a^2 = \frac{2}{\beta^2}\left[(h\beta - \alpha^2) + \sqrt{\Delta}\right], \quad b^2 = \frac{2}{\beta^2}\left[(h\beta - \alpha^2) - \sqrt{\Delta}\right] \qquad (9)$$

and $\Delta = (h\beta - \alpha^2)^2 + \beta^2(\alpha^2 - h^2) = \alpha^2(\alpha^2 + \beta^2 - 2h\beta)$.

Thus, we can use (8) to obtain all parametric representations of periodic orbits and homoclinic orbits of $(5)_{\gamma=0}$ in the different parameter regions of the parameter space (α, β). We shall use the Jacobian elliptic function $\mathrm{cn}(u, k)$, $\mathrm{dn}(u, k)$ et al (see [Byrd, P.F. et al, 1971]).

1. The case $\beta > 2\alpha$. In this case, we have $h_0 = -\alpha < 0 < h_\pi = \alpha < h_2 = \frac{\beta}{2} < h_1$. On the bounded phase cylinder $S^1 \times (-1, 1)$, there exist three families of oscillating periodic orbits $\{\Gamma_1^h\}$, $\{\Gamma_{2\pm}^h\}$ and two families of rotating periodic orbits $\{\Gamma_{r\pm}^h\}$.

(1) Corresponding to $H(\phi, \eta) = h$, $h \in (-\alpha, \alpha)$, an oscillating periodic orbit Γ_1^h has the parametric representation

$$\eta(t) = a\,\mathrm{cn}(\Delta^{\frac{1}{4}}t, k_1), \qquad (10)$$

where $k_1 = \frac{a}{\sqrt{a^2 - b^2}}$, a^2 and b^2 are given by (9) and $b^2 < 0$. The period of Γ_1^h is $T_1(k_1) = \frac{4K(k_1)}{\Delta^{\frac{1}{4}}}$, where $K(k)$ is the complete elliptic integral of the first kind. It is easy to show that $T_1(k_1)$ is monotonous as k_1 increases from 0 to 1.

(2) Corresponding to $H(\phi, \eta) = \alpha$, there are two homoclinic orbits $\Gamma^\alpha_{H\pm}$ to the saddle point $(\pm\pi, 0)$(which are identified as a point) with the parametric representations

$$\eta(t) = \pm\frac{2\Omega}{\beta}\operatorname{sech}(\Omega t). \tag{11}$$

where $\Omega = \sqrt{\alpha(\beta - \alpha)}$.

(3) Corresponding to $H(\phi, \eta) = h$, $h \in (\alpha, \frac{\beta}{2})$, the rotating periodic orbits $\Gamma^h_{r\pm}$ have the parametric representations

$$\eta(t) = \pm a\operatorname{dn}\left(\frac{\beta}{2}at, k_r\right), \tag{12}$$

where $k_r = \sqrt{\frac{a^2 - b^2}{a^2}} = \frac{1}{k_1}$, $b^2 > 0$. The period of $\Gamma^h_{r\pm}$ is $T_r(k_r) = \frac{4K(k_r)}{\beta a}$ which is monotonous as k_r increases.

(4) Corresponding to $H(\phi, \eta) = \frac{\beta}{2}$, there are two orbits connecting the two straight lines $y = \pm 1$.

(5) Corresponding to $H(\phi, \eta) = h$, $h \in \left(\frac{\beta}{2}, \frac{\alpha^2 + \beta^2}{2\beta}\right)$, the two oscillating periodic orbits $\Gamma^h_{2\pm}$ have the same parametric representations as (12).

2. The case $\beta = 2\alpha$. In this case, there is not two families $\{\Gamma^h_{r\pm}\}$ of the rotating orbits. The other orbits are the same as **1**.

3. The case $\alpha < \beta < 2\alpha$.

(1) Corresponding to $H(\phi, \eta) = h$, $h \in (-\alpha, \frac{\beta}{2})$, there exists a family $\{\Gamma^h_1\}$ of oscillating periodic orbits enclosing the equilibrium point $(0, 0)$ which has the same parametric representation as (10).

(2) Corresponding to $H(\phi, \eta) = h$, $h \in (\frac{\beta}{2}, \alpha)$, there exists a family $\{\Gamma^h_2\}$ of oscillating periodic orbits enclosing three equilibrium points $P_\pi(\pm\pi, 0)$ and $P^\pm_\pi(\pm\pi, \pm\eta_1)$ which has the same parametric representation as (10).

(3) Corresponding to $H(\phi, \eta) = \alpha$, there are two homoclinic orbits $\Gamma^\alpha_{2H\pm}$ with "figure-eight" to the saddle point $P_\pi(\pm\pi, 0)$(which are identified as a point) enclosing the equilibrium points $P^+_\pi(\pm\pi, \eta_1)$ and $P^-_\pi(\pm\pi, -\eta_1)$, respectively, which have the same parametric representations as (11).

(4) Corresponding to $H(\phi, \eta) = h$, $h \in \left(\alpha, \frac{1}{2\beta}(\alpha^2 + \beta^2)\right)$, there exist two families $\{\Gamma^h_{3\pm}\}$ of oscillating periodic orbits enclosing the equilibrium points $P^\pm_\pi(\pm\pi, \pm\eta_1)$, respectively, which have the same parametric representations as (12).

4. The case $0 < \beta \le \alpha$. In this case, the equilibrium point $P_\pi(\pi, 0)$ becomes a center.

Corresponding to $H(\phi, \eta) = h$, $h \in (-\alpha, \frac{\beta}{2})$ and $h \in (\frac{\beta}{2}, \alpha)$, respectively, there exist two families of periodic orbits which have the same parametric representations as (10).

3. The Melnikov analysis for the perturbed system (6) and numerical examples

In this section, we consider system (6). We suppose that $\beta_0 > 2\alpha_0$ or $\alpha_0 < \beta_0 < 2\alpha_0$ in (6). Namely, in the phase cylinder, there exist two homoclinic orbits with "figure-eight", having the parametric representation (11). It is well known that the Melnikov technique can be used to find the condition leading to the non-empty transversal intersection of stable and unstable manifolds of a saddle point and the conditions of subharmonic bifurcations (see [Guckenheimer and Holmes, 1983], [Li, J.B. et al, 1989]). For the perturbed system (6), the Melnikov function of homoclinic bifurcations evaluated along the unperturbed homoclinic orbits $\Gamma_{H\pm}^{\alpha}$ of $(5)_{\gamma=0}$ has the form:

$$
\begin{aligned}
M(t_0) &= \int_{-\infty}^{\infty} \{[-\alpha_0\sqrt{1-\eta^2}\sin\phi][\gamma_0 + \beta_1\eta \\
&\quad + (\gamma_1 + \alpha_1\eta(1-\eta^2)^{-\frac{1}{2}}\cos\phi)\sin(\omega(t+t_0))] \\
&\quad + [\beta_0\eta + \alpha_0\eta(1-\eta^2)^{-\frac{1}{2}}\cos\phi][\alpha_1\sqrt{1-\eta^2}\sin\phi\sin(\omega(t+t_0))]\}dt \\
&= \left(\int_{-\infty}^{\infty}\left(\frac{\alpha_1\beta_0}{\alpha_0}\eta^2 - \gamma_1\eta\right)\sqrt{\Omega^2 - \frac{\beta_0^2}{4}\eta^2}\sin\omega t\, dt\right)\cos\omega t_0 \\
&= \frac{2\pi\omega\Omega}{\beta_0}\left[\frac{\alpha_1\omega}{\alpha_0\sinh\frac{\pi\omega}{2\Omega}} - \frac{\gamma_1}{\cosh\frac{\pi\omega}{2\Omega}}\right]\cos\omega t_0 \equiv J\cos\omega t_0.
\end{aligned}
$$

We next only assume that $\alpha_0 > 0, \beta_0 > 0$. Then, system $(6)_{\epsilon=0}$ has different families of periodic orbits (see Fig. 4), which have the parametric representations given by (10) and (12), respectively. For example, we investigate the Fig. 4 (4-1), i.e., we consider the following resonant conditions for five families of periodic orbits $\{\Gamma_1^h\}$, $\{\Gamma_{2\pm}^h\}$ and $\{\Gamma_{r\pm}^h\}$:

$$
T_i(k_i) = \frac{2m\pi}{\omega n} = \frac{mT}{n}, \quad i = 1, 2, r. \tag{14}
$$

These relations determine $k_i = k_i(m)$ for $n = 1$. Along the unperturbed orbits Γ_i^h with $h = h(k_i(m))$ to calculate subharmonic Melnikov functions, we have

$$
M_i^m(t_0) = \int_0^{mT}(\alpha_1\beta_0\eta_i - \alpha_0\gamma_1)\sqrt{1-\eta_i^2}\sin\phi\sin\omega(t+t_0)dt
$$

$$
= \left[\int_0^{mT}\left(\frac{\alpha_1\beta_0}{2\alpha_0}\eta_i^2 - \gamma_1\eta_i\right)\cos\omega t\, dt\right]\cos\omega t_0. \tag{15}
$$

By using the parameter representations of $\{\Gamma_1^h\}$ and $\{\Gamma_{r\pm}^h\}$ (i.e.$\{\Gamma_{2\pm}^h\}$), we obtain the results as follows:

$$
M_1^m(t_0) = I_1^m\cos\omega t_0, \tag{16}
$$

where $I_1^m = \frac{\alpha_1\beta_0 a_0^2 m\pi}{2k_1^2\alpha_0 K(k_1)}\text{csch}\left(\frac{m\pi K'(k_1)}{K(k_1)}\right) - \frac{2\gamma_1\pi}{k_1}\text{sech}\left(\frac{(2m+1)\pi K'(k_1)}{2K(k_1)}\right).$

$$
M_r^m(t_0) = M_2^m(t_0) = I_r^m\cos\omega t_0, \tag{17}
$$

where $I_r^m = \frac{\alpha_1 \beta_0 a_0^2 m \pi^2}{2\alpha_0 K(k_1)} \mathrm{csch} \left(\frac{m \pi K'(k_1)}{K(k_1)} \right) - 2\gamma_1 \pi \mathrm{sech} \left(\frac{m \pi K'(k_1)}{K(k_1)} \right)$.

Thus, we obtain the following conclusion.

Theorem 1

(i) Suppose that $\beta_0 > 2\alpha_0$ or $\alpha_0 < \beta_0 < 2\alpha_0$ in (6). Making the parameters α_1, γ_1 and ω such that $J \neq 0$, then $M(t_0)$ defined by (13) has simple zeros. It implies that system (6) has Smale horseshoe dynamical behavior, i.e, the solutions of (6) are chaotic.

(ii) Suppose that $\alpha_0 > 0, \beta_0 > 0$ in (6). Making the parameters α_1, γ_1 such that $I_1^m \neq 0$ and $I_r^m \neq 0$, then $M_1^m(t_0)$ and $M_2^m(t_0)$ defined by (16) and (17) have simple zeros. It means that system (6) has $m-$order subharmonic periodic solutions bifurcated from the periodic families $\{\Gamma_1^h\}$, $\{\Gamma_{2\pm}^h\}$ and $\{\Gamma_{r\pm}^h\}$, respectively.

We consider the dynamical behavior for some orbits near two homoclinic orbits $\Gamma_{H\pm}^{\alpha_0}$ of $(6)_{\epsilon=0}$. As two numerical examples, Fig. 5 (5-1) gives the phase portrait of an orbit of (6) with the initial condition $\phi(0) = 0$, $\eta(0) = 0.88$, under the parameter conditions $\alpha_0 = 1, \beta_0 = 3, \epsilon = 0.1, \gamma_1 = 1.2, \omega = 1.25, \beta_1 = 1, \alpha_1 = \gamma_0 = 0$ in (6). Fig. 5 (5-2) is the corresponding Poincare map (4000 points are plotted); Fig. 5 (5-3) gives the phase portrait of an orbit of (6) with the initial condition $\phi(0) = 0$, $\eta(0) = 0.87$, under the parameter conditions $\alpha_0 = 1, \beta_0 = 3, \epsilon = 0.1, \gamma_1 = 1.15, \omega = 1.25, \beta_1 = 0.88, \alpha_1 = \gamma_0 = 0$ in (6). Fig. 5 (5-4) is the corresponding Poincare map (6000 points are plotted). We see from Fig. 5 (5-2) and (5.4) that it is different from the Duffing attractor, two Poincare maps give rise to a new type of strange attractor.

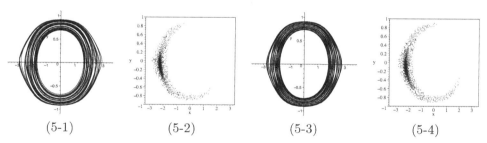

(5-1)	(5-2)	(5-3)	(5-4)

Fig. 5. The phase portraits of two orbits system (6) and their Poincare maps

Finally, we consider the dynamical behavior for some orbits near two homoclinic orbits $\Gamma_{2H\pm}^{\alpha}$ with "figure-eight" of $(6)_{\epsilon=0}$. As two numerical examples, Fig. 6 (6-1) gives the phase portrait of an orbit of (6) with the initial condition $\phi(0) = \pi$, $\eta(0) = 0.2$, under the parameter conditions $\alpha_0 = 5, \beta_0 = 5.25, \epsilon = 0.1, \gamma_1 = 1.2, \omega = 1.25, \beta_1 = 1, \alpha_1 = \gamma_0 = 0$ in (6). Fig. 6 (6-2) gives the phase portrait of an orbit of (6) with the initial condition $\phi(0) = \pi$, $\eta(0) = 0.2$, under the parameter conditions $\alpha_0 = 5, \beta_0 = 5.5, \epsilon = 0.1, \gamma_1 = 1.2, \omega = 1.25, \beta_1 = 1, \alpha_1 = \gamma_0 = 0$ in (6).

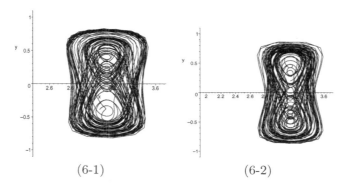

(6-1)   (6-2)

Fig. 6. The phase portraits of two orbits system (6)

References

1. Byrd, P.F. and Fridman, M.D.,[1971] Handbook of Elliptic Integrals for Engineers and Sciensists. Springer, Berlin.
2. Guckenheimer, J. and Holmes, P.J.,[1983] Nonlinear Oscillations, Dynamical Systems and Bifurcations of Vector Fields, Springer-Verlag, Berlin.
3. Marino, I., Raghavan, S., Fantoni, S., Shenoy, S.R., Smerzi, A.,[1999] "Bose-condensate tunneling dynamics: momentum-shortened pendulum with damping," Physical Review A, **60**, 1: 487-493.
4. Li, J.B., and Wan, B.H.,[1989] "Chaos and subharmonic bifurcations in the periodically forced system of phase-locked loops," Ann. Diff. Eqns **5**, 407-426.
5. Raghavan, S., Smerzi, A., Fantoni, S., Shenoy, S.R.,[1999] "Coherent oscillations between two weakly coupled Bose-Einstein condensates: Josephson effects, π oscillations, and macroscopic quantum self-trapping," Physical Review A, **59**, 1: 620-633.
6. Xia, B.L., Hai, W.H., Chong G.S.,[2006] "Stability and chaotic behavior a two-component Bose-Einstein condensate," Physics Letter A, **361**, 136-142.
7. Xie, Q.T., Hai, W.H. and Chong, G.S.,[2003] "Chaotic atomic tunneling between two periodically driven Bose-Einstein condensates," Chaos, **13**, 3:801-805.

Perspectives in Mathematical Sciences
Interdisciplinary Mathematical Sciences, Volume 9, 2009
pp. 145–180

Chapter 8

Recent Developments and Perspectives in Nonlinear Dynamics

Zengrong Liu

Institute of Systems Biology, Shanghai University, Shanghai 200444, China.
E-mail: zrongLiu@online.sh.cn

Dedicated to Professor Youzhong Guo on the occasion of his 75th birthday

We summarize three progresses in nonlinear dynamics in the recent decade: chaos control, chaos synchronization and complex networks, and also introduce our work on these aspects. Finally, we give some perspectives on the development of nonlinear dynamics combining with complex systems theory.

1. Introduction

Historically speaking, the development of nonlinear dynamics has close relationship with the research on scientific complexity. Novel prize winner Prigogine proposed that science should transform from studying simple to complexity in the late 1960s. For the understanding of how to investigate theoretically the complexity in scientific phenomenon was also gradually deepened. According to our present understanding, people have at least realized that the following several aspects should be considered: the system consisting of two interacting simple units, the system consisting of a large number of interacting units, and how the effects of decisiveness and randomness harmoniously embody in the system and so on.

Since the second half of the 1970s, chaos has been found from the interacting units. Thereafter chaos became a main research topic in nonlinear dynamics and had a quick development. After endeavoring for about twenty years, the basic framework of chaos theory has been constituted. At this time, how the nonlinear dynamics develop further became a scientific problem cared about by many researchers.

According to the authors' viewpoint, investigating the production mechanism of scientific complexity by utilizing the ideas and methods of nonlinear dynamics

Key words and phrases: Nonlinear dynamics, chaos control, chaos synchronization, complex networks.

This project was jointly supported by NNSF grants 10672093, 10832006 and Shanghai Leading Academic Discipline Project, Project Number: S30104.

should be one of the motive powers of the development of nonlinear dynamics theory. In view of such analysis, based on the framework of chaos theory getting constructed, we proposed the concepts and methods of chaos control and chaos synchronization combining applied basic research. The researches on this two problems were the important advances of nonlinear dynamics in recent ten years. Meanwhile, due to considering the complexity induced by the system consisting of a large number of interacting units, thus caused the formation of another research hotspot of nonlinear dynamics — complex networks theory.

Here we want to point out that chaos control, chaos synchronization and complex networks, this three different progresses directions in nonlinear dynamics, have the trend of integration at present due to affected by the research thought of nonlinear dynamics. They provide a powerful tool for the behavioral study of open system consisting of a large number of interaction (decisiveness and randomness) units. According to the current stage, it should begin to enter the stage of studying the behavior of such system for scientific complexity exploration.

This paper is organized as follows. The main methods and scientific thoughts of three progresses are introduced in Section 1. Section 2 reports our results obtained on this aspect. Finally, we propose some suggestions for the perspective on the development of nonlinear dynamics in Section 3.

2. Three progresses in nonlinear dynamics

2.1. *Chaos control*

The discovery of chaos phenomenon shows that there exists pseudo-similar random dynamical behavior in a deterministic system. This behavior is ubiquitous. It sometimes plays a beneficial role in the concrete practice problems, while sometimes plays a harmful role in that, therefore the problems of chaos control and chaotifying (chaos anti-control) are proposed from the view of application.

(a): In 1990, Ott, Grebogi, and Yorke published a paper in PRL, which firstly presented the concepts and methods of chaos control, and this method is now called OGY method for short.[4]

Consider a two-dimensional chaotic iteration

$$\begin{cases} \xi_{n+1} = F(\xi_n, p_n), \\ \xi_i \in R^2, \qquad p \in (-p_{max}, p_{max}), \end{cases} \tag{2.1}$$

where p is a controllable parameter, p_{max} is the maximal adjustment range. Let $\xi_F^* = 0$ be the saddle fixed-point of (2.1) with $p = 0$. According to continuity, we can assume that $\xi^*(p)$ with p as parameter is the saddle fixed-point coordinate. Moreover, ξ_n can be assumed to fall into the neighborhood of ξ_F^* due to the ergodicity of chaos, then we can choose a suitable p_n such that $\xi_{n+1} = F(\xi_n, p_n)$ falls on the tangent of stable manifold. Thus, under the linear approximation, we can obtain

$$\xi_{n+1} - gp_n \approx A(\xi_n - gp_n), \tag{2.2}$$

where $g = \frac{\partial \xi^*(p)}{\partial p}| \neq 0$ is in order to make $\xi_{n+1} \cdot f_u \approx 0$ (f_u is a vector perpendicular to the stable direction), that is, ξ_{n+1} approximately falls on the tangent of stable manifold. From (2.2), we can get

$$p_n = \lambda_u(\lambda_u - 1)^{-1}\frac{(\xi_n \cdot f_u)}{(g \cdot f_u)}, \tag{2.3}$$

where λ_u is a unstable eigenvalue. In such way, we can adjust the parameter step by step to make the orbit fall on the tangent of stable manifold, once it is achieved, the chaos orbit can then be forced to the saddle fixed-point due to the attractability of stable manifold.

It can be seen from the above that the main features of OGY method are as follows: 1) The control quantity is small, it is a only small perturbations method. 2) The control method does not change the existence of global chaotic attractor of the system, which is different from the general control theory. 3) Due to there existing infinite many saddle period points similar to this kind of saddle fixed-point, and all of them are embedded into chaos attractors, therefore the control method can in principle be applied to realize controlling a system to infinite kinds of stationary states. For the work related to this method and idea, can see references [2-20].

(b): In 1998, Chen and Lai proposed the principle and method for making a discrete-time dynamical system chaotic by using linear feedback method [21].

Consider a discrete-time nonlinear dynamical system

$$\begin{cases} x_{k+1} = f_k(x_k) + u_k, & x_k \in R^n \\ x_0 \text{ is given.} \end{cases} \tag{2.4}$$

Choose $u_k = B_k x_k$, where B_k is a $n \times n$ constant matrix. Let

$$J_j(z) = f'_j(z) + B_j, \tag{2.5}$$

be the system (2.4) Jacobian matrix, evaluated at z, $j = 0, 1, 2, \cdots$, and let

$$T_j = T_j(x_0, x_1, \cdots, x_j)$$
$$= J_j(x_j)J_{j-1}(x_{j-1})\cdots J_1(x_1)J_0(x_0).$$

Moreover, let $u_i^j = u_i(T_j)$ be the i-th eigenvalue of T_j, then the Lyapunov exponents of the system (2.4), starting from x_0, are that

$$l_i(x_0) = \overline{\lim_{j \to 0}}\frac{\ln |u_i^j|}{j}, \quad i = 1, 2, \cdots, n. \tag{2.6}$$

The objective of selecting the control scheme is to make all the above Lyapunov exponents exist and strictly positive, and, meanwhile, to ensure the system orbit be bounded, this will guarantee making controllable system chaotic. Therefore, suppose that there is a constant $N > 0$ such that

$$\sup_{0 \leq k < \infty} \|f_k'\| < N. \tag{2.7}$$

Then we can select some constant $c > 0$, and let

$$B_k = (N + e^c)I_n. \tag{2.8}$$

It follows that $J_j(z) = f_j'(z) + (N + e^c)I_n$, and for any given j, z, all eigenvalues of the matrix $J_j(z)$ are between e^c and $2N + e^c$, then we can obtain

$$l_i(x_0) > c, \quad i = 1, 2, \cdots, n. \tag{2.9}$$

Furthermore, in order to ensure the system orbit be bounded, we apply the mod-operation to the system,

$$x_{k+1} = f_k(x_k) + (N + e^c)x_k \quad \mathrm{mod}\, 1, \tag{2.10}$$

and thereby guarantee realizing chaotifying. Particularly, we can use the following linear system for simplicity

$$\begin{cases} x_{k+1} = Ax_k + u_k \quad \mathrm{mod}\, 1 \\ u_k = (N + e^c)x_k. \end{cases} \tag{2.11}$$

This method is very simple, and can be rigorously proved in the theory that it satisfies all requirements of chaos [22-24]. For the further developments, can see references [25-38].

The above theories and methods of chaos control and chaotifying got very great development in the following a period of time, and became a research hotspot in nonlinear dynamics in recent ten years.

2.2. *Chaos synchronization*

Besides proposing chaos control, people also considered the potential application brought by the long-term unpredictable of dynamical system due to the existence of chaos. This idea leaded to the research on the problem of chaos synchronization.

The pioneer paper about chaos synchronization problem was the paper published in PRL by Pecara and Carroll in 1990, and this method is now called the driving-response method [39].

Consider a n-dimensional chaotic dynamical system

$$\dot{U} = f(U), \quad U \in R^n. \tag{2.12}$$

Decompose it into a stable subsystem

$$\dot{V} = f_1(V, W), \tag{2.13}$$

and a unstable system

$$\dot{W} = f_2(V, W), \tag{2.14}$$

where $V \in R^p, W \in R^q, p + q = n$. Equation (2.12) is called the driving system, duplicate a corresponding system which is the same as the driving subsystem (14) by using V as a driving signal

$$\dot{W}' = f_2(V, W'). \tag{2.15}$$

If, for the same driving signal V and arbitrary different response initial values W_0, W_0', the following condition is satisfied

$$\lim_{t \to \infty} \|W - W'\| = 0,$$

then we can say that the trajectories of the systems (2.14) and (2.15) asymptotically approach to synchronize. In [39], the authors pointed out that, if all Lyapunov exponents (generally called conditional Lyapunov exponent) of the respond system under the effect of the driving signal are negative, then synchronization was realized.

Since the concept of chaos synchronization was proposed, the theories and methods of chaos synchronization have got a great development. More importantly, after the discovery of synchronization phenomenon, a large number of various synchronization phenomenon have been found, such as complete synchronization, phase synchronization, lag synchronization and generalized synchronization. The discovery of these synchronization phenomenon showed that the dynamical behaviors of two dynamical systems can exhibit various relations through coupling. This idea, reflecting in the complex system consisting of a large number of interaction basic units, may make them exhibit spatial correlation interaction. This opened a new direction for the research on complex system, and thus urged synchronization phenomenon to become an important research hotspot in nonlinear dynamics in the past decade [40-68].

2.3. *Complex networks*

After developing a relatively complete systematic results about the chaotic phenomenon induced by interaction units, people started to turn the attention to the analysis of the system consisting of interaction subsystems gradually. For such system, the first thing to solve was to how to model, and network was found to be the best tool to model such system after many years exploration. From a theoretical perspective, it is imperative first to understand the topological structure of the constructed model before carrying out theoretical analysis for the system behavior.

From the viewpoint of graph theory, the topological structure of network can be described by the average path length L, the clustering coefficient C, and the degree distribution $P(k)$. Before 1998, the theory result of ER random graph one of the most studied, showed that it had small L, small C, and its $p(k)$ followed the feature of exponential distribution. However, the measured results of network consisting of a large number of realistic complex systems showed that they often had the characters different from the above features, and then naturally attracted people research interests.

In 1998, Watt and Strogntz first published a paper on Nature [69]. They constructed a new network by using deterministic and stochastic hybrid method to establish connections among nodes, and the numerical examples showed that such network had the characteristics of small L and large C, which was consistent with that of many measured networks. Such network is now called small-world network.

In 1999, Barnbnsi and Albert published another paper about network' topological structure on Science [70]. They obtained another network by using the network growth and the preferential attachment among nodes. The degree distribution of such network was descried by a power law. This feature was also consistent with that of many measured networks. Such network is now called scale-free network.

These results showed that the realistic system network had a sequence of topological characteristics. When investigating the system consisting of a large number of interaction subsystems, we first must correctly construct network model reflecting the nature of system from the real and objective mechanism of the formation of system. According to recent opinions, such network may have many interesting topological properties, and we use complex network as generic term of them.

The research on complex network opened a completely new research direction in nonlinear dynamics, which told us that it perhaps may model a class of complex systems by using scientific methods, and then analyzed their behaviors. This direction has been drawn great attention since it was proposed. At present, the research objects mainly focus on the network' topological structure, information transmission on the network and the study of network dynamics [71-139].

3. Related work introduction

Over the past ten years, according to the above international latest research idea, we also have developed a series of research around the three progresses, and gained some results. Now, we introduce the results as follows:

3.1. *Chaos control*

On this aspect, controlling hyperchaos and chaotifying a continuous system, this two work are introduced mainly in this section.

(a) The OGY method can not be used to control hyperbolic fixed point without stable manifold, but in the two-dimensional plane hyperchaos map, the embedded fixed point has such property, so how to control it? Therefore, we proposed straightline stabilization method to solve this problem [140].

Consider a planar map

$$\xi_{n+1} = F_\varepsilon(\xi_n) \tag{3.1}$$

where $\xi = (x, y) \in R^2$, $\varepsilon \in (p, q) \in R^2$ is a small control parameter vector, and $F_\varepsilon(\xi_n)$ is a vector-valued function of ξ_n with ε as parameter. Let ξ_*^0 be the fixed point of map (3.1) with $\varepsilon = 0$. Without loss of generality, it is assumed that proper coordinate changes have been made so that ξ_*^0 is the origin of the two-dimensional space. Let J be the Jacobian matrix of the map with $\varepsilon = 0$ evaluated at the fixed point, i.e.,

$$J = \left(\frac{\partial F_0}{\partial \xi_n}\right)_{\xi_n = \xi_*^0}. \tag{3.2}$$

When $\mathrm{Det}(J - I) \neq 0$, the implicit function theorem can be used to obtain a small neighborhood V in the ε planar. When $\varepsilon \in V$ and $\varepsilon \neq 0$, the map has a corresponding fixed point $\xi_*(\varepsilon) = (x_*(\varepsilon), y_*(\varepsilon))$. Define the following matrix:

$$J_1 = \left(\frac{\partial \xi_*(\varepsilon)}{\partial \varepsilon}\right)_{\varepsilon = 0}. \tag{3.3}$$

In the neighborhood of $\xi_*(\varepsilon)$, $\xi_*(\varepsilon)$ and ξ_*^0 have the same stability. Their linear approximation is

$$\xi_{n+1} - \xi_*(\varepsilon) \approx \bar{J}(\xi_n - \xi_*(\varepsilon)), \tag{3.4}$$

where $\bar{J}$ is the Jacobian matrix of the map with parameter ε evaluated at $\xi_*(\varepsilon)$. For small enough ε, $\bar{J}$ can be approximated by matrix J in Eq. (3.2). Hence, for $\varepsilon \to 0$, we have

$$\xi_{n+1} - \xi_*(\varepsilon) = J(\xi_n - \xi_*(\varepsilon)). \tag{3.5}$$

Thus, by means of the instability of $\xi_*(\varepsilon)$, it can derived that ξ_n make ξ_{n+1} become more close to ξ_*^0.

To stabilize the unstable orbit, we propose to require the straight-line stabilization method,

$$\xi_{n+1} = k\xi_n, \quad -1 < k < 1. \tag{3.6}$$

Clearly, for $\varepsilon \to 0$, we have

$$\xi_*(\varepsilon) = J_1 \varepsilon. \tag{3.7}$$

It is easy to know that J_1 is invertible since $(J - I)$ is invertible. Substituting (3.6) and (3.7) to (3.5) gives

$$\varepsilon_n = J_1^{-1}(J - I)^{-1}(J - kI)\xi_n. \tag{3.8}$$

Here ε has been re-denoted by ε_n to indicate that the parameter adjustment is in the nth iteration of the map.

Fig. 8.1 is the idea of straight-line stabilization method. Apply this method to the following example

$$x_{n+1} = 1 - 2(x_n^2 + y_n^2) + p, \quad y_{n+1} = -4x_n y_n + q,$$

and the obtained result can see Fig. 8.2. Both theoretical analysis and numerical results indicate the effectiveness of the proposed methodology, and this method can be applied to a wide range of problems. For the detailed theory on this aspect and the related work, can see references [141-143].

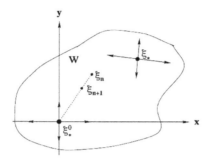

Fig. 8.1.   The idea of straight-line stabilization method.

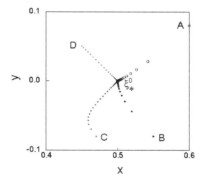

Fig. 8.2.   Example of straight-line stabilization method.

(b) For the problem of chaotifying a dynamical system, we extended the method used to chaotify a discrete-time system to continuous dynamical system.

Consider a n-dimensional continuous autonomous system

$$\dot{x} = f(x), \tag{3.9}$$

where $x \in R^n$ and $f : R^n \to R^n$ is a continuous function. Assume that the system (3.9) has a stable limit orbit $\gamma : P(t)$, of period T. Let matrix $C = \int_0^T \Phi^{-1}(T-S)dS,$

where $\Phi(t)$ is the fundamental matrix of the system $\dot{x} = A(t)x$, and $A(t)$ is the Jacobian matrix of f along γ. Divide the tangent space at a selected point $p \in \gamma$ into orthogonal sum of two linear subspaces $\pi_1 \oplus \pi_2$, where π_1 is the image space and π_2 is the kernel space of matrix C. Suppose that the dimensions of π_1 and π_2 are m and $n - m$ $(1 \leq m \leq n)$, respectively.

Now, the task of chaotifying a continuous system is transformed into making the system exhibit a chaotic map in the π_1 planar, that is, make the system

$$\dot{x} = f(x) + \sum_{k=1}^{\infty} g(x(kT))H(\frac{t}{T} - k), \quad x_0 \in \widetilde{\omega}, \tag{3.10}$$

where $\widetilde{\omega}$ is a small neighborhood of p, $g(x(kT)) \in R^n$ is a constant vector determined by $x(kT)$, and H is a Heaviside function, show a chaotic map in the π_1 planar.

Due to the map in π_1 is m-dimension, thus according to the result in the previous section, we can get that, for any arbitrary $m \times m$ bounded matrix B, the map

$$q_{k+1} = Bq_k + u_k, \quad \mod(r), \tag{3.11}$$

can be chaotic if an appropriate control input sequence $\{u_k\}_{k=0}^{\infty}$ is applied, where r is a positive constant. Indeed, one may simple use

$$u_k = (N + e^c)q_k, \tag{3.12}$$

to drive the map (3.11) chaotic, where $N \geq \|B\|$. In particular, one may use $B = 0_{m \times m}$, and $N = 0$, which will reduce (3.11) to

$$q_{k+1} = e^c q_k, \quad \mod(r), \tag{3.13}$$

Denote this system as $q_{k+1} = U_{c,r}(q_k)$, it can realize the map chaotifying.

The linearization of (3.10) is given by

$$Z((k+1)T) = \Phi(T)Z(kT) + Cg(x(kT)), \tag{3.14}$$

where $Z(t) = D_x(x(t) - p(t))$. Let $E_1 = [I_1, I_2, \cdots, I_m]$, and $E_2 = [I_{m+1}, \cdots, I_n]$, which are mutual orthogonal unit vectors in π_1 and π_2, respectively. Moreover, let q_k be the projection of $Z(kT)$ onto π_1, then we have

$$\begin{aligned}
q_{k+1} &= E_1^\top Z((k+1)T) \\
&= E_1^\top \left[\Phi(T)Z(kT) + Cg(x(kT)) \right].
\end{aligned} \tag{3.15}$$

The chaotifying can be realized provided that the map satisfies

$$q_{k+1} = U_{c,r}(E_1^\top Z(kT)) = e^c q_k(\mod r), \tag{3.16}$$

and, meanwhile, the trajectory is prevented from diverging in the π_2 direction. We can prove that $g(x(kT))$ in (30) can be obtained as follows

$$g(x(kT)) = \begin{cases} -\alpha[E_1, E_2][0, E_2]^\top (x(t) - p(t)), \\ \qquad\qquad k = 0, 2, 4, \ldots, \\ E_1 M^{-1}\{U_{c,r}E_1^\top (x((k-1)T) - p)) \\ -E_1^\top \phi(T)(x(kT) - p)\}, \, k = 1, 3, 5, \ldots \end{cases} \tag{3.17}$$

where $\alpha > -\ln\left(\min\{\frac{1}{4}, \frac{1}{4\lambda^2}\}\right)/T$, with $\lambda = \sup\limits_{x \in R^n} \dfrac{\|[E_1, E_2]\phi(T)x\|}{\|[E_1, E_2]^\top x\|}$, $M = E_1^\top \subset E_1$

(E_1 is invertible).

We apply the above method to the non-chaotic system $\dot{x} = x - y - x(x^2 + y^2)$, $\dot{y} = x + y - y(x^2 + y^2)$, $\dot{z} = -z$, which has a limit cycle. The numerical results show the chaotifying is achieved, and its Poincare map can see Fig. 8.3.

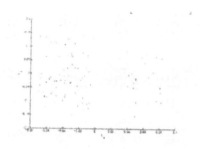

Fig. 8.3.   Chaotifying limit cycle system.

This method is simple and operationable. For detailed work about chaotifying a continuous system, can see references [144-145].

The above are our two main work about chaos control, we also did some related work, the interested readers can see references [146-153].

3.2. *Chaos synchronization*

On this aspect, synchronization classification, synchronization in the delayed neural network, and cluster synchronization in the network, this three work are introduced mainly in this section.

(a) The synchronization phenomenon discovered in two systems now have various forms. Generally speaking, this two systems themselves can be same or difference. Four kinds of the most common synchronization phenomenon are complete synchronization, phase synchronization, lag synchronization and generalized synchronization. For complete synchronization in two systems, it has been obtained many theoretical results. What relationships existing among the remaining seven kinds of synchronization phenomenon, should be our concerned problem. We can proved that all the remaining seven kinds of synchronization phenomenon could be reduced to the generalized synchronization of two identical systems from the viewpoint of differential homeomorphism between dynamical systems in mathematical sense.

According to this theoretical results, we furthermore gave a general framework for investigating generalized synchronization by using differential homeomorphism. These theoretical results were correct in mathematical sense. However, their rationality in application are still required further research. The interested readers can see references [154-159].

(b) In recent years, coupled delayed neural networks system has been successfully applied to image processing, information transmission and secure communication, etc. Thus, the study on synchronization dynamic of coupled delayed neural networks system is indispensable to implement and design artificial neural network technology. We consider the following coupled delayed neural networks system [160-161]:

$$\dot{x}_i(t) = -Cx_i(t) + Af(x_i(t)) + A^\tau f(x_i(t-\tau)) + u(t)$$

$$+ \sum_{j=1}^{N} b_{ij}\Gamma x_j(t), \quad i = 1, 2, \cdots, n, \qquad (3.18)$$

where $x_i(t) = (x_{i1}(t), x_{i2}(t), \ldots, x_{in}(t))^\top \in R^n$ are the state variables of the ith delayed neural network, $C = \mathrm{diag}(c_1, c_2, \ldots, c_n)$ is a diagonal matrix with positive diagonal entries, $c_r > 0\,(r = 1, 2, \ldots, n)$, $A = (a_{rs}^0)_{n\times n}$ is a weight matrix, $A^\tau = (a_{rs}^\tau)_{n\times n}$ is a delayed weight matrix, $u(t) = (u_1(t), u_2(t), \ldots, u_n(t))^\top \in R^n$ is the input vector function, $\tau(r) = (\tau_{rs})$ with the delays $\tau_{rs} \geq 0$, $r, s = 1, 2, \ldots, n$, and $f(x_i(t)) = \left[f_1(x_{i1}(t)), f_2(x_{i2}(t)), \ldots, f_n(x_{in}(t))\right]^\top$. $\Gamma = \mathrm{diag}(\gamma_1, \gamma_2, \ldots, \gamma_n) \in R^{n\times n}$ represents the inner linking matrix between nodes, and $B = (b_{ij})_{n\times n} \in R^{n\times n}$ represents coupling symmetric matrix with Laplacian structure of the whole networks. We assume that each of the activation functions $f_r(x)$ is globally Lipschitz continuous, i.e., the following conditions are satisfied: there exist constants $k_r > 0\,(r = 1, 2, \ldots, n)$, such that

$$|f_r(x_1) - f_r(x_2)| \leq k_r|x_1 - x_2|, \quad r = 1, 2, \ldots, n,$$

for any two different $x_1, x_2 \in R$.

Based on the Lyapunov functional method and Hermitian matrices theory, it can be proved that the sufficient condition for global synchronization of delayed dynamical network (3.18) is as follows: Let the eigenvalues of coupling symmetric matrix B be ordered as $0 = \lambda_1 > \lambda_2 \geq \lambda_3 \cdots \geq \lambda_N > -\infty$. If there exist n positive numbers $p_1, \cdots, p_n$ and two positive numbers $r_1 \in [0,1]$, $r_2 \in [0,1]$, and the following conditions are satisfied for all $i = 1, 2, \ldots, n$

$$|a_{ii}^0|p_ik_i + \alpha_i + p_i\gamma_i\lambda(\gamma_i) < 0, \quad i = 1, \cdots, n, \qquad (3.19)$$

where

$$\lambda(\gamma_i) = \begin{cases} \lambda_2, & \text{if } \gamma_i > 0, \\ 0, & \text{if } \gamma_i = 0, \\ \lambda_N, & \text{if } \gamma_i < 0. \end{cases}$$

and

$$\alpha_i \overset{\text{def}}{=} -c_i p_i + \frac{1}{2} \sum_{j=1, j \neq i}^{n} \left(p_i |a_{ij}^0| k_j^{2r_1} + p_j |a_{ji}^0| k_i^{2(1-r_1)} \right)$$
$$+ \frac{1}{2} \sum_{j=1}^{n} \left(p_i |a_{ij}^\tau| k_j^{2r_2} + p_j |a_{ji}^\tau| k_i^{2(1-r_2)} \right).$$

The above results show that the synchronizability of the coupled neural network system still is determined by the second largest eigenvalue of the coupling matrix. It can be concluded that the global synchronization of such coupled delayed neural networks can be achieved by a suitable design of the coupling matrix and the inner linking matrix. Fig. 8.4 is the case of the synchronization error of neural networks system consisting of three CNN cellular change with time.

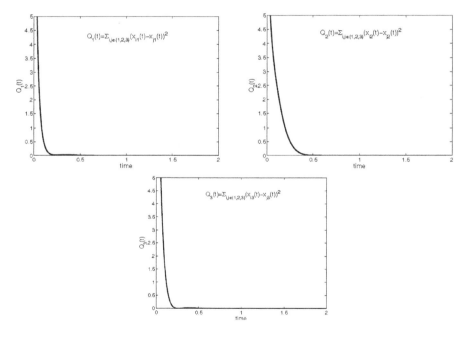

Fig. 8.4.   Global synchronization of the coupled network consisting of three CNN.

Moreover, we investigated the adaptive control and synchronization problems of coupled delayed neural networks system, and derived a sequence of simple yet generic criteria for robust synchronization of chaotic delayed neural networks with all the parameters unknown [162-165].

(c) Many numerical results have shown that the network, in the process of realizing complete synchronization, was often gradually developed through cluster synchronization. For a network with N nodes, it is said to realize cluster synchronization if the N nodes split into n clusters, the number of nodes in each cluster are denoted as $m_1, m_2, \ldots, m_n$, where $m_1 + m_2 + \cdots + m_n = n$, and the nodes in the same cluster synchronization with each other, while the nodes in the different cluster are desynchronization. Up to now, there not only exist difficulty in the theory on this problem, but also lack method in practical operation.

We consider the following network

$$\dot{x}_i = f(x_i) + \epsilon \sum_{j=1}^{N} c_{ij} P x_j, \quad i = 1, \ldots, N, \tag{3.20}$$

where $x_i = (x_i^1, \ldots, x_i^m)^\top$ is the m-dimensional state variable of the $i$the node and N is the total number of the nodes in the network. $C = (c_{ij})$ is a $N \times N$ real symmetric matrix reflecting the network topology, satisfying $\sum_{j=1}^{N} C_{ij} = 0, i = 1, 2, \ldots, N$ (its non-diagonal element is 1 or 0). The nonzero elements of the $m \times m$ matrix P determine the coupling components among the states of nodes. Here consider $P = \text{diag}(p_1, p_2, \ldots, p_m)$, where $p_h > 0$ for $h = 1, \ldots, s$ and $p_h = 0$ for $h = s + 1, \ldots, m$. $\epsilon > 0$ is the coupling strength.

In order to make the network achieve n-cluster synchronization, we construct a matrix C of the form

$$C = \begin{pmatrix} b_1 c_{11} & c_{12} & 0 & \cdots & 0 & 0 \\ c_{21} & b_2 c_{22} & c_{23} & \cdots & 0 & 0 \\ 0 & c_{32} & b_3 c_{33} & \cdots & 0 & 0 \\ \cdots & \cdots & \cdots & \cdots & \cdots & \cdots \\ 0 & 0 & 0 & \cdots & b_{n-1} c_{n-1,n-1} & c_{n-1,n} \\ 0 & 0 & 0 & \cdots & c_{n,n-1} & b_n c_{n,n} \end{pmatrix}$$

where $b_1 \geq 1 + \frac{2m_2}{m_1}$, $b_i \geq 1 + \frac{2(m_{i-1} + m_{i+1})}{m_i}$, for $1 < i < n$, $b_n \geq m_n \times \{2, 1 + \frac{2m_{n-1}}{m_n}\}$, $C_{i,i} \in R^{m_i \times m_i}$, $C_{i+1,i}^\top = C_{i,i+1} \in R^{m_i \times m_{i+1}}$, the structure of $m_i \times m_i$ matrix $C_{i,i}$ is required to have the same form as matrix C. It can be proved that the cluster synchronization manifold $M^* = \{x_1 = \cdots = x_{m_1}, \cdots, x_{m_1+m_2+\cdots+m_{n-1}+1} = \cdots = x_N\}$ is an invariant manifold of the network (3.20) in this case.

We assume that the individual system $\dot{x}_i = f(x_i)$ is eventually dissipative, then it can be proved that the following inequality holds:

$$\epsilon \sum_{i=1}^{N} \sum_{j \sim i} \bar{G}_{\bar{j}} X_{ji}^\top B P X_{ji} \geq \sum_{i=1}^{N} \sum_{j \sim i} X_{ji}^\top B A X_{ji}, \tag{3.21}$$

where $G_{\bar{j}}$ denotes the set of subscripts of the cluster in which the jth node is, i.e., $j \in G_{\bar{j}}$, and $\bar{G}_{\bar{j}}$ denotes the number of elements in $G_{\bar{j}}$. The means of matrix A and B can see reference [166]. The related corresponding results of the globally asymptotically stable of cluster invariant manifold M^* can be derived from the above inequality, that is, the desired cluster synchronization can be realized [166].

We apply the above method to the network consisting of 8 coupled Lorenz systems.

$$\dot{x}_i = \sigma(y_i - x_i) + \epsilon \sum_{j=1}^{8} c_{ij}x_j, \quad \dot{y}_i = rx_i - y_i - x_iz_i,$$

$$\dot{z}_i = -bz_i + x_iy_i, \quad i = 1, 2, \ldots, 8.$$

Require realizing three-cluster synchronization in the above network, where the sets of subscripts of three clusters are $G_1 = \{1, 2\}$, $G_2 = \{3, 4, 5\}$, and $G_3 = \{6, 7, 8\}$. The numerical results show that the three-cluster synchronization is achieved, (for example, Fig. 8.5 shows node 1 and node 2 are synchronous, node 2 and node 8 are synchronous), but the dynamical behaviors are difference among different clusters (for example, Fig. 8.5 shows the dynamical behaviors of node 1 and node 6 are difference), the coupling matrix of the network in this case are cooperative and competitive, and weighted.

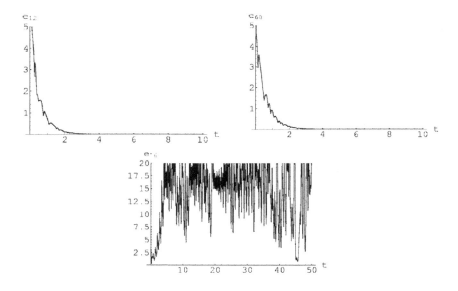

Fig. 8.5.   Cluster synchronization of the network.

The above three results are our representational results obtained in chaos synchronization. In addition, we also obtained many interesting results on impulsive synchronization, partial synchronization, and adaptive synchronization, etc., and on the synchronization theory in life sciences application, etc.[167-173].

3.3. *Complex networks*

On this aspect, we mainly introduce biological networks model construction, the periodic response of delayed neural networks with periodically varying external stimuli, synchronization in complex dynamical networks via impulsive control, epidemic spreading on uncorrelated heterogeneous networks with non-uniform transmission, synchronization mechanisms of circadian rhythms in the suprachiasmatic nucleus, the dynamics of MicroRNA-mediated motifs, and information transmission in scale-free or degree association networks.

For the complex networks model construction, we mainly considered the modeling of biological networks. Based on the basic principles of biology evolution—duplication and divergence, we tried to construct network models with the basic characteristics of observed biological networks. Our existing first work showed that duplication would make biological networks evolve towards the networks with disassortative mixing property, and if introducing the proper anti-preference property, this trend would be strengthened. The further work is in progress [174].

For the other work about network model construction, can see references [175-177].

(b) Considering the evolution process of biological systems in the real world and the evolution law of artificial intelligence system, it is necessary to investigate the dynamic attractor of artificial neural networks. Thus, the following delayed neural networks with periodically varying external stimuli and network parameters is proposed [178-180].

$$\dot{x}_i(t) = -c_i(t)x_i(t) + \sum_{j=1}^{n} a_{ij}^0 f_j(x_j(t))$$

$$+ \sum_{j=1}^{n} a_{ij}^\tau f_j(x_j(t - \tau_{ij})) + u_i(t), i = 1, \cdots, n, \tag{3.22}$$

where $x_i(t) = (x_{i1}(t), x_{i2}(t), \ldots, x_{in}(t))^\top \in R^n$ are the state variables of the delayed neural network, $C(t) = \text{diag}(c_1(t), \ldots, c_2(t))$, $A(t) = (a_{rs}^0(t))_{n \times n}$, $A^\tau(t) = (a_{rs}^\tau(t))_{n \times n}$, and $u(t) = (u_1(t), \ldots, u_2(t))^\top$ are the continuous ω- periodic matrix-valued functions and ω-periodic functions with respect to the time variable t, respectively. Generally, the model (3.22) is called the periodic delayed neural networks model (PDNNs).

Assume that each of the activation functions $f_i(x)$ satisfies the following condition: there exist real numbers $k_i \geq 0$ and $d_i \geq 0$, such that

$$|f_i(x)| \leq k_i|x| + d_i, \quad i = 1, 2, \ldots, n. \tag{3.23}$$

Without assuming the smoothness, monotonicity and boundedness of the activation functions, by using the Mawhin coincidence degree theory, we presents a sufficient condition for the existence of dynamic periodic attractor for delayed Hopfied neural networks and delayed cellular neural networks (CNNs) descried by the infinite-dimensional nonautonomous dynamical system (3.22): There exist n positive numbers $p_1, \ldots, p_n$ and two positive numbers $r_1 \in [0, 1]$, $r_2 \in [0, 1]$, and the following conditions are satisfied for all $i = 1, 2, \ldots, n$,

$$\left| \max_{[0,\omega]} a_{ii}^0(t) \right| p_i k_i + \alpha_i < 0, \quad i = 1, \cdots, n, \tag{3.24}$$

where

$$\alpha_i \overset{\text{def}}{=} -\min_{[0,\omega]} c_i(t) p_i + \frac{1}{2} \sum_{j=1, j \neq i}^{n} \max_{[0,\omega]} \left(p_i |a_{ij}^0(t)| k_j^{2r_1} \right.$$
$$\left. + p_j |a_{ji}^0(t)| k_i^{2(1-r_1)} \right) + \frac{1}{2} \sum_{j=1}^{n} \max_{[0,\omega]} \left(p_i |a_{ij}^\tau(t)| k_j^{2r_2} \right.$$
$$\left. + p_j |a_{ji}^\tau(t)| k_i^{2(1-r_2)} \right).$$

We also obtained some sufficient conditions for the global exponential stability of the above delayed neural networks model. Fig. 8.6 is the case of the dynamic periodic attractor of the above model with non-monotonic, non-differentiable and unbounded activation function. Moreover, we discussed the global dynamics behavior of delayed bidirectional associative memory (BAM) neural networks [181].

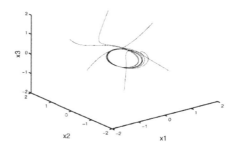

Fig. 8.6.   Dynamic attractor of neural networks with periodically varying external stimuli.

(c) In order to investigate the possibility of realizing network synchronization by utilizing impulses, we consider the following network

$$\dot{x}_i = f(x_i(t)) + g_i(x_1, x_2, \cdots, x_N), i = 1, 2, \ldots, N. \tag{3.25}$$

where $x_i = (x_1^1, x_i^2, \ldots, x_i^n)^\top \in R^n$ are the state variables of node i, $f : R^n \to R^n$ is vector function that reflect the isolated node dynamics, and $g_i : R^{nN} \to R^n$ reflect the coupling influence in the network.

By adding the impulsive control input to the above network (3.25), then the network can be written as

$$\dot{x}_i = f(x_i(t)) + g_i(x_1, x_2, \cdots, x_N)$$
$$+ \sum_{k=1}^{\infty} B_k \left[x_i(t) - \sum_{j=1}^{N} \xi_j x_j(t) \right] \delta(t - t_k), \quad i = 1, 2, \ldots, N. \tag{3.26}$$

where the impulsive instant sequence $\{t_k\}_{k=1}^{\infty}$ satisfies $0 \leq t_1 < t_2 < \cdots < t_k < t_{k+1} < \cdots$, $\lim_{k\to\infty} t_k = \infty$, $B_k \in R^{n \times n}$ denote the control gains, $\xi_j \in R$, $j = 1, 2, \ldots, N$ denote the weight of node j in the network. Assume that the coupling function $g_i(x_1, \ldots, x_N)$ and the vector function $f(x)$ satisfy the following conditions:

$$\|g_i(x_1, x_2, \cdots, x_N)\| \leq \sum_{j=1}^{N} d_{ij} \|x_j(t) - x_i(t)\|,$$
$$\|f(x) - f(y)\| \leq L\|x - y\|.$$

It can be proved that, if there exist a set of positive-definite matrices P_{ij}, $i, j = 1, 2, \ldots, N$, and a constant $\xi > 1$ such that the following condition holds:

$$\ln \xi \gamma \lambda_k + \alpha(t_k - t_{k-1}) \leq 0, \quad \forall k = 1, 2, \ldots, \tag{3.27}$$

where $\gamma = \max_{i,j=1,2,\ldots,N} \left\{ \lambda_{max}(P_{ij})/\lambda_{min}(P_{ij}) \right\}$, $\lambda_{max}(P_{ij})$ and $\lambda_{min}(P_{ij})$ are the largest and the smallest eigenvalues of P_{ij}, respectively, λ_k is the largest eigenvalue of $(I_n + B_k)^{\top}(I_n + B_k)$, $\alpha = \max_{i,j=1,2,\ldots,N} \left\{ \frac{\beta_{ij}}{\lambda_{\min}(P_{ij})} + \lambda_{max}(P_{ij}) \right\}$, $\beta_{ij} = L^2 + \lambda_{max}(P_{ij}) \sum_{k=1}^{N} d_{ik} + d_{ij} \sum_{k=1}^{N} \lambda_{max}(P_{ik}) + \lambda_{max}(P_{ij}) \sum_{k=1}^{N} d_{jk} + d_{ji} \sum_{k=1}^{N} \lambda_{max}(P_{kj})$, then the network (3.25) achieves synchronization under the impulsive control.

This theory tells us that the realizing of impulsive synchronization is determined by several factors. Most interestingly, we can propose various schemes to make the network achieve complete synchronization through changing the weighted sequence. Therefore, achieving synchronization by using impulses is a very effective scheme [188]. Fig. 8.7 shows the synchronization process of the network consisting of 100 coupled chaotic Chua system, where synchronization is achieved through adjusting the impulsively controlled weights of the nodes in the network.

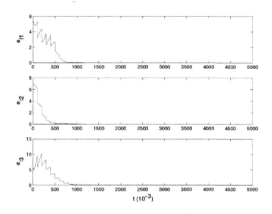

Fig. 8.7. Realizing complete synchronization of the network by changing impulsive control weights

(d) As one of the typical cases of interacting particle systems on networks, epidemic spreading has been studied intensively. Many recent works on this topic mainly focus in how the combination of the properties of the disease and the topology structure of network determine the dynamics of spreading by assuming the uniform transmission of edges. Considering the only consideration structure may make previous works not valid in many realistic cases [189], we developed a method that combines the distribution of transmissions and network structure in framework of mean-field rate equation; here the transmissions of links are simply set to be related to the nodes degree. At the mean-field level, the density of susceptible, infected, and removed nodes of connectivity k at time t, $s_k(t)$, $\rho_k(t)$, $r_k(t)$ satisfy the coupled differential equations differentiated by connectivity classes:

$$\begin{cases} \dfrac{ds_k(t)}{dt} = -ks_k(t)\Theta(t), \\ \dfrac{d\rho_k(t)}{dt} = ks_k(t)\Theta(t) - \mu\rho_k(t), \\ \dfrac{dr_k(t)}{dt} = \mu\rho_k(t), \end{cases} \tag{3.28}$$

where μ is the recovery rate, and the factor $\Theta(t)$ represents the probability that any given link will transmit diseases. By analyzing the quantity $\Theta(t)$, we obtain (for details see [189])

$$\Theta(t) = \frac{\sum_{k'} \lambda(k')(k'-1)P(k')\rho_{k'}(t)}{<k>}, \tag{3.29}$$

where $<k>$ the normalization factor. Unlike previous works, here we introduce the transmission distribution as one of the functional factors of the network, and this factor brings different modality of how topology structure affects the epidemic spreading. It is shown that different kinds of $P(k)$ (degree distribution) and $\lambda(k)$ (the transmission), will affect the threshold of λ_0, which can be seen in Figs. 8.8-8.9.

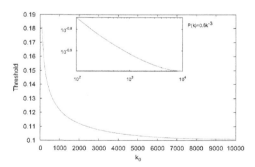

Fig. 8.8.   The dependence of epidemic threshold λ_0^c on infectivity bound k_0, the inset is log scale.

Our results show that in epidemiology, the only knowledge of the connectivity structure is not enough. To thoroughly understand the properties of epidemics,

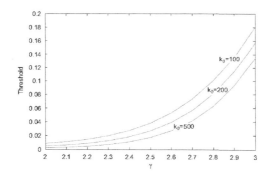

Fig. 8.9. The dependence of epidemic threshold λ_0^c on topological structure γ.

the particular studying of functional factors such as transmissions of edges and infectivities of nodes is indispensable.

(e) It has been shown that isolated single neurons are able to produce circadian oscillations, with periods ranging from 20 to 28 hours. Less well understood is how individual cells are assembled to create a whole tissue pacemaker that can govern behavioral and physiological rhythmicity and be reset by environmental light. In order to study the synchronization mechanisms and the circadian rhythm generation in mammals, we proposed the following heterogeneous network of circadian oscillators in which individual oscillators are self-sustained, based on the structural and functional heterogeneity of the suprachiasmatic nucleus (SCN) of the hypothalamus (for details see [190]):

$$
\begin{aligned}
\dot{x}_i(t) = {}& (A_x + \delta A_i(t))x_i + (B_{1x} + \delta B_{1i}(t))f(x_i) \\
& -(B_{2x} + \delta B_{2i}(t))g(x_i) + (B_{2x} + \delta B_{2i}(t))I \\
& + \sum_{j=1}^{N_1} c_{ij}\Gamma H(x_j, x_i), \ i = 1, \ldots, N_1,
\end{aligned}
\tag{3.30}
$$

$$
\begin{aligned}
\dot{y}_i(t) = {}& (A_y + \delta A_i(t))y_i + (B_{1y} + \delta B_{1i}(t))f(y_i) \\
& -(B_{2y} + \delta B_{2i}(t))g(y_i) + (B_{2y} + \delta B_{2i}(t))I \\
& + \sum_{j=1}^{N_1} d_{ij}\Gamma H(x_j, y_i), \ i = N_1 + 1, \ldots, N_1 + N_2,
\end{aligned}
$$

$$
\tag{3.31}
$$

where $x_i = (x_i^{(1)}, \cdots, x_i^{(n)})^\top \in R^n$, $y_i = (y_i^{(1)}, \cdots, y_i^{(n)})^\top \in R^n$, are the state vectors of the ith oscillator in VL and DM parts respectively, and $\delta A_i(t)$, $\delta B_{1i}(t)$ and $\delta B_{2i}(t)$ are the mismatch matrices which can be time varying. $\sum_{j=1}^{N_1} c_{ij}\Gamma H(x_j, x_i)$ shows the coupling among the oscillators in the VL part of the SCN. $\sum_{j=1}^{N_1} d_{ij}\Gamma H(x_j, y_i)$ shows the coupling between the VL and DM parts. $\Gamma = (\gamma_{ij}) \in R^{n \times n}$ determines by which variables the oscillators are coupled.

In our model, all the entries of Γ are zero except for $\gamma_{1n} = 1$. $H(x_j, x_i) = (h_1(x_j^{(1)}, x_i^{(1)}), \ldots, h_n(x_j^{(n)}, x_i^{(n)}))^\top$ is a function of each oscillator's elements that describes the direct coupling between oscillators, where $h_k(x_j^{(k)}, x_i^{(k)}) : R^2 \to R$. In the model, $h_k(x_j^{(k)}, x_i^{(k)}) \equiv 0$ for $k = 1, \ldots, n-1$.

Based on the Lyapunov method, we derived the sufficient conditions for the synchronization of the coupled nonidentical oscillators (3.30) in the VL region under some assumptions: If the following two conditions are satisfied, the network of Eq.(3.30) is asymptotically synchronized.

1) There exist positive real constants $a > 0$ and $\gamma > 0$, and matrices $P > 0$ with the first column $(0,0,0,1)^\top$, $\Lambda_1 = \mathrm{diag}(\lambda_{11}, \cdots, \lambda_{1n}) > 0$, $\Lambda_2 = \mathrm{diag}(\lambda_{21}, \cdots, \lambda_{2n}) > 0$, such that the following matrix inequality holds:

$$M = \begin{pmatrix} M_1 & M_2 & M_3 \\ M_2^\top & -2\Lambda_1 & 0 \\ M_3^\top & 0 & -2\Lambda_2 \end{pmatrix} < 0, \qquad (3.32)$$

where $M_1 = PA_x + A_x^\top P - 2\Lambda + \gamma E$, $M_2 = PB_{1x} + K_1\Lambda_1$, $M_3 = -PB_{2x} + K_2\Lambda_2$, $\Lambda = \mathrm{diag}(0,0,0, ah_{min}) \in R^{n \times n}$, $E \in R^{n \times n}$ is the unit matrix.

2)

$$\sum_{k=1}^{d} c_k \bar{h}_n'(x_{jk}^{(n)}, x_{ik}^{(n)}) X_{i_k j_k}^{(n)^2} \geq \frac{a}{N_1} \sum_{k=1}^{d} \bar{h}_n'(x_{jk}^{(n)}, x_{ik}^{(n)}) X_{i_k j_k}^{(n)^2}, \qquad (3.33)$$

where d is the number of non-zero elements in the coupling matrix C, and $X_{i_k j_k} = (X_{i_k j_k}^{(1)}, \cdots, X_{i_k j_k}^{(n)}) = (x_{jk}^{(1)} - x_{ik}^{(1)}, \ldots, x_{jk}^{(n)} - x_{ik}^{(n)})^\top \in R^n$ $(k = 1, \ldots, d)$ are defined by links. Here the coupling coecients are present and c_k are the corresponding coupling strength.

We denote the synchronization manifold of (3.30) after it achieves synchronization as $\bar{x}$, which can be described as follows by appropriate mathematical manipulations

$$\dot{\bar{x}} = \bar{A}\bar{x} + \bar{B}_1 f(\bar{x}) - \bar{B}_2 g(\bar{x}) + \bar{B}_2 I, \qquad (3.34)$$

where $\bar{A}, \bar{B}_1$, and $\bar{B}_2$ are appropriate parameter matrices dened as those in the model of a single oscillator and $I = (1,1,1,1)^\top$. It should be emphasized that the period of the synchronization solution $\bar{x}$ approaches the average period of the oscillators coupled and may not be 24h, which is also consistent with the fact that the average periods of dispersed neurons (uncoupled) and SCN slices (coupled) are not significantly different. Thus, intercellular communication seems to adjust the periods of individual oscillators toward the average period.

By applying an external forcing, we consider the circadian rhythm entrained by LD cycle. The neurons of the VL part receiving the LD cycle signal are described by the following equation

$$\dot{\bar{x}} = \bar{A}\bar{x} + \bar{B}_1 f(\bar{x}) - \bar{B}_2 g(\bar{x}) + \bar{B}_2 I + L(t) E_1, \qquad (3.35)$$

where $L(t)$ denotes the effect of the LD cycle and $E_1 = [1, 0, 0, 0]^\top$. $L(t)$ is a square-wave function switching from $L = 0$ in dark phase to $L = L_0$ in light phase. In our previous work [171], we proved that the system Eq.(3.35) is a 24h- periodic oscillator under several conditions, i.e., the neurons in the VL part have a 24h-period rhythm by the intercellular coupling and the LD cycle signal. For the dynamical mechanism of the effect of the LD cycle signal, one can consult Ref. [171]. Similarly, the 24-period oscillator of Eq. (3.35) can be rewritten as follows:

$$\dot{x} = Ax + B_1 f(x) - B_2 g(x) + B_2 I, \qquad (3.36)$$

where A, B_1, and B_2 are appropriate parameter matrices defined as those in the model of a single oscillator. Hence, the transmission of the circadian rhythm, i.e., the coupling between the VL part and the DM part can be simplified to the coupling between Eq.(3.36) and the oscillators in the DM part. Therefore, we obtain

$$\begin{cases} \dot{x} = Ax + B_1 f(x) - B_2 g(x) + B_2 I, \\ \dot{y}_i = (A + \delta A_i(t))y_i + (B_1 + \delta B_{1i}(t))f(y_i) - (B_2 \\ \quad + \delta B_{2i}(t))g(y_i) + (B_2 + \delta B_{2i}(t))I - \varepsilon_i \Gamma H(y_i, x), \end{cases} \qquad (3.37)$$

where ε_i is the coupling strength. It is easy to derive the following sufficient conditions for the neurons in the DM part to achieve the circadian rhythm: If the following two conditions are satisfied, the network of Eq.(3.37) is asymptotically synchronized for $\forall i = N_1 + 1, \cdots, N_1 + N_2$.

1) There exist positive real constants $a > 0$ and $\gamma > 0$, and matrices $P > 0$ with the first column $(0, 0, 0, 1)^\top$, $\Lambda_1 = \mathrm{diag}(\lambda_{11}, \cdots, \lambda_{1n}) > 0$, $\Lambda_2 = \mathrm{diag}(\lambda_{21}, \cdots, \lambda_{2n}) > 0$, such that the following matrix inequality holds:

$$M = \begin{pmatrix} \tilde{M}_1 & \tilde{M}_2 & \tilde{M}_3 \\ \tilde{M}_2^\top & -2\Lambda_1 & 0 \\ \tilde{M}_3^\top & 0 & -2\Lambda_2 \end{pmatrix} < 0, \qquad (3.38)$$

where $\tilde{M}_1 = PA + A^\top P - 2\Lambda + \gamma E$, $\tilde{M}_2 = PB_1 + K_1\Lambda_1$, $\tilde{M}_3 = -PB_2 + K_2\Lambda_2$, $\Lambda = \mathrm{diag}(0, 0, 0, ah_{min}) \in R^{n \times n}$, $E \in R^{n \times n}$ is the unit matrix.

2) $\varepsilon_i \geq a$.

For simplicity, we consider a small size of network with $N1 = 10$ and $N2 = 20$ neurons to show that after the signal of circadian rhythm is transmitted from the VL part to the DM part, the wave form of the circadian rhythm becomes smoother, which can be viewed as the adaptability of organisms for responding the fluctuations of the environment. Figs. 8.10 and 8.11 give the time evolution of mRNA of the uncoupled oscillators in the VL and DM parts respectively. It is obvious that the self-sustained oscillators of neurons have different periods. When an appropriate coupling of the synchronized oscillators with 24h-period in the VL part is added to the DM part, the oscillators in the DM achieve synchronization with 24h-period, as shown in Fig. 8.12.

From the above theoretical analysis, we obtain the following conclusions:

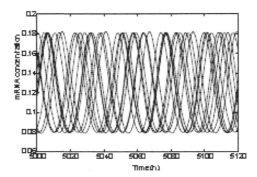

Fig. 8.10. The time evolution of the mRNA concentrations of the uncoupled oscillators in the VL part, which have different periods ranging from 23.8h to 27.1h.

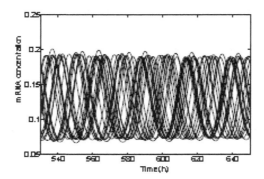

Fig. 8.11. The time evolution of the mRNA concentrations of the uncoupled oscillators in the DM part, which have different periods ranging from 22.1h to 26.9.

(i) One major effect for coupling of the neurons in the VL part is to increase their synchronizability. Accordingly, with the dense coupling, every neuron in the VL part is able to easily achieve circadian rhythm entrained by the 24h LD cycle. Even if there is a discrete jumping for the periodic LD cycle, the periodic response is robust. On the other hand, comparing with the number of DM parts, the number of the neurons in the VL part is generally small. As a result, the overall biological network is still kept to be sparse.

(ii) Under the precondition that every neuron in the VL part has the same wave form, all neurons can be regarded as a single equivalent neuron, coupling to the neurons in the DM part, thereby making them achieve circadian rhythm rapidly. At the same time, the steep wave form of LD cycle becomes smoother due to the structure of DM and VL.

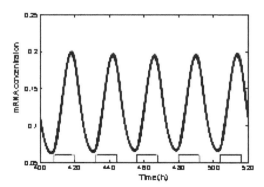

Fig. 8.12. The time evolution of the mRNA concentrations of the coupled oscillators in the VL and the DM parts. In this case, the synchronized oscillators have the period of 24h, and the waveform becomes smoother.

(F) Recently, as more and more cases of post-transcriptional regulation manifested by small non-coding RNAs are being uncovered, it is recognized that post-transcriptional regulation also plays a prominent role in the regulation of cellular processes. MicroRNAs are a kind of post-transcriptional regulatory non-coding RNAs recently discovered in animals and plants. Therefore, we analyze the dynamics and functions of microRNA-mediated motifs. According to the external input which can be classified into the four categories: (1) the same external input regulates the expression of microRNA and mRNA; (2) different external inputs regulate the expression of microRNA and mRNA, respectively; (3) external input acts only on mRNA; and (4) external input acts only on microRNA, as shown in Fig. 8.13.

The rate equations of this four kinds of microRNA-mediated motifs are as follows (for details see [191]):

$$
\begin{cases}
\dot{s}(t) = \dfrac{\alpha_1 u_1(t)^m}{1 + u_1(t)^m} - \gamma s(t)m(t) - \beta_1 s(t), \\[2mm]
\dot{m}(t) = \dfrac{\alpha_2 u_2(t)^n}{1 + u_2(t)^n} - \gamma s(t)m(t) - \beta_2 m(t), \\[2mm]
\dot{p}(t) = \alpha_3 m(t) - \beta_3 p(t), \\[1mm]
\dot{u}_1(t) = \alpha_4 \omega_1 - \beta_4 u_1(t), \\[1mm]
\dot{u}_2(t) = \alpha_5 \omega_2 - \beta_5 u_2(t).
\end{cases}
\tag{3.39}
$$

$$
\begin{cases}
\dot{s}(t) = \dfrac{\alpha_1 u(t)^m}{1 + u(t)^m} - \gamma s(t)m(t) - \beta_1 s(t), \\[2mm]
\dot{m}(t) = \dfrac{\alpha_2 u(t)^n}{1 + u(t)^n} - \gamma s(t)m(t) - \beta_2 m(t), \\[2mm]
\dot{p}(t) = \alpha_3 m(t) - \beta_3 p(t), \\[1mm]
\dot{u}(t) = \alpha_4 \omega - \beta_4 u(t).
\end{cases}
\tag{3.40}
$$

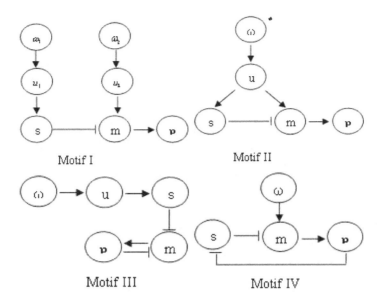

Fig. 8.13. The four kinds of microRNA-mediated motifs and one example, where $\omega, \omega_1, \omega_2$ are external inputs, u, $u1$, $u2$, s, and m denote the upstream factors, microRNA, and the target gene, respectively.

$$\begin{cases} \dot{s}(t) = \dfrac{\alpha_1 u(t)^m}{1 + u(t)^m} - \gamma s(t)m(t) - \beta_1 s(t), \\[2mm] \dot{m}(t) = \dfrac{\alpha_2}{1 + p(t)^n} - \gamma s(t)m(t) - \beta_2 m(t), \\[2mm] \dot{p}(t) = \alpha_3 m(t) - \beta_3 p(t), \\[2mm] \dot{u}(t) = \alpha_4 \omega - \beta_4 u(t). \end{cases} \tag{3.41}$$

$$\begin{cases} \dot{s}(t) = \dfrac{\alpha_1}{1 + p(t)^m} - \gamma s(t)m(t) - \beta_1 s(t), \\[2mm] \dot{m}(t) = \alpha_2 \omega - \gamma s(t)m(t) - \beta_2 m(t), \\[2mm] \dot{p}(t) = \alpha_3 m(t) - \beta_3 p(t). \end{cases} \tag{3.42}$$

We analyze the dynamical behavior of the above four motifs, and conclude that four motifs all have a unique equilibrium in the first quadrant when there is no time delay and noise influence. Moreover, for motif I and II, this property is robust to external input qualitatively and quantitatively.

As an example, we just consider the case $\omega_1 = \omega_2 = \omega$, and $m = n = 4$ for motif I. Fig. 8.14 is the steady state of p versus ω, which shows that when $\omega > 0.1$ the stable state of target protein remain almost unchanged. This result shows qualitatively and quantitatively that microRNA regulation make the expression of target protein robust to the change of the inducer ω. Fig. 8.15 is the steady state of p versus γ, which shows that the expression level of p decreases while the base pairing rate increases. The introduction of microRNA regulation plays an important role in repressing the expression of target gene.

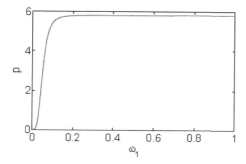

Fig. 8.14.   The steady state of p versus ω.

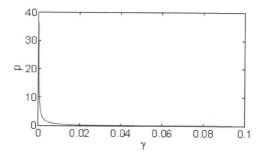

Fig. 8.15.   The steady state of p versus γ.

We also investigate how time delays and noise influence system's dynamics, respectively. Theoretical analysis and numerical simulations proved that all the four motifs have a unique equilibrium in the first quadrant except that motif III and IV exhibit Hopf bifurcation when time delays satisfy certain conditions, and the position of the equilibrium almost remain unchanged in a wide range of parameters. The above properties are preserved under time delay and are almost uninfluenced by the stochastic cellular environment. On the other hand, We also proved that the occurrence of Hopf bifurcation is only related to time delay, not the external input. So Hopf bifurcation is determined by the inner attributes and has nothing to do with external changes. In conclusion, the systems dynamic behaviors arise from inner but not external changes. All the four motifs exhibit strong robustness to external and internal stochastic perturbations which is indispensable to biological system.

(G) We investigated the effects of degree distribution and degree association of the node in the network on information transmission, the finished work were mainly to analyze the spread of infectious disease through scale-free networks and the strategy problem in the marketing, the results can see references [192-194]. The further important work on the node degree association networks is in progress.

In addition to the above seven representative results, we also did many other work on the complex networks, the readers having interest in them can see references [195-198].

3.4. *The analysis of the problems related to emergence*

We think that the research on nonlinear dynamics should be closely related to the representation form of dynamic complexity—emergence. The researches on synchronization and complex networks provided possibilities for investigating emergence in the theory, thus we also made some explorative research on this aspect. The obtained results mainly show in the following two aspects:

(a) We consider two different systems, the representation forms of their dynamical behaviors are

$$\begin{cases} \dot{x} = f_1(x), \\ \dot{y} = f_2(y), \end{cases} \tag{3.43}$$

and their final dynamical behaviors are different. By using the synchronization research method, we can make this two different systems both synchronize to a same system through feedback and structure adaptive. The same system is of the form

$$\dot{z} = \varepsilon_1 f_1(z) + \varepsilon_2 f_2(z), \tag{3.44}$$

where $\varepsilon_1 + \varepsilon_2 = 1$.

This mathematical expression explicitly expresses that, through cooperative and competitive ways, two different systems can mutual absorb the other part advantage and reserve own specialty, and reach consensus in the end. Of course, due to there existing two parameters ε_1, ε_2 and a restrictive condition, then a parameter is free, thus the final synchronization system can fall into simple parameter system family, and the synchronized behavior may be diversity, this is just the requirement of emergence [155].

(b) We believe that the nodes in the network can realize various synchronization behaviors through interaction, particularly generalized synchronization. This kind of generalized synchronization reflects that the nodes distributed in space exist spatial interrelated relations, thus establish the pattern, and achieve emergence.

Therefore, we designed a class of Map Lattices on the plane, which form a network. We then proved that these lattices could achieve lag synchronization in the theory, thus formed various patterns which appear in the form of wave. The following figure shows the types of pattern evoked by the lag synchronization in spatiotemporal networks, see [199-201] for details. This work illustrated the rationality of our conjecture. Certainly should say, this aspect work just start, and have many work to do.

Fig. 8.16. Pattern formation evoked by the lag synchronization in spatio-temporal networks.

4. Some perspectives

We think that the development of the theory of nonlinear dynamics will depend on the deepening research of science of complexity. The exploration of the science of complexity is based on the deepening of the research on various systems. Nowadays, rather typical complex systems must be open systems composing of a large number of interaction units. For such systems, people at least have realized the following two points up to now:

(1) Network is its best modeling way, however this model generally has both deterministic components and stochastic components, which is different from purely deterministic or purely stochastic in the past.

(2) Its common behavior is emergence.

According to scientific history, there was only Statistical Physics that could deal with the emergence phenomenon comparatively well. However, this theory is to discuss many closed system with purely random interaction, so it can not be completely used to deal with the present problems. Actually, it should be admitted that we do not have a complete knowledge of the topological characteristics of existing network models. Due to this situation, I think that the most important task is to construct a network with the characteristics (at least reflects the measured topological characteristics) according to the fundamental principle followed the study system, and must also strengthen the research on the network characteristics in this process.

After establishing the correct model, we should analyze its behavior, and pay attention to the relation between the analysis and the emergence characteristics. The following aspects should be noticed nowadays: (a) The effect of the network topological characteristics on the dynamical behaviors; (b) How the decisiveness and randomness affect the network behaviors through harmoniously coexisting; (c) How to investigate various cooperative behaviors of the nodes in the network combining synchronization.

Another point which I want to propose is that we should pay more attention to how environments respond the network behaviors in the theory due to this class of systems generally all possess adaptability. Such adaptability should be represented through the openness of the network. To the best of our knowledge, about how to deal with this problem in the theory, few has been done. It is not favorable for the whole research. We think this topic will surely become a direction which the nonlinear dynamics theory must attach enough importance to.

References

1. T. Aoki, *Calcul exponentiel des opérateurs microdifferentiels d'ordre infini.* I, Ann. Inst. Fourier (Grenoble) **33** (1983), 227–250.
2. R. Brown, *On a conjecture of Dirichlet*, Amer. Math. Soc., Providence, RI, 1993.
3. R. A. DeVore, *Approximation of functions*, Proc. Sympos. Appl. Math., vol. 36, Amer. Math. Soc., Providence, RI, 1986, pp. 34–56.
4. E. Ott, C. Grebogi, J.A. Yorke, *Controlling chaos*, Phys. Rev. Lett., **64** (1990), 1196-1199.
5. W.L. Ditto, S.N. Rauseo, M.L. Spano, *Experimental control of chaos*, Phys. Rev. Lett., **65** (1990), 3211-3214.
6. G. Chen, X. Dong, *From Chaos to Order: Perspectives, Methodologies, and Applications*, World Scientific Pub., 1998.
7. T. Kapitaniak, *Controlling Chaos*, Academic Press, New York, 1996.
8. K. Pyragas, *Continuous control of chaos by self-controlling feedback*, Phys. Lett. A, **170** (1992), 421-428.
9. M. Basso, R. Genesio, A. Tesi, *Stabilizing periodic orbits of forced systems via generalized Pyragas controller*, IEEE Trans. Circ. Syst. I, **170** (1992), 421-428.
10. P. Cellka, *Experimental verification of Pyragas's chaos control method applied to Chua's circuit*, Int. J. of Bifurcation and Chaos, **4** (1994), 1703-1706.
11. T. Ushio, *Limitation of delayed feedback control in nonlinear discrete-time systems*, IEEE Trans. Circuits and Systems I, **43** (1994), 815-816.
12. T. Ushio, S. Yamamoto, *Delayed feedback control with nonlinear estimation in chaotic discrete-time systems*, Phys. Lett. A, **247** (1998), 112-118.
13. R. Lima, M. Pettini, *Suppression of chaos by resonant parametric perturbations*, Phys. Rev. A, **41** (1990), 726-733.
14. Y. Braiman, I. Goldhirsch, *Taming chaotic dynamics with weak periodic perturbations*, Phys. Rev. Lett., **66** (1991), 2545-2548.
15. R. Roy, T.W. Murphy, T.D. Maier, et al, *Dynamical control of a chaotic laser: Experimental stabilization of a globally coupled system*, Phys. Rev. Lett., **68** (1992), 1259-1262.
16. E.A. Jackson, *On the control of complex dynamic systems*, Physica D, **50** (1991), 341-366.
17. T. Kapitaniak, *Targeting unstable stationary states of Chua's circuit*, J. of Circ. Sys. Comput., **3** (1993), 195-199.

18. K. Pyragas and A. Tamasevicius, *Experimental control of chaos by delayed self-controlling feedback*, Phys. Lett. A, **180** (1993), 99-102.

19. Y. Liu, N. Kikuchi, J. Ohtsubo, *Controlling dynamical behavior of a semiconductor laser with external optical feedback*, Phys. Rev. E, **51** (1995), 2697-2700.

20. T. Yang, C.W. Wu, L.O. Chua, *Cryptography based on chaotic systems*, IEEE on Trans. Circuits and Systems I, **44** (1997), 469-472.

21. A.T. Parker, K.M. Short, *Reconstructing the key stream from a chaotic encryption scheme*, IEEE Trans. on Circuits and Systems I, **48** (2001), 624-630.

22. S. Hayes, C. Grebogi, E. Ott, *Communication with chaos*, Phys. Rev. Lett., **67** (1993), 3021-3034.

23. E. Bollt, Y.C. Lai, *Dynamics of coding in communicating with chaos*, Phys. Rev. E, 58 (1998), 1724-1736.

24. G. Chen, D. Lai, Feedback anti-control of discrete chaos, Int. J. Bifurcation and Chaos, **8** (1998), 1585-1590.

25. X.F. Wang, G. Chen, *On feedback anti-control of discrete chaos*, Int. J. Bifurcation and Chaos, **9** (1999), 1435-1442.

26. X.F. Wang, G. Chen, *Chaotifying a stable LTI system by tiny feedback control*, IEEE Trans. on Circuits and Systems I, **47** (2000), 410-415.

27. X.F. Wang, G. Chen, *Chaotification via arbitrarily small feedback controls: Theory, method and applications*, Int. J. Bifurcation and Chaos, **10** (2000), 549-570.

28. X.F. Wang, G. Chen, X.H. Yu, *Anti-control of chaos in continuous-time system via time-delay feedback*, Chaos, **10** (2000), 771-779.

29. X.F. Wang, G. Chen, K.F. Man, *Making a continous-time minimum-phase system chaotic by using time-delay feedback*, IEEE Trans. on Circuits and Systems I, **48** (2001),641-645.

30. G. Chen, S.B. Hsu, J. Zhou, et al., *Chaotic vibrations of the one-dimensional wave equation due to a self-excitation boundary condition, Part I, controlled hysteresis*, Trans. Amer. Math. Soc., **350** (1998), 4265-4311.

31. G. Chen, S.B. Hsu, J. Zhou, *Chaotic vibrations of one-dimensional wave equation due to a self-excitation boundary condition, Part II, energy injection, period doubling and homoclinic orbits*, Int. J. Bifurcation and Chaos, **8** (1998), 423-446.

32. G. Chen, S.B. Hsu, J. Zhou, *Chaotic vibrations of one-dimensional wave equation due to a self-excitation boundary condition, Part III, Natural hysteresis memory effects*, Int. J. Bifurcation and Chaos, **8** (1998), 447-470.

33. D.S. Chen, H.O. Wang, G. Chen, *Anti-control of Hopf bifurcation, IEEE Trans. on Circuits and Systems I*, **48** (2001), 661-672.

34. S. Codreanu, *Desynchronization and chaotification of nonlinear dynamical systems*, Chaos, Solitons, and Fractals, **13** (2002), 839-843.

35. A. Potapov, M.K. Ali, *Chaotic neural control*, Phys. Rev. E, **63** (2001), 046215.

36. E.N. Sanchez, J.P. Perez, G. Chen, *Using dynamic neural network to generate chaos: A inverse optimal control approach*, Int. J. Bifurcation and Chaos, **11** (2001), 857-863.

37. X. Li, Z.Q. Chen, Z.Z. Yuan, et al., *Generating chaos by an Elman network, IEEE Trans. on Circuits and Systems I*, **48** (2001), 1126-1131.

38. K. Yagasaki, *Homoclinic and heteroclinic behavior in an infinite-degree-of-freedom Hamiltonian system: Chaotic free vibrations of an undamped buckled beam*, Phys. Lett. A, **285** (2001), 55-62.

39. K.S. Tang, G.Q. Zhong, G. Chen, et al., *Generation of scroll attractor via sine function*, IEEE Trans. on Circuits and Systems I, **48** (2001), 1369-1372.

40. K.S. Tang, K.F. Man, G.Q. Zhong, et al., *Generating chaos via x|x|, IEEE Trans. on Circuits and Systems,* **48** (2001), 636-641.

41. G.Q. Zhong, K.F. Man, G. Chen, *Generating chaos via a dynamical controller,* Int. J. Bifurcation and Chaos, **11** (2001), 865-869.

42. L.M. Pecora, T.L. Carroll, *Synchronization in chaotic systems,* Phys. Rev. Lett., **64** (1990), 821-824.

43. L.M. Pecora, T.L. Carroll, *Driving systems with chaotic signals,* Phys. Rev. A, **44** (1991), 2374-2383.

44. S. Boccalettia, J. Kurths, G. Osipov, et al., *The synchronization of chaotic systems,* Physics Reports, **366** (2002), 1-101.

45. H. Fujisaka, T. Yamada, *Stability Theory of Synchronized Motion in Coupled-Oscillator Systems,* Prog. Theor. Phys., **69** (1983), 32-47.

46. M.G. Rosenblum, A.S. Pikovsky, J. Kurths, *Phase synchronization of chaotic oscillators,* Phys. Rev. Lett., **76** (1996), 1804-1807.

47. E.R. Rosa, E. Ott, M.H. Hess, *Transition to phase synchronization of chaos,* Phys. Rev. Lett., **80** (1998), 1642-1645.

48. M.G. Rosenblum, A.S. Pikovsky, J. Kurths, *From phase to lag synchronization in coupled chaotic oscillators,* Phys. Rev. Lett., **78** (1997), 4193-4196.

49. N.F. Rulkov, M.M. Sushchik, L.S. Tsimring, et al., *Generalized synchronization of chaos in directionally coupled chaotic systems,* Phys. Rev. E, **51** (1995), 980-994.

50. L. Kocarev, U. Parlitz, *Generalized synchronization, predictability, and equivalence of unidirectionally coupled dynamical systems,* Phys. Rev. Lett., **76** (1996), 1816-1819.

51. S. Boccaletti, D.L. Valladares, *Characterization of intermittent lag synchronization,* Phys. Rev. E, **62** (2000), 7497-7500.

52. H.D.I. Abarbanel, N.F. Rulkov, M.M. Sushchik, *Generalized synchronization of chaos:The auxiliary system approach,* Phys. Rev. E, **62** (2000), 7497-7500.

53. V. Afraimovich, A. Cordonet, N.F. Rulkov, *Generalized synchronization of chaos in non-invertible maps,* Phys. Rev. E, **66** (2002), 016208.

54. M.A. Zaks, E.-H. Park, M.G. Rosenblum, et al,*Alternating locking ratios in imperfect phase synchronization,* Phys. Rev. Lett., **82** (1999), 4228-4231.

55. P.K. Moore, W. Horsthemke, *Localized patterns in homogeneous networks of diffusively coupled reactors,* Physica D, **206** (2005), 121-144.

56. Z. Zeng, J. Wang, *Complete stability of cellular neural networks with time-varying delays,* IEEE Trans. Circuits Syst.I, **53** (2006), 944-955.

57. C. Li, G. Chen, X. Liao, et al., *Chaos quasisynchronization induced by impulses with parameter mismatches,* Chaos, **16** (2006), 023102.

58. J. Cao, P. Li, W. Wang, *Global synchronization in arrays of delayed neural networks with constant and delayed coupling,* Phys. Lett. A, **353** (2006), 318-325.

59. W. Lu, T. Chen, *Synchronization of coupled connected neural networks with delays,* IEEE Trans. Circuits Syst. I, **51** (2004), 2491-2503.

60. X. Wang, G. Chen, *Synchronization in small-world dynamical networks,* Int. J. Bifurcation and Chaos, **12** (2002), 187-192.

61. X. Wang, G. Chen, *Synchronization in scale-free dynamical networks: Robustness and fragility,* IEEE Trans. Circuits Syst. I, **49** (2002), 54-62.

62. P.M. Gade, C-K. Hu, *Synchronous chaos in coupled map lattices,* Phys. Rev. E, **65** (2002), 6409-6413.

63. L. Kocarev, U. Parlitz, *General approach for chaotic synchronization with applications to communication,* Phys. Rev. Lett., **74** (1995), 5028-5031.

64. U. Parlitz, L. Kocarev, T. Stojanovski, et al, *Encoding messages using chaotic synchronization,* Phys. Rev. E, **53** (1996), 4351-4361.

65. W. Lu, T. Chen, *Synchronization analysis of linearly coupled networks of discrete time systems*, Physica D, **198** (2004), 148-168.

66. Y. Wang, Z. Guan, H. Wang, *Feedback and adaptive control for the synchronization of Chen system via a single variable*, Phys. Lett. A, **312** (2003), 34-40.

67. H.N. Agiaz, *Chaos synchronization of Lü dynamical system*, Nonlinear Analysis, **58** (2004), 11-20.

68. E.M. Elabbasy, H.N. Agiaz, M.M. EI-Dessoky, *Adaptive synchronization of Lü system with uncertain parameters*, Chaos, Solitons and Fractals, **21** (2004), 657-667.

69. T. Yang, L. Yang, C. Yang, *Impulsive synchronization of Lorenz systems*, Phys. Lett. A, **226** (1997), 349-354.

70. L. Yang, T. Yang, *Impulsive synchronization of nonautonomous chaotic systems*, Acta Physica Sinica, **49** (2000), 33-37.

71. J. Sum, Y. Zhang, *Impulsive control and synchronization of Chua's oscillators*, Mathematics and Computers in Simulation, **66** (2004), 499-508.

72. D.J. Watts, S.H. Strogatz, *Collective dynamics of small-world*, Nature, **393** (1998), 440-442.

73. Barabási A.L., Albert R., *Emergence of scaling in random networks*, Science, **286** (1999), 509-512.

74. P. Erdös, A. Rényi, *On the evolution of random graphs*, Publications of the Mathematical Institute of the Hungarian Academy of Sciences, **5** (1960), 17-61.

75. M.E.J. Newman, D.J. Watts, *Renormalization group analysis of the small-world network model*, Phys. Lett. A, **263** (1999), 341-346.

76. R. Albert, A.L. Barabasi, *Statistical mechanics of complex networks*, Rev. Modern Physics, **74** (2002), 48-97.

77. Newman M.E.T., *The structure and function of complex networks*, SIAM Review, **45** (2003), 167-256.

78. S. Boccaletti, V. Latora, Y. Moreno, et al., *Complex networks: Structure and dynamics*, Physics Report, **424** (2006), 175-308.

79. K.I. Goh, E. Oh, B. Kahng, et al, *Between centrality correlation in social networks*, Phys. Rev. E, **67** (2003), 017101.

80. Newman M.E.J., *Mixing pattern in networks*, Phys. Rev. E., **67** (2003), 026126.

81. R. Monasson, *Diffusion, localization and dispersion relation on small-world lattices*, Eur. Phys. J. B., **12** (1999), 555-567.

82. M.E.J. Newman, C. Moore, D.J. Watt, *Mean-field solution of the small-world network model*, Phys. Rev. Lett.,**84** (2000), 3201-3204.

83. A. Barrat, M. Weigt, *On the properties of small-world networks*, Eur. Phys. J. B., **13** (2000), 547-560.

84. R. Albert, A.L. Barabasi, *Dynamics of complex systems: Scaling laws for the period of Boolean networks*, Phys. Rev. Lett., **84** (2000), 5660-5663.

85. R. Albert, A.L. Barabasi, *Topology of evolving networks: local events and universality*, Phys. Rev. Lett., **85** (2000), 5234-5237.

86. P.L. Krapinsky, S. Redner, F. Leyvraz, *Connectivity of growing random networks*, Phys. Rev. Lett., **85** (2000), 4692-4632.

87. S.N. Dorogovtsev, J.F.F. Mendes, *Effect of the accelerating growth of communication networks on their structure*, Phys. Rev. E, **63** (2001), 025101.

88. S.N. Dorogovtsev, J.F.F. Mendes, *Scaling behavior of developing and decaying networks*, Europhys. Lett., **52** (2000),33-39.

89. P.L. Krapivsky, S.A. Redner, *statistical physics perspective on Web growth*, Computer Networks, **39** (2002),261-276.

90. B. Tadic, *Dynamics of directed graphs: The World-Wide Web*, Physica A, **293** (2001), 273-284.

91. G. Bianconi, A.L. Barabasi, *Bose-Einstein condensation in complex networks*, Phys. Rev. Lett, **86** (2001), 5632-5635.

92. A. Vazquez, et al., *Modeling of protein interaction networks*, ComPlexUs, **1** (2003), 38-46.

93. R. Solo, et al, *A model of large scale proteome evolution*, Advances in Complex Systems, **5** (2002), 43-54.

94. F. Chung, L. Lu, *Coupling online and off-line analyses for random power graphs*, Internet Mathematics, **1** (2004), 409-461.

95. J. Berg, M. Lassing, A. Wangner, *Structure and evolution of protein interaction networks: a statistical model for linking dynamics and gene duplications*, BMC Evolutionary Biology, **4** (2004), 51.

96. R. Milo, et al., *Network motifs: simple building blocks of complex networks*, Science, **298** (2002), 824-827.

97. M. Newman, *Fast algorithm for detecting community structure in networks*, Phys. Rev. E, **69** (2004), 066133.

98. M. Newman, M. Girvan, *Finding and evaluating community structure in networks*, Phys. Rev. E, **69** (2004), 026113.

99. R. Milo, et al., *Superfamilies of evolved and designed networks*, Science, **303** (2004), 1538-1542.

100. F. Radicchi, et al, *Defining and identifying communities in networks*, Proc. Nat. Acad. Sci., **101** (2004), 2658-2663.

101. E. Ziv, M. Middendorf, C. Wiggins, *An information-theoretic approach to network modularity*, Phys. Rev. E, **71** (2005), 046117.

102. S. Yook, Z. Oltvai, A. Barabasi, *Functional and topological characterization of protein interaction networks*, Protemics, **4** (2004), 928-942.

103. S. Wuchty, Z. OLtvai, A. Barabasi, *Evolutionary conservation of motif constituents in the yeast protein interaction network*, Nature Genetics, **35** (2003), 176-179.

104. E. Segal, et al., *Module networks: identifying regulatory modules and their condition-specific regulators from gene expression data*, Nature Genetics, **34** (2003), 166-176.

105. N. Przulj, D. Wigle, I. Jurisica, *Functional topology in a network of protein interactions*, Bioinformatics, **20** (2004), 340-348.

106. N. Kashtan, et al, *Topological generalizations of network motifs*, Phys. Rev. E, **70** (2004), 031909.

107. S. Fortunato, V. Latora, M. Marchiori, *Method to find community structures based on information centrality*, Phys. Rev. E, **70** (2004), 056104.

108. M. Girvan, M. Newman, *Community structure in social and biological networks*, Proc. Nat. Acad. Sci., **99** (2002), 7821-7826.

109. Z. Enright, S.V. Dongen, C. Ouzounis, *An efficient algorithm for large-scale detection of protein families*, Nucleic Acids Research, **30** (2002), 1575-1584.

110. Z. Bar-Joseph, et al, *Computational discovery of gene modules and regulatory networks*, Nature Biotechnology, **21** (2003), 1337-1342.

111. C. Li, G. Chen, X. Liao, et al, *Chaos quasisynchronization induced by impulses with parameter mismatches*, Chaos, **16** (2006), 023102.

112. C.W. Wu, L.O. Chua, *Synchronization in an array of linearly coupled dynamical systems*, IEEE Trans. Circuits and Systems-I, **42** (1995), 430-447.

113. C.W. Wu, L.O. Chua, *Synchronization in networks of nonlinear dynamical systems via a directed graph*, Nonlinearity **18** (2005), 1057-1064.

114. X. Li, G.R. Chen, *Synchronization and desynchronization of complex dynamical networks: an engineering viewpoint*, IEEE Trans. Circuits and Systems-I, **50** (2003), 1381-1390.

115. J.H. Lu, X.H. Yu, G.R. Chen, et al, *Characterizing the synchronizability of small-world dynamical networks*, IEEE Trans. Circuits and Systems-I, **51** (2004), 787-796.

116. Z. Li, and G.R. Chen, *Global synchronization and asymptotic stability of complex dynamical networks*, IEEE Trans. Circuit and Systems-II, **53** (1998), 28-33.

117. L.M. Pecora, T.L. Carroll, *Master stability functions for synchronized coupled systems*, Phys. Rev. Lett., **80** (1998), 2109.

118. C.G. Li, G.R. Chen, *Synchronization in general complex dynamical networks with coupling delays*, Physica A, **343** (2004), 236-278.

119. J.A. Acebron, L.L. Bonilla, C.J.P. Vicente, et al, *The Kuramoto model: A simple paradigm for synchronization phenomena*, Rev. Mod. Phys., **77** (2005), 137-185.

120. S.H. Strogatz, *From Kuramoto to Grawford: exploring the onset of synchronization in populations of coupled oscillators*, Physica D, **143** (2002), 1-20.

121. H. Hong, Parkand H., Choi M.Y., *Collective synchronization in spatially extended systems of coupled oscillators with random frequencies*, Phys. Rev. E, **72** (2005), 036217.

122. H. Hong, B.J. Kim, M.Y. Choi, et al, *Factors that predict better synchronizability on complex networks*, Phy. Rev. E., **69** (2004), 067105.

123. H. Hong, M.Y. Choi, B.J. Kim, *Synchronization on small-world networks*, Phys. Rev. E., **65** (2002), 026139.

124. Y.M. Vega, M. Vásquez-Prada, A.F. Pacheco, *Fitness of synchronization of network motifs*, Physica A, **343** (2004), 279-287.

125. M. Chen, D. Zhou, *Synchronization in uncertain complex networks*, Chaos, **16** (2006), 013101.

126. X. Li, G.R. Chen, *Synchronization in general complex dynamical networks with coupling delays*, Physica A, **343** (2004), 263-278.

127. F. Liljeros, et al., *The Web of Human sexual contact*, Nature, **411** (2001), 907.

128. R. Pastor-Satorras, et al, *Epidemic spreading in scale-free networks*, Phys. Rev. Lett., **86** (2002), 3200-3203.

129. R.M. May, A.L. Lloyd, Infection dynamics on scale-free networks, Phys. Rev. E., **64** (2001), 066112.

130. R. Pastor-Satorras, A. Vespignani, *Epidemic dynamics in finite size scale-free networks*, Phys. Rev. E., **65** (2002), 035108.

131. M. Boguna, R. Pastor-Satorras, A. Vespignani, *Absence of epidemic threshold in scale-free networks with degree correlations*, Phys. Rev. Lett., **90** (2003), 028701.

132. R.M. Anderson, R.M. May, *Infectious Disease of Humans: Dynamics and Control*, Oxford University Press, 1991.

133. H. Hethcote, *The mathematics of infectious diseases*, SIAM. Review, **42(3)** 2002, 599-653.

134. D. Hwang, et al., *Thresholds for epidemic outbreaks in finite scale-free networks*, Mathematical Bioscience and Engineering, **2** (2005), 317-327.

135. M. Boguna, R. Pastor-Satorras, *Epidemic spreading in corrected complex networks*, Phys. Rev. E.,., **66** (2002), 047104.

136. Z. Dezso, A. Barabasi, *Halting viruses in scale-free networks*, Phys. Rev. E., **65** (2002), 055103.

137. R. Pastor-Satorras, A. Vespignani, *Immunization of complex networks*, Phys. Rev. E., **65** (2002), 036104.

138. S. Eubank, et al., *Modelling disease outbreak in realistic urban social networks*, Nature, **429** (2004), 180-182.

139. N. Becker, et al, *Controlling emerging infectious disease like SARS*, Mathematical Bioscience, **193** (2005), 205-221.

140. S. Shen-Orr, R. Milo, S. Mangan, et al., *Network motifs in the transcriptional regulatory network of Escherichia coli*, Nature Genetics, **31** (2002), 64-68.

141. I. Belykh, V. Belykh, M. Hasler, *Blinking model and synchronization in small-world networks with a time-varying coupling*, Physica D, **195** (2004), 188-206.

142. D.J. Stilwell, E.M. Bollt, D.G. Roberson, *Sufficient Conditions for Fast Switching Synchronization in Time-Varying Network Topologies*, SIAM Journal on Applied Dynamical Systems, **5** (2006), 140-156.

143. L. Yang, Z.R. Liu, J.M. Mao, *Controlling hyperchaos*, Phys. Rev. Lett., **67(1)** (2000), 67-71.

144. Mao J.M., Liu Z.R., Yang L., *Straight-line stabilization*, Phy. Rev. E, **62(4)** (2000), 4846-4849.

145. L. Yang, Z.R. Liu, J.M. Mao, *Controlling hyperchaos in planar systems by adjusting parameters*, Appl. Math and Mech., **24(2)** (2003), 351-356.

146. Z.R. Liu, J.M. Mao, *Control of unstable flows*, Chin. Phys. Lett, **20(2)** (2003), 351-356.

147. L. Yang, Z.R. Liu, G. Chen, *Chaotifying a continuous-time system via impulse input*, Inter. J. Bifurcation and Chaos, **12(5)** (2002), 1121-1128.

148. G. Chen, L. Yang, Z.R. Liu, *Anti-control of chaos for continuous-time systems*, IEICE, E85-A, (2002), 1333-1335.

149. Chen L.Q., Liu Z.R., *Control of a hyperchaotic discrete system*, Appl. Math and Mech., **22(7)** (2001), 741-746.

150. Y.T.L. Andrew, Z.R. Liu, *Suppressing chaos for some nonlinear oscillators*, Inter. J. Bifur. and Chaos, **14(4)** (2004), 1455-1465.

151. Y.T.L. Andrew, Z.R. Liu, *Some new methods to suppress chaos for a kind of nonlinear oscillators*, Inter. J.. Bifur. and Chaos, **14(8)** (2004), 2955-2961.

152. Z.R. Liu, K.W. Chung, *Hybrid control of bifurcation in continuous nonlinear dynamical systems*, Inter. J. Bifur. and Chaos, **15(2)** (2005), 3895-3903.

153. Z.R. Liu, G. Chen, *On the relationship between parametric variation and state feedback in chaos control*, Inter. J. Bifurcation and Chaos, **12(6)** (2002), 1411-1415.

154. L. Yang, Z.R. Liu, Y.A. Zheng, *"Middle" periodic orbit and its application to chaos control*, Inter. J. Bifurcation and Chaos, **12(8)** (2002), 1869-1876.

155. Y.A. Zheng, Y.B. Nian, Z.R. Liu, *Impulsive control for the stabilization of discrete chaotic system*, Chin. Phy. Lett., **19(9)** (2002), 1252-1253.

156. Y.A. Zheng, G. Chen, Z.R. Liu, *On Chaotification of discrete systems*, Inter. J. Bifur. and Chaos, **13(11)** 2003, 3443-3447.

157. Z.R. Liu, *Several Theoretic Problems for Synchronization*, Nature Magazine (in Chinese), **26(5)** (2004), 298-300.

158. Z.R. Liu, *Using Structure Adaptive to Realize Complete Synchronization between Different Systems*, Communication on Applied Mathematics and Computation, **18(2)** (2004), 68-72.

159. Liu Z.R., Zhang G., Ma Z.J., *Several results to realize generalized synchronization in dynamical systems*, DCDIS. B, (2005), 790-794.

160. Z.R. Liu, J.G. Luo, *Realization of complete synchronization between different systems by using structure adaptation*, Chin. Phys. Lett., **23(5)** (2006), 1118-1121.

161. G. Zhang, Z.R. Liu, Z.J. Ma, *Generalized synchronization of continuous dynamical system*, Appl. Math. and Mech., **28(2)** (2007), 157-162.

162. Z.J. Ma, Z.R. Liu, G. Zhang, *Generalized synchronization of discrete systems*, Appl. Math. And Mech., **28(5)** (2007), 609-614.
163. G.R. Chen, J. Zhou, Z.R. Liu, *Global synchronization of coupled delayed neural networks and applications to chaotic CNN models*, Inter J. Bifurcation and Chaos, **14(7)** (2004), 2229-2240.
164. J. Zhou, T.P. Chen, L. Xiang, *Adaptive synchronization of delayed neural networks based on parameters identification*, Lecture Note in Computer Science, **3496** (2005), 308-313.
165. J. Zhou, T.P. Chen, L. Xiang, *Robust synchronization of delayed neural networks based on adaptive control and parameters identification*, Chaos, Solitons and Fractals, **27(4)** (2006), 905-913.
166. G.R. Chen, J. Zhou, S. Celikovsky, *On LaSalle's invariance principle and its application to robust synchronization of general vector Liénard equations*, IEEE Trans on Automat. Control, **50(6)** (2005), 869-874.
167. J. Zhou, T.P. Chen, L. Xiang, *Adaptive synchronization of coupled chaotic delayed systems based on parameter identification and its applications*, Inter J. Bifurcation and Chaos, **16(10)** (2006), 2923-2933.
168. J. Zhou, T.P. Chen, L. Xiang, *Chaotic lag synchronization of coupled delayed neural networks and its applications in secure communication*, Circuits Systems and Signal Processing, **24(5)** (2005), 599-613.
169. Z.J. Ma, Z.R. Liu, G. Zhang, *A new method to realize cluster synchronization in connected chaotic network*, Chaos, **16** (2006), 023103.
170. y. A. Zheng, Y.B. Nian, Z.R. Liu, *Impulsive synchronizatiion of discrete chaotic systems*, Chin. Phys. Lett., **20(2)** (2003), 199-201.
171. Z.R. Liu, G. Chen, *On a possible mechanism of the brain for responding to dynamical features extracted from input signals*, Chaos, Solitons and Fractals, **18** (2003), 785-794.
172. J. Chen, Z.R. Liu, *A method of controlling synchronization in different systems*, Chin. Phys. Lett., **20(9)** (2003), 1441-1443.
173. J. Chen, Z.R. Liu, *Partial synchronization between different systems*, Appl. Math. and Mech., **26(9)** (2005), 1132-1137.
174. Y. Li, Z.R. Liu, J.B. Zhang, *Circadian oscillator and phase synchronization under a light-dark cycle*, Inter. J. Nonlinear Science, **1(3)** (2006), 131-138.
175. D.J. Zhao, Z.R. Liu, *Lag synchronization in nonlinear systems based on adaptive control*, J. Shanghai University, **8(1)** (2004), 34-37.
176. G. Zhang, Z.R. Liu, J.B. Zhang, *Adaptive synchronization of a class of continuous chaotic systems with uncertain parameters*, Phys. Lett. A, **372** (2008), 447-450.
177. D. Zhao, Z.R. Liu, J.Z Wang, *Duplication: a mechanism producing disassortive mixing networks in biology*, Chin. Phys. Lett., **24(10)** (2007), 2766-2768.
178. C.D. Dong, Z.R. Liu, *An ideal disassortative network and synchronization*, Inter. J. Bifur. and Chaos, **16(10)** (2006), 3039-3102.
179. C.D. Dong, Z.R. Liu, *An ideal assortative network and synchronization*, Commun. Theor. Phys.,**47(1)** (2007), 186-192.
180. Z.R. Liu, C.D. Dong, Q.D. Fan, *Multicenter network and synchronization*, Inter J. Bifur. and Chaos, **17(6)** (2007), 2109-2115.
181. J. Zhou, Z.R. Liu, G.D. Chen, *Dynamics of periodic delayed neural networks*, Neural Networks, **17(1)** (2004), 87-101.
182. J. Zhou, Z.R. Liu, G. Chen, *Global dynamics of periodic delayed neural networks models*, Dynamics of Continuous Discrete and Impulsive Systems-Series, B-Applications and Algorithms, **12(5-6)** (2005), 689-699.

183. L. Xiang, J. Zhou, Z.R. Liu, et al, *On the asymptotic behavior of Hopfield neural network with periodic inputs*, Appl. Math. and Mech., **23(12)** (2002), 1367-1373.

184. J. Zhou, Z.R. Liu, L. Xiang, *Global dynamics of delayed bidirectional associative memory (BAM) neural networks*, Appl. Math and Mech., **26(3)** (2005), 327-335.

185. J. Zhou, T.P. Chen, *Synchronization in general complex delayed dynamical networks*, IEEE Trans. on Circuits and Systems-I, **53(3)** (2006), 733-744.

186. J. Zhou, L. Xiang, Z.R. Liu, *Global Synchronization in general complex delayed dynamical networks and its applications*, Physica A., **385(2)** (2007), 729-742.

187. J. Zhou, L. Xiang, Z.R. Liu, *Synchronization in complex delayed dynamical networks with impulsive effects*, Physica A., **384(2)** (2007), 684-692.

188. J. Zhou, T.P. Chen, L. Xiang, et al, *Global synchronization of impulsive coupled delayed neural networks*, Lecture Notes in Computer Science, **3971** (2006), 303-308.

189. L. Xiang, Z.R. Liu, J. Zhou, *Robust impulsive synchronization of coupled delayed neural networks*, Lecture Notes in Computer Science, **4492** (2007),16-23.

190. L. Xiang, Z.R. Liu, J. Zhou, *Robust impulsive synchronization of delayed dynamical networks and its applications*, Lecture Notes in Operation Research, **7** (2007), 170-177.

191. G. Zhang, Z.R. Liu, Z.J. Ma, *Synchronization of complex dynamical networks via impulsive control*, Chaos **17** (2007), 043126.

192. J.Z. Wang, Z.R. Liu, J.H. Xu, *Epidemic spreading on uncorrelated heterogeneous networks with non-uniform transmission*, Physica A, **382** (2007), 715-721.

193. Y. Li, Z.R. Liu, J.B. Zhang, R.Q. Wang, L.N. Chen, *Synchronization mechanisms of circadian rhythms in the suprachiasmatic nucleus*, IET Systems Biology, **3(2)** (2009), 100-112.

194. F.D. Xu, Z.R. Liu, J.W. Shen, R.Q. Wang, *The dynamics of MicroRNA-mediated motifs*, IET Systems Biology (accepted).

195. W.Q. Duan, Z. Chen, Z.R. Liu, et al, *Efficient target strategies for contagion in scale free networks*, Phys. Rev. E, **72** (2005), 026133.

196. W.Q. Duan, Z. Chen, Z.R. Liu, *Epidemic spreading in contact network based on exposure level*, Chin. Phys. Lett., **23(5)** 2006, 1347-1350.

197. W.Q. Duan, Z. Chen, Z.R. Liu, *Phase transition dynamics of collective decision in scale free networks*, Chin. Phys. Lett., **22(8)** 2005, 2137-2139.

198. Y.S. Wan, Z. Chen, Z.R. Liu, *Modelling the two power-law degree distribution of banking networks*, DCDIS, Series B Appl. and Algorithm, **13(3-4)** 2006, 441-449.

199. B. Yang, Z. Chen, Z.R. Liu, *Research on structure evolution and pattern emergence of socioeconomic complex networks based on individual choices*, DCDIS, Series B, Appl. and Algorithm, **13(3-4)** (2006), 387-394.

200. J.B. Zhang, Z.R. Liu, Y. Li, *An approach to analyze phase synchronization in oscillator networks with weak coupling*, Chin. Phys. Lett., **24(6)** (2007), 1494-1497.

201. Y. Li, Z.R. Liu, J.B. Zhang, *Dynamics of network motifs in genetic regulatory network*, Chin. Phys., **16(9)** (2007), 2587-2594.

202. Z.R. Liu., Luo J.G., *From lag synchronization to pattern formation in one-dimensional open flow models*, Chaos, Solitons and Fractals, **30** (2006), 1198-1205.

203. Z.Y. Zhang, J.G. Luo, Z.R. Liu, *From lag synchronization to pattern formation in networked dynamics*, Physica A, **378** (2007), 537-549.

204. J.G. Luo, Z.R. Liu, *From Synchronization to Emergence*, Complex systems and complex science, **2(1)** (2005), 29-32.

Perspectives in Mathematical Sciences
Interdisciplinary Mathematical Sciences, Volume 9, 2009
pp. 181–210

Chapter 9

Mathematical Aspects of the Cold Plasma Model

Thomas H. Otway

Department of Mathematics, Yeshiva University,
New York, NY 10033, USA
otway@yu.edu

A simple model for electromagnetic wave propagation through zero-temperature plasma is analyzed. Many of the complexities of the plasma state are present even under these idealized conditions, and a number of mathematical difficulties emerge. In particular, boundary value problems formulated on the basis of conventional electromagnetic theory turn out to be ill-posed in this context. However, conditions may be prescribed under which solutions to the Dirichlet problem exist in an appropriately weak sense. In addition to its physical interest, analysis of the cold plasma model illuminates generic difficulties in formulating and solving boundary value problems for mixed elliptic-hyperbolic partial differential equations.

1. Introduction

Among the many equations of mathematical physics which change from elliptic to hyperbolic type along a smooth curve, only the equations for transonic flow have received sustained attention. In this brief review we consider elliptic-hyperbolic equations originating in a simple model for the propagation of electromagnetic waves through zero-temperature plasma. Solutions to such equations are likely to have significantly weaker regularity than solutions to the linearized equations of transonic flow. Recognizing the interdisciplinary nature of the topic, we assume a familiarity with physics but not necessarily plasma physics, and analysis but not necessarily elliptic-hyperbolic equations. However, the physics is confined to a review of fundamental results in Sec. 2, whereas the mathematical results of Sec. 3 are somewhat more technical. There we consider the extent to which problems formulated primarily for linearized equations of gas dynamics possess analogies for equations arising from a different physical problem. Continuing such investigations in various physical and geometric contexts (*c.f.* Ref. 29), one may hope to obtain eventually a natural theory for linear elliptic-hyperbolic partial differential equations.

1.1. *Physical background*

The plasma state is characterized by the dominance of long-range, nonlinear effects. For matter in such a state, it is particularly difficult to obtain mathematical problems which can be stated with a satisfactory degree of rigor, and for which solutions can be shown to exist. Without a proof of the existence and uniqueness of solutions — which, in particular, specifies the function spaces in which solutions lie — it is hard to place appropriate boundary conditions on numerical experiments and to gauge the reliability of the results obtained.

If one hopes to obtain a tractable mathematical problem, it is usually necessary to impose harsh assumptions on both the plasma and the applied field. Perhaps the harshest of these fixes the temperature of the plasma to be zero. This permits one to neglect altogether the fluid properties of the medium, which is then treated as a linear dielectric. Somewhat surprisingly, the assumption of zero plasma temperature is a useful first approximation to the products of tokamaks: low-density plasmas which are remarkably free of expected high-temperature phenomena such as collisions and wall effects. See the remarks in the introduction to Ref. 36 and the more detailed discussions in Ref. 39. More generally, the cold plasma model approximates the effects of small-amplitude electromagnetic waves, propagating with phase velocities which are sufficiently large in comparison to the thermal velocity of the particles.

We note that the term *cold plasma* is highly ambiguous. Although we take this to imply zero temperature, in the astrophysics literature interstellar plasmas on the order of 10^4 K to 10^5 K are typically referred to as "cold" (see, *e.g.*, Ref. 11). Very recently, "ultracold" neutral plasmas, having electron temperatures in the range from 1 K to 10^3 K and ion temperatures ranging from 10^{-3} K to 1 K, have been created experimentally. The cold plasma model explored in this paper is apparently too simple to yield quantitative insight into those plasmas. In particular, the fluid dynamics aspects of experimental ultracold plasmas appear to be non-negligible (*c.f.* Sec. 3 of Ref. 17).

The other physical hypotheses imposed in this review are also quite restrictive: Although the plasma is not assumed to be homogeneous, the inhomogeneity is taken to be two-dimensional, so the governing equations for the model are also essentially two-dimensional. Moreover, the applied magnetic field is assumed in Sec. 2.4 to be longitudinal and the resulting wave motion confined to electrostatic oscillations. In Sec. 2.5 we consider electromagnetic waves, but we find (after reviewing a detailed analysis by Weitzner[39]) that elliptic-hyperbolic equations arising in the electrostatic case retain their validity as a qualitative model for the general case.

For the most part, the outstanding mathematical problems relevant to the cold plasma model are boundary value problems for Maxwell's equations. The dielectric tensor for these equations will render them of elliptic type on one part of their domain and of hyperbolic type on the remainder, except for a smooth curve (the *parabolic line*) separating the two regions. Little is known about the formulation of well-posed boundary value problems for equations which change type in this way, especially as the equations that arise in the cold plasma model appear to have certain fundamental differences from those that arise in gas dynamics.

Careful reasoning about the mathematical properties of plasma models is not needed merely in order to prevent "mathematicians' nightmares." An example is known[26] in which the boundary conditions suggested by physical reasoning about the plasma lead to a mathematically ill-posed problem in the expected function space. Moreover, numerical experiments tend to confirm the difficulties that arise when the model equations are subjected to classical analytic techniques; see Ref. 26 and various remarks in Ref. 39.

High-frequency waves can be modelled via geometrical optics. (Any propagating electromagnetic field will tend to have high frequency relative to the characteristic plasma frequencies; see, *e.g.*, Sec. 2.4 of Ref. 23.) Mathematical problems that arise in the geometrical optics approximation are quite different from those that arise from applying Maxwell's equations directly, and we do not pursue the geometrical optics approach in this review. The complexity of the geometrical optics approximation is due to significant difference in magnitude among the terms of the plasma conductivity tensor at lower hybrid frequencies; see Ref. 34 and the references therein.

The physics presented in Secs. 2.1 and 2.2 essentially goes back to the work of Tonks and Langmuir[38] in the late 1920s. The results of Sec. 2.3 were already well known in the 1950s;[2,3,35] those of Secs. 2.4 and 2.5 date from the 1970s[20,33] and 1980s, respectively. In particular, Sec. 2.5 derives some fundamental analytic formulas introduced by Weitzner in Refs. 39 and 40; see also Ref. 16. Section 3 is based on recent results,[27,28] which extend analogous research on equations of Tricomi type — particularly Refs. 21 and 22; see also Ref. 41, an earlier paper which is based on Ref. 24.

In the sequel, a subscripted variable denotes (usually partial) differentiation in the direction of the variable, whereas subscripted numbers denote components of a matrix, vector, or tensor. Differentiation of vector or matrix components in the direction of a variable is indicated by preceding the subscripted variable by a comma. Unless otherwise stated, a cartesian coordinate system is assumed in which the subscript 1 denotes a component projected onto the x-axis, the subscript 2 denotes a component projected onto the y-axis, and the subscript 3 denotes a component projected onto the z-axis. In particular, $\mathbf{x} = (x_1, x_2, x_3) = (x, y, z)$ and we denote the canonical cartesian basis by $\left(\hat{\imath}, \hat{\jmath}, \hat{k}\right)$.

2. The cold plasma model

A *plasma* is a fluid composed of electrons and one or more species of ions. Because it is a fluid, its evolution must satisfy the equations of fluid dynamics. But because the particles of the fluid are charged, they act as sources of an electromagnetic field,

which is governed by Maxwell's equations. The presence of this intrinsic field leads
to highly nonlinear behavior. Indeed, the dominance of long-range electromagnetic
interactions over the short-range interatomic or intermolecular forces is often cited
as the defining characteristic of the plasma state.

If the plasma is at zero temperature, then Amontons' Law implies that the
pressure term in the equations for fluid motion will also be zero, and the laws of fluid
dynamics will enter only through the conservation laws for mass and momentum.
In fact, because collisions can be neglected, the fluid aspect of the medium can
be virtually ignored. The plasma is then represented as a static dielectric through
which electromagnetic waves propagate.

In particular, *zero-order quantities* — the plasma density, proportions of ions
to electrons, and the background magnetic field — can all be considered static in
time and uniform in space. *First-order quantities* — the electric field $\mathbf{E}$ and particle
velocities $\mathbf{v}$, are assumed to be expressible as plane waves: sinusoidal waves propor-
tional to functions having the form $\exp\left[i\left(\mathbf{k}\cdot\mathbf{r}-\omega t\right)\right]$, where $\mathbf{k}$ is the propagation
vector of the wave (not to be confused with the notation for the cartesian basis
vector $\hat{k}$); $\mathbf{r}$ is the radial coordinate in space; ω is angular frequency; $i^2=-1$. Thus
in cartesian coordinates, $\mathbf{k}\cdot\mathbf{r}=k_1 x+k_2 y+k_3 z$.

In the following we review basic elements of the physical theory that results
from these assumptions. The material in Secs. 2.1–2.3 is completely standard and
can be found in many sources. The classical reference is Ch. 1 of Ref. 36; see also
Refs. 1, 8, 13, and Sec. 2 of Ref. 40. More recent surveys include nodes 43–45 of
Ref. 12 and Ch. 2 of Ref. 37. A recent review of theoretical plasma physics can be
found in Ref. 6. We employ *SI* units except where other units are specified.

2.1. *Equations of motion*

Consider a single particle of mass m, having charge $q=Z\delta e$, where Z is a positive
integer, δ equals 1 or -1, and e is the charge on an electron. Let the particle be
subjected only to the Lorentz force

$$\mathbf{F}_L=q\left(\mathbf{E}+\mathbf{v}\times\mathbf{B}\right),$$

where

$$\mathbf{B}=B_0\hat{k}. \tag{2.1}$$

Equation (2.1) implies that the applied magnetic field is *longitudinal*: its only
nonzero component is directed along the positive z-axis. (In fact, there is little
harm in assuming, somewhat more generally, that

$$\mathbf{B}=B_0\hat{k}+\tilde{\mathbf{B}}\left(x,y,z\right)\exp\left[i\left(\mathbf{k}\cdot\mathbf{r}-\omega t\right)\right],$$

with $|\tilde{\mathbf{B}}|<<|B_0|$.)

The equation of motion for the particle is given by Newton's Second Law of Motion, that is,

$$m\frac{d\mathbf{v}}{dt} = \mathbf{F}_L.\tag{2.2}$$

In accordance with our assumption about first-order quantities, we write

$$\mathbf{v}(x, y, z, t) = \tilde{\mathbf{v}}(x, y, z)\exp\left[i\left(\mathbf{k}\cdot\mathbf{r} - \omega t\right)\right],$$

or

$$\frac{d\mathbf{v}}{dt} = -i\omega\mathbf{v}.$$

Substituting this result into (2.2) yields

$$-im\omega\tilde{\mathbf{v}} = q\left(\tilde{\mathbf{E}} + \tilde{\mathbf{v}}\times\mathbf{B}\right),\tag{2.3}$$

where

$$\mathbf{E}(x, y, z, t) = \tilde{\mathbf{E}}(x, y, z)\exp\left[i\left(\mathbf{k}\cdot\mathbf{r} - \omega t\right)\right].\tag{2.4}$$

Initially we will take $\tilde{\mathbf{E}}$ to be a constant vector:

$$\tilde{\mathbf{E}}(x, y, z) = E_1\hat{\imath} + E_2\hat{\jmath} + E_3\hat{k},\tag{2.5}$$

where E_1, E_2, and E_3 are constants, and similarly for $\tilde{\mathbf{v}}$.

Defining the *cyclotron frequency*

$$\Omega = \left|\frac{qB_0}{m}\right|,$$

Eq. (2.3) has solutions $\mathbf{v} = (v_1, v_2, v_3)$ satisfying

$$v_1 = \frac{iq}{m\left(\omega^2 - \Omega^2\right)}\left(\omega E_1 + i\delta\Omega E_2\right);\tag{2.6}$$

$$v_2 = \frac{iq}{m\left(\omega^2 - \Omega^2\right)}\left(\omega E_2 - i\delta\Omega E_1\right);\tag{2.7}$$

$$v_3 = \frac{iq}{m\omega}E_3.\tag{2.8}$$

2.2. *The dielectric tensor*

Although the above relations were derived for an individual particle, they also hold, in our simplified linear model, for each species of particle in a plasma consisting of electrons and $N-1$ species of ions. In particular, the plasma current can be written as the sum

$$\mathbf{j} = \sum_{\nu=1}^{N} n_\nu q_\nu\mathbf{v}_\nu,\tag{2.9}$$

where n_ν is the density of particles having charge magnitude $|q_\nu| = Z_\nu e$.

In the sequel we will only consider the aggregate of particles, in which Eqs. (2.1)–(2.8) pertain with the quantities $\mathbf{v}$, m, q, Z, δ, and Ω indexed by ν, where $\nu = 1, ..., N$. Introduce the *electric displacement vector*

$$\mathbf{D} = \text{vacuum displacement} + \text{plasma current} = \epsilon_0 \mathbf{E} + \frac{i}{\omega}\mathbf{j}, \qquad (2.10)$$

where ϵ_0 is the permittivity of free space. It will be convenient to express (2.10) in the form

$$\mathbf{D} = \epsilon_0 \mathbf{K} \mathbf{E}, \qquad (2.11)$$

where

$$D_i = \epsilon_0 \sum_{j=1}^{3} K_{ij} E_j \qquad (2.12)$$

and $\mathbf{K} = (K_{ij})$ is the *dielectric tensor* (also called the *cold plasma conductivity tensor*). The tensorial nature of this quantity reflects the anisotropy of the plasma due to the presence of an applied magnetic field. (Note that in the sequel the reader will be expected to distinguish between the notation $\mathbf{K}$ for the dielectric tensor, the notation K_{ij} for the scalar element of its ith row and j^{th} column, and the notation $\mathcal{K}$ for the type-change function of an elliptic-hyperbolic equation.) Equations (2.6)–(2.11) imply that

$$\mathbf{K} = \begin{pmatrix} s & -id & 0 \\ id & s & 0 \\ 0 & 0 & p \end{pmatrix}, \qquad (2.13)$$

where s, d, and p are defined in terms of:

  i) the *plasma frequency*, which for particles of the νth species is given by

$$\Pi_\nu^2 = \frac{n_\nu q^2}{\epsilon_0 m_\nu};$$

  ii) the *permittivities* R or L of a right- or left-circularly polarized wave travelling in the direction $\hat{k}$; these are given by

$$R = 1 - \sum_{\nu=1}^{N} \frac{\Pi_\nu^2}{\omega^2}\left(\frac{\omega}{\omega + \delta_\nu \Omega_\nu}\right)$$

and

$$L = 1 - \sum_{\nu=1}^{N} \frac{\Pi_\nu^2}{\omega^2}\left(\frac{\omega}{\omega - \delta_\nu \Omega_\nu}\right).$$

In terms of these quantities,

$$s = \frac{1}{2}\left(R + L\right),$$

$$d = \frac{1}{2}\left(R - L\right),$$

and

$$p = 1 - \sum_{\nu=1}^{N} \frac{\Pi_\nu^2}{\omega^2}.$$

The mass of an electron is considerably smaller than the mass of any ion; so the squared ion cyclotron frequencies obtained from combining fractions in R and L can be neglected, leading to the approximate formulas

$$R \approx 1 - \sum_{\nu=1}^{N-1} \frac{\Pi_e^2}{\omega^2 + \omega\Omega_e + \Omega_e\Omega_{i_\nu}} \qquad (2.14)$$

and

$$L \approx 1 - \sum_{\nu=1}^{N-1} \frac{\Pi_e^2}{\omega^2 - \omega\Omega_e + \Omega_e\Omega_{i_\nu}}. \qquad (2.15)$$

In these formulas, the subscripted e denotes the value of the relevant quantity for the electrons and the subscripted i_ν denotes that value for the νth species of ion. Note that, by the same reasoning, the ion plasma frequencies can be neglected in the definition of p.

2.3. *The plasma dispersion relation*

The field equations for the system described in Sec. 2.2 are Maxwell's equations,

$$\nabla \times \mathbf{E} = -\frac{\partial \mathbf{B}}{\partial t}, \qquad (2.16)$$

$$\nabla \times \mathbf{B} = \mu_0 \left(\mathbf{j} + \epsilon_0 \frac{\partial \mathbf{E}}{\partial t} \right), \qquad (2.17)$$

where μ_0 is the permeability of free space.

From the form of Eq. (2.4), it is clear that whenever $\mathbf{E}$ and $\mathbf{B}$ are plane waves, Eqs. (2.16) and (2.17) reduce to the simpler form

$$\mathbf{k} \times \mathbf{E} = \omega\mathbf{B} \qquad (2.18)$$

and

$$\mathbf{k} \times \mathbf{B} = -i\mu_0\mathbf{j} - \omega\mu_0\epsilon_0\mathbf{E}. \qquad (2.19)$$

We can rewrite Eq. (2.19) to read

$$\mathbf{k} \times \mathbf{B} = -\omega\mu_0 \left(\frac{i\mathbf{j}}{\omega} + \epsilon_0\mathbf{E} \right)$$
$$= -\omega\mu_0\mathbf{D} = -\epsilon_0\mu_0\omega\mathbf{K}\mathbf{E}, \qquad (2.20)$$

where we have used (2.10) and (2.11) in deriving the last identity.

Now using (2.18), (2.20), and the elementary identity $\mu_0\epsilon_0 = c^{-2}$, where c is the speed of light *in vacuo*, we obtain

$$\mathbf{k} \times (\mathbf{k} \times \mathbf{E}) = \mathbf{k} \times (\omega\mathbf{B}) = \omega\,(\mathbf{k} \times \mathbf{B})$$

$$= -\omega^2\epsilon_0\mu_0\mathbf{KE} = -\left(\frac{\omega}{c}\right)^2\mathbf{KE},$$

implying that

$$\mathbf{k} \times (\mathbf{k} \times \mathbf{E}) + \left(\frac{\omega}{c}\right)^2\mathbf{KE} = 0. \tag{2.21}$$

Define the *index of refraction vector*

$$\mathbf{n} = \frac{c}{\omega}\mathbf{k}.$$

With this construction, the scalar index of refraction acquires a direction: that of the wave propagation vector $\mathbf{k}$. In terms of $\mathbf{n}$, Eq. (2.21) reads

$$\mathbf{n} \times (\mathbf{n} \times \mathbf{E}) + \mathbf{KE} = 0. \tag{2.22}$$

Conventionally, $\mathbf{k}$ (and thus $\mathbf{n}$) lies in the xz-plane. Denote by θ the angle subtended by the vectors $\mathbf{k}$ and $\mathbf{B}$. Then (2.22) can be written as the matrix equation

$$\begin{pmatrix} s - n^2\cos^2\theta & -id & n^2\cos\theta\sin\theta \\ id & s - n^2 & 0 \\ n^2\cos\theta\sin\theta & 0 & p - n^2\sin^2\theta \end{pmatrix} \begin{pmatrix} E_1 \\ E_2 \\ E_3 \end{pmatrix} = 0.$$

This matrix equation has a nontrivial solution precisely when the determinant of the 3×3 matrix vanishes. The condition for the vanishing of that determinant, the *cold plasma dispersion relation* is, geometrically, the equation for the wave-normal surface:

$$An^4 - Bn^2 + C = 0, \tag{2.23}$$

where the coefficients satisfy

$$A = s\sin^2\theta + p\cos^2\theta, \tag{2.24}$$

$$B = \left(s^2 - d^2\right)\sin^2\theta + ps\left(1 + \cos^2\theta\right), \tag{2.25}$$

and

$$C = p\left(s^2 - d^2\right). \tag{2.26}$$

Because the left-hand side of Eq. (2.23) is a quadratic polynomial in n^2, we obtain from the quadratic formula the solutions

$$n^2 = \frac{B \pm F}{2A}$$

for F satisfying $F^2 = B^2 - 4AC$. Using (2.24)–(2.26) to write

$$F^2 = (RL - ps)^2\sin^4\theta + 4p^2d^2\cos^2\theta,$$

we obtain

$$\tan^2 \theta = -\frac{p\left(n^2 - R\right)\left(n^2 - L\right)}{\left(sn^2 - RL\right)\left(n^2 - p\right)}.$$

These equations yield criteria for *cutoff*, where $n = 0$, or *resonance*, where $n \to \infty$.

Physically, cutoffs and resonances correspond to a change in the behavior of the wave from possible propagation to evanescence. Mathematically, we will identify certain resonances with a change in type of the governing field equation, from hyperbolic (implying wave propagation) to elliptic (implying evanescence). These transitions may be accompanied, under certain conditions, by reflection and/or absorption of the wave.

Sufficient conditions for cutoff are $p = 0$, $R = 0$, or $L = 0$ — that is, a sufficient condition is $C = 0$. A sufficient condition for resonance is $A = 0$ which, given Eq. (2.24), can be written

$$\tan^2 \theta = -\frac{p}{s}. \tag{2.27}$$

Particular cases of interest are $\theta = 0$ (propagation parallel to the magnetic field) and $\theta = \pi/2$ (propagation perpendicular to the magnetic field). We will be particularly interested in the *hybrid resonances* at $\theta = \pi/2$, which occur at frequencies for which $s = 0$.

2.4. *Electrostatic waves*

The electric field is said to be *electrostatic* if it approximately satisfies

$$\mathbf{E} = -\nabla \Phi, \tag{2.28}$$

where Φ is a scalar potential. Equation (2.28) is satisfied locally by all time-independent electric fields and in an ordinary dielectric, the converse is also true. However in cold plasma there also exist time-dependent solutions of (2.28). This recalls the characterization of cold plasma as a linear dielectric through which, in distinction to ordinary dielectrics, electromagnetic waves propagate — including as a special case, electrostatic oscillations.

We write Φ in the form

$$\Phi(x, y, z; t) = \varphi\left(x, y, z\right) \exp\left[\mathbf{k} \cdot \mathbf{r} - i\omega t\right],$$

and add to Maxwell's equations (2.16), (2.17) the additional equation

$$div\,\mathbf{D} = 0, \tag{2.29}$$

which follows from Gauss' law for electricity.

Equation (2.28) implies immediately that

$$\nabla \times \mathbf{E} = 0. \tag{2.30}$$

This is most easily seen if we use differential forms, and note that in terms of the exterior derivative, $\mathbf{E} = d\Phi$, so (2.30) is just the well-known property that

$$d\mathbf{E} = d^2\Phi = 0.$$

(Here and below we will switch from vectors to forms whenever the calculation is made more transparent thereby; but we will not change notation for the underlying geometric object, making the convention that the argument of the exterior derivative is always assumed to be a differential form.) In either the vector or form notation, identity (2.30) follows from the equality of mixed partial derivatives. Applying the arguments relating (2.16) to (2.18) and (2.17) to (2.19) (*translation into Fourier modes*), we rewrite (2.30) in the form

$$\mathbf{k} \times \mathbf{E} = 0.$$

This implies, by the properties of the cross product, the geometric fact that the vectors $\mathbf{k}$ and $\mathbf{E}$ are parallel. We say that electrostatic waves are *longitudinal*. Physically, they appear as oscillations along the axis of the magnetic field.

Thus we conclude that *transverse* waves, which propagate in a direction perpendicular to the magnetic field, must satisfy

$$\mathbf{k} \cdot \mathbf{E} = 0.$$

Again, differential forms are illuminating: The above identity becomes $\delta d\Phi = 0$, where δ is the formal adjoint of the exterior derivative d. But Φ is a 0-form, so $\delta\Phi = 0$ automatically, and we find that transverse waves satisfy

$$\delta d\Phi + d\delta\Phi \equiv \Delta\Phi = 0,$$

that is, transverse waves are necessarily harmonic.

2.4.1. *Plane-layered media*

If we allow a plane-layered inhomogeneous medium (parameterized by x), the electrostatic potential has the form

$$\Phi(x, y, z) = \varphi(x) \exp\left[i\left(k_2 y + k_3 z\right)\right],$$

where k_j is the jth component of the wave vector $\mathbf{k}$ for $j = 1, 2, 3$. Substitution of this form for the electric potential into Eq. (2.29) yields, using (2.12), the single scalar equation[20]

$$K_{11}\varphi_{xx} + \left(K_{11,x} + i\sigma_0\right)\varphi_x = 0, \qquad (2.31)$$

where

$$\sigma_0 = k_3\left(K_{13} + K_{31}\right) + k_2\left(K_{12} + K_{21}\right),$$

and zero-order terms in φ have been neglected. This equation has a power-series solution except where K_{11} vanishes.

Explicit solutions of the model equation (2.31) under various physical assumptions are given in Sec. 1 of Ref. 33, the Appendix to Ref. 20, and Sec. C of Ref. 16.

It is easy to believe that inhomogeneities may develop in a plasma. For example, if the temperature is not exactly zero, the difference in velocity between electrons and ions can be expected to destabilize an initially homogeneous distribution. However, it is difficult to imagine a force that will restrict these inhomogeneities to a 1-parameter foliation, which would be necessary in order to arrive at Eq. (2.31). Formally, an electromagnetic potential leading to Eq. (2.31) could be induced by applying a driving potential to the metallic plates of a condenser. But in practice, this plasma geometry has little application either in the laboratory or in nature.

2.4.2. *A two-dimensional inhomogeneity*

Suppose instead that the medium is a cold, anisotropic plasma with a two-dimensional inhomogeneity parameterized by two variables, x and z. Then the field potential has the form

$$\Phi(x, y, z) = \varphi(x, z) \exp[ik_2 y].$$

The electric field **E** is then given by

$$E = -\nabla \Phi = (E_1, E_2, E_3) = -\left(\varphi_x e^{ik_2 y}, ik_2 \varphi e^{ik_2 y}, \varphi_z e^{ik_2 y}\right).$$

Maxwell's equations for the electric displacement vector $\mathbf{D} = (D_1, D_2, D_3)$ take the form

$$0 = \nabla \cdot \mathbf{D} = D_{1,x} + D_{2,y} + D_{3,z}. \tag{2.32}$$

We continue to neglect those terms which do not contain derivatives of φ, as φ is assumed to oscillate rapidly.

Because neither φ nor K_{ij} have any dependence on y, the problem is effectively two-dimensional. Applying Eq. (2.12), the surviving terms of Eq. (2.32) are (setting ϵ_0 equal to 1)

$$D_{1,x} = -\left[K_{11}\varphi_{xx} + K_{11,x}\varphi_x + K_{12}\varphi_x ik_2 + K_{13}\varphi_{zx} + K_{13,x}\varphi_z\right] e^{ik_2 y};$$

$$D_{2,y} = -\left[K_{21}\varphi_x ik_2 + K_{23}\varphi_z ik_2\right] e^{ik_2 y};$$

$$D_{3,z} = -\left[K_{31}\varphi_{xz} + K_{31,z}\varphi_x + K_{32}ik_2\varphi_z + K_{33}\varphi_{zz} + K_{33,z}\varphi_z\right] e^{ik_2 y}.$$

Collecting terms, we find that[33]

$$K_{11}\varphi_{xx} + 2\sigma\varphi_{xz} + K_{33}\varphi_{zz} + \alpha_1\varphi_x + \alpha_2\varphi_z = 0, \tag{2.33}$$

where

$$2\sigma = K_{13} + K_{31};$$

$$\alpha_1 = K_{11,x} + ik_2 \left(K_{12} + K_{21} \right) + K_{31,z};$$

$$\alpha_2 = K_{13,x} + ik_2 \left(K_{23} + K_{32} \right) + K_{33,z}.$$

Two-dimensional inhomogeneities of the kind represented by Eq. (2.33) can be expected to arise in toroidal fields, such as those created in tokamaks.

The entries of the matrix K under our assumptions on $\mathbf{B}_0$ imply that $\sigma = 0$, so we can write Eq. (2.33) in the form

$$K_{11}\varphi_{xx} + K_{33}\varphi_{zz} + \text{lower-order terms} = 0. \tag{2.34}$$

Equation (2.34) is of either elliptic or hyperbolic type, depending on whether the sign of the product

$$K_{11} \cdot K_{33} = \left(1 - \sum_{\nu=1}^{N} \frac{\Pi_\nu^2}{\omega^2 - \Omega_\nu^2} \right) \cdot \left(1 - \sum_{\nu=1}^{N} \frac{\Pi_\nu^2}{\omega^2} \right) \tag{2.35}$$

is, respectively, positive or negative.

The sign of K_{11} changes at the *cyclotron* resonances $\omega^2 = \Omega_\nu^2$. The cold plasma model breaks down at these resonances, as three terms of the dielectric tensor become infinite. The sign of K_{11} also changes at the *hybrid* resonances, at which

$$1 = \sum_{\nu=1}^{N} \frac{\Pi_\nu^2}{\omega^2 - \Omega_\nu^2}. \tag{2.36}$$

(These resonances, which have both a low-frequency and a high-frequency solution, are hybrid in that they involve both plasma and cyclotron frequencies.) In particular, the sign changes at the *lower* hybrid resonance,

$$1 + \frac{\Pi_e^2}{\Omega_e^2} = \frac{\Pi_i^2}{\omega^2}, \tag{2.37}$$

where as before, the subscript e denotes electron frequency, and the subscript i denotes ion frequency. At the hybrid resonance frequencies the cold plasma model retains its validity.

The sign of K_{33} changes on the surface

$$1 = \sum_{\nu=1}^{N} \frac{\Pi_\nu^2}{\omega^2}, \tag{2.38}$$

the resonance at which the frequency of the applied wave equals the plasma frequency of the medium. We may suppose that an electromagnetic wave propagating through a plasma does so at a much higher frequency than any of the characteristic frequencies of the plasma. Otherwise, the plasma magnetic field would prevent the waves from propagating very far (*c.f.* Ref. 18). Thus in evaluating (2.35) and in the sequel we will take K_{33} to be strictly positive.

Borrowing the terminology of fluid dynamics, we will refer to resonances such as (2.36)–(2.38) as *sonic conditions* on Eq. (2.34).

2.4.3. *The type of the governing equation*

In order to understand the possible variants of Eq. (2.34), we consider the coordinate transformation $(x, z) \rightarrow (\xi(x, z), \eta(x, z))$, where

$$\xi = K_{11}(x, z).$$

In these coordinates, the higher-order terms of Eq. (2.34) assume the form

$$K_{11}\varphi_{xx} + K_{33}\varphi_{zz} = \left(\xi\xi_x^2 + K_{33}\xi_z^2\right)\varphi_{\xi\xi} +$$
$$\left(\xi\xi_x\eta_x + K_{33}\xi_z\eta_z\right)\varphi_{\xi\eta} + \left(\xi\eta_x^2 + K_{33}\eta_z^2\right)\varphi_{\eta\eta}. \qquad (2.39)$$

In order that the transformation $(x, z) \rightarrow (\xi, \eta)$ be nonsingular, we require that its Jacobian be nonvanishing, *i.e.*,

$$\xi_x\eta_z - \xi_z\eta_x \neq 0. \qquad (2.40)$$

Because we want the coefficients of the cross term $\varphi_{\xi\eta}$ to be zero in the new coordinates, we impose the condition that

$$\xi\xi_x\eta_x + K_{33}\xi_z\eta_z = 0. \qquad (2.41)$$

Both ξ and K_{33} are given, and it is easy for the two first derivatives of η to satisfy (2.40) and (2.41) simultaneously.

Two possibilities arise. Either

i) ξ and ξ_z never vanish simultaneously, or

ii) there exist one or more points (x, z) on the domain at which

$$\xi(x, z) = \xi_z(x, z) = 0. \qquad (2.42)$$

In case *i)*, the condition $\xi = 0$ implies, via (2.41) and the assumption that K_{33} is positive, the accompanying condition $\eta_z = 0$. But if ξ and η_z both vanish, then the coefficient of $\varphi_{\eta\eta}$ in (2.39) vanishes; that is,

$$\xi\eta_x^2 + K_{33}\eta_z^2 = 0$$

whenever $\xi = 0$. Again using (2.41), we obtain from Eqs. (2.34) and (2.39) an equation with higher-order terms having the form

$$\varphi_{\xi\xi} + \frac{\xi\eta_x^2 + K_{33}\eta_z^2}{\xi\xi_x^2 + K_{33}\xi_z^2}\varphi_{\eta\eta} = 0. \qquad (2.43)$$

The denominator in the coefficient of $\varphi_{\eta\eta}$ cannot be zero: ξ and ξ_z cannot vanish simultaneously, and if ξ_x vanishes, then ξ_z must be nonzero in order to preserve condition (2.40). So Eq. (2.43) is of the form

$$\varphi_{\xi\xi} + \mathcal{K}(\xi, \eta)\varphi_{\eta\eta} = 0, \qquad (2.44)$$

where $\mathcal{K}(\xi, \eta) = 0$ if and only if $\xi = 0$, an equation of *Tricomi type*.

In case *ii)*, condition (2.40) prevents η_z from vanishing when ξ_z vanishes. Thus if ξ and ξ_z vanish together, the coefficient of $\varphi_{\eta\eta}$ in (2.39) will not vanish at that point. Thus in case *ii)* we obtain from (2.34), (2.39), and (2.41) an equation with higher-order terms having the form

$$\frac{\xi\xi_x^2 + K_{33}\xi_z^2}{\xi\eta_x^2 + K_{33}\eta_z^2}\varphi_{\xi\xi} + \varphi_{\eta\eta} = 0, \tag{2.45}$$

where the numerator in the coefficient of $\varphi_{\xi\xi}$, but not the denominator, is zero whenever ξ is zero. That is, Eq. (2.45) is an equation of the form

$$\mathcal{K}(\xi, \eta)\varphi_{\xi\xi} + \varphi_{\eta\eta} = 0, \tag{2.46}$$

where $\mathcal{K}(\xi, \eta) = 0$ if and only if $\xi = 0$, an equation of *Keldysh type*.

See Sec. 1.2 of Ref. 5, Ref. 9, Sec. 1 of Ref. 26, and Eqs. (75)–(78) of Ref. 39 for arguments of this kind.

The regularity of weak solutions to equations of Tricomi type can be established by microlocal arguments; see Refs. 14 and 15 and, especially, Refs. 30 and 31. These arguments appear to fail for equations of Keldysh type, and one expects weaker regularity for weak solutions to such equations.

The crucial question is: does condition (2.42) occur in our physical model? The answer to that question is "yes."

2.4.4. *Geometry of the resonance curve (after Piliya and Fedorov)*

Returning to our original xz-coordinates, we set the elements K_{11} and K_{22} of the dielectric tensor equal to $\mathcal{K}$, and the element K_{33} equal to η. The coefficients of the only other nonzero elements, K_{12} and K_{21}, are zero in Eq. (2.34), so only K_{11} and K_{33} play a direct role in the analysis. The sonic condition is equivalent to the alternative:

$$\mathcal{K} = 0 \tag{2.47}$$

or

$$\mathcal{K}\sin^2\theta + \eta\cos^2\theta = 0, \tag{2.48}$$

where $\theta(x, z)$ is the angle between the direction of $\mathbf{B_0}$ and the xz-plane; *c.f.* (2.27).

The singular points on the *sonic line* (2.47) are the points at which this curve (which is not a generally a line in standard coordinates) is tangent to the projection of the force lines of $\mathbf{B_0}$ in the xz-plane — that is, the flux lines of the magnetic field. The singular points of the graph Γ of Eq. (2.48) are the points at which the flux lines of $\mathbf{B_0}$ are normal to Γ.

This motivates the placement of the origin at a singular point of the sonic line, with the z-axis (the axis along which $\mathbf{B_0}$ is directed) tangent to the sonic line. The x-axis is directed along the inward normal to the sonic line, relative to the hyperbolic region of Eq. (2.33). Then K_{11} and σ both vanish at the origin. Taking both x and z to be small, one can write

$$K_{11} = \frac{x}{a} + \frac{z^2}{b} \tag{2.49}$$

and

$$K_{33} = -\eta_0, \tag{2.50}$$

where η_0 is a positive constant. Scale x and z, via

$$x \to \tilde{x} = x/a \tag{2.51}$$

and

$$z \to \tilde{z} = z/a\sqrt{\eta_0}, \tag{2.52}$$

in order to obtain dimensionless variables $\tilde{x}$ and $\tilde{z}$. In this way, one obtains in place of (2.33) the equation

$$-\left(\tilde{x} + A\tilde{z}^2\right)\varphi_{\tilde{x}\tilde{x}} + \varphi_{\tilde{z}\tilde{z}} - \varphi_{\tilde{x}} = 0, \tag{2.53}$$

where A is a constant, for the simple case in which the coefficients of cross terms vanish identically.[33]

2.5. *Analytic difficulties in the electromagnetic case* (after H. Weitzner)

In this section we suppose that the electric field satisfies Eqs. (2.4) and (2.5), but no longer assume that the electric field satisfies condition (2.28). Closely following Ref. 39, we attempt to study the resulting field equations using conventional analytic tools, in order to see what difficulties arise.

Repeating the calculations of Eqs. (2.16)–(2.21) in greater detail, we compute

$$\nabla \times (\nabla \times \mathbf{E}) = \nabla \times \left(-\frac{\partial B}{\partial t}\right) = \nabla \times (i\omega\mathbf{B}) =$$

$$i\omega\left(\nabla \times \mathbf{B}\right) = i\omega\left[\mu_0\left(\mathbf{j} + \epsilon_0\frac{\partial \mathbf{E}}{\partial t}\right)\right] =$$

$$i\omega\mu_0\mathbf{j} + i\omega\mu_0\epsilon_0\left(-i\omega\mathbf{E}\right) = i\omega\mu_0\mathbf{j} + \omega^2\mu_0\epsilon_0\mathbf{E}$$

$$= \mu_0\omega^2\left(\frac{i}{\omega}\mathbf{j} + \epsilon_0\mathbf{E}\right) = \mu_0\omega^2\mathbf{D} = \mu_0\epsilon_0\omega^2\mathbf{KE}. \tag{2.54}$$

Now

$$\nabla \times \mathbf{E} = (E_{3,y} - E_{2,z})\,\hat{\imath} + (E_{1,z} - E_{3,x})\,\hat{\jmath} + (E_{2,x} - E_{1,y})\,\hat{k}.$$

It is obvious that this quantity vanishes identically in the electrostatic case: apply (2.28) and equate mixed partial derivatives. But if $\nabla \times \mathbf{E}$ is itself a gradient, then the quantity $\nabla \times (\nabla \times \mathbf{E})$ vanishes for the general case as well. We will understand the seriousness of this latter difficulty once we evaluate the left-hand side of Eq. (2.54). Explicitly,

$$\nabla \times (\nabla \times \mathbf{E}) = (E_{2,xy} - E_{1,yy} - E_{1,zz} + E_{3,xz})\,\hat{\imath}+$$

$$(E_{3,yz} - E_{2,zz} - E_{2,xx} + E_{1,yx})\,\hat{\jmath} + (E_{1,zx} - E_{3,xx} - E_{3,yy} + E_{2,zy})\,\hat{k}.$$

Applying (2.4), (2.5) to the right-hand side, we obtain the algebraic expression

$$\left[k_1k_2E_2 - \left(k_2^2 + k_3^2\right)E_1 + k_1k_3E_3\right]\hat{\imath} + \left[k_2k_3E_3 - \left(k_3^2 + k_1^2\right)E_2 + k_{21}E_1\right]\hat{\jmath}$$
$$+ \left[k_{31}E_1 - \left(k_1^2 + k_2^2\right)E_3 + k_{32}E_2\right]\hat{k}. \quad (2.55)$$

This object can be written as the matrix operator

$$L\mathbf{E} = \begin{pmatrix} -\left(k_2^2 + k_3^2\right) & k_1k_2 & k_1k_3 \\ k_2k_1 & -\left(k_3^2 + k_1^2\right) & k_2k_3 \\ k_3k_1 & k_3k_2 & -\left(k_1^2 + k_2^2\right) \end{pmatrix} \begin{pmatrix} E_1 \\ E_2 \\ E_3 \end{pmatrix}.$$

The system (2.54) is uniquely solvable if and only if the operator L can be inverted — that is, if and only if

$$\det \begin{pmatrix} -\left(k_2^2 + k_3^2\right) & k_1k_2 & k_1k_3 \\ k_2k_1 & -\left(k_3^2 + k_1^2\right) & k_2k_3 \\ k_3k_1 & k_3k_2 & -\left(k_1^2 + k_2^2\right) \end{pmatrix} \neq 0.$$

But it is easy to check that this determinant vanishes identically for all (k_1, k_2, k_3). Of course (2.55) is just a translation of $\nabla \times (\nabla \times \mathbf{E})$ into Fourier mode. Because the symbol of a differential operator is a natural generalization of the idea of Fourier modes, we can interpret the foregoing computation to mean that the symbol of the differential operator $\nabla \times (\nabla \times)$ vanishes identically. This is a serious obstacle to understanding (2.54). As Weitzner notes in Ref. 40, neither the type of Eq. (2.54) (which is given by the sign of the symbol) nor the order of the equation (which is given by the degree of the symbol) are determined by standard analytic methods.

2.5.1. *Choices of potential and gauge*

It is therefore necessary to impose an additional hypothesis. A natural one is that an electromagnetic potential exists. But in distinction to the electrostatic case, we do not assume that $\mathbf{E}$ can be derived by simply taking the negative gradient of a scalar field.

In order to compare our computations with the extensive expositions in Refs. 39 and 40 we adopt, only in Secs. 2.5.1 and 2.5.2, the convention that the time-harmonic dependence is of the form $\exp[i\omega t]$ in units of c/ω. (This is in distinction to (2.4).) Because our time derivatives usually end up being taken twice, this only has an effect on the sign in a few intermediate calculations. However, with this sign convention, Maxwell's equations for plane waves assume the slightly different form

$$\nabla \times \mathbf{E} = -i\mathbf{B}, \quad (2.56)$$

$$\nabla \times \mathbf{B} = i\mathbf{D} = i\mathbf{KE}. \quad (2.57)$$

The first choice of potentials is to let the vector $\mathbf{A}$ denote the magnetic potential and to introduce a second, scalar potential, Φ. We then write

$$\mathbf{B} = \nabla \times \mathbf{A} \quad (2.58)$$

and

$$\mathbf{E} = -i\mathbf{A} - \nabla\Phi. \tag{2.59}$$

Taking the curl of the second equation, we obtain

$$\nabla \times \mathbf{E} = -i\nabla \times \mathbf{A} - \nabla \times \nabla\Phi. \tag{2.60}$$

Evaluating the last term on the right-hand side of (2.60) using differential forms, Φ is a zero-form and

$$\nabla \times \nabla\Phi = d^2\Phi = 0. \tag{2.61}$$

Equations (2.58) and (2.61) imply that (2.56) is satisfied under condition (2.59) for any smooth choice of $\mathbf{A}$ and Φ.

Notice that we automatically obtain from hypothesis (2.58) an extra condition

$$\nabla \cdot \mathbf{B} = \nabla \cdot (\nabla \times \mathbf{A}),$$

which is to say, in terms of differential forms, that the 2-form $\mathbf{B}$ and the 1-form $\mathbf{A}$ satisfy

$$d\mathbf{B} = d^2\mathbf{A} = 0.$$

In order to evaluate (2.57), we notice, continuing to interpret the magnetic potential $\mathbf{A}$ as a 1-form and Φ as a zero-form, that if g is defined by

$$g(\mathbf{A}) = \mathbf{A} + df,$$

where f is a smooth 0-form, then

$$d(g(\mathbf{A})) = d(\mathbf{A} + df) = d\mathbf{A} + d^2f = d\mathbf{A} = \mathbf{B},$$

so the magnetic field remains invariant under the *gauge transformation g*. Moreover, because $\delta f = 0$ for any zero-form f, we have

$$\Delta f = -(\delta d + d\delta) f = -\delta df.$$

Thus, given any smooth potential $\mathbf{A}$, we can choose f to satisfy the Poisson equation

$$\Delta f = \delta\mathbf{A},$$

in which case

$$\delta(g(\mathbf{A})) = \delta(\mathbf{A} + df) = \delta\mathbf{A} - \Delta f = 0.$$

We call the gauge produced by such a g a *Coulomb* (*transverse, radiation,* or *Hodge*) gauge. In vector notation,

$$i\nabla \cdot g(\mathbf{A}) = 0.$$

Computing (2.57) in the Coulomb gauge, we obtain[39]

$$\Delta g(\mathbf{A}) - i\mathbf{K}\nabla\Phi + \mathbf{K}g(\mathbf{A}) = 0.$$

Computing the symbol σ of this operator by the same method that was applied to the operator $\nabla \times (\nabla \times \mathbf{E})$, we find that $\sigma = -|\mathbf{k}|^4 (\mathbf{Kk}) \cdot \mathbf{k}$, a polynomial of degree six in $\mathbf{k}$. That the corresponding system is of order six is an expected result for a system of two first-order equations for vectors in $\mathbb{R}^3$.

Replacing the Coulomb gauge by a slightly more complicated gauge in which

$$i\nabla \cdot (\mathbf{K}^*\mathbf{A}) = 0,$$

where the superscripted asterisk denotes the adjoint matrix, we obtain a self-adjoint operator in (2.57). This more complicated gauge can be constructed by the same general method that led to the Coulomb gauge, provided that we solve a slightly more complicated Poisson problem. The symbol corresponding to this self-adjoint operator can also be calculated by the methods introduced earlier, and that symbol is also a sixth-degree polynomial in $\mathbf{k}$.

However, we can obtain a fourth-order system, which is more convenient for analysis, if we impose an additional hypothesis: that the plasma has axisymmetric geometry. While this is a very strong hypothesis, it is satisfied by plasmas produced in tokamaks.

In order to motivate the choice of potential in this case, we make a few preliminary calculations. Only for the remainder of this section, the subscripts r, θ, and z when affixed to a vector are to be interpreted as the radial, angular, and axial components of the vector unless preceded by a comma; if preceded by a comma, they are to be considered partial derivatives in the direction of the component. (The subscripted-variable notation for partial derivatives of scalar functions remains unchanged.) Adopting the basis

$$\mathbf{u}_r = \cos\theta \hat{\imath} + \sin\theta \hat{\jmath},$$

$$\mathbf{u}_\theta = -\sin\theta \hat{\imath} + \cos\theta \hat{\jmath},$$

$$\mathbf{u}_z = -\hat{k},$$

we recall that in the axisymmetric case we can write

$$\mathbf{E} = \left(E_r(r,z)\,\mathbf{u}_r + E_\theta(r,z)\,r\mathbf{u}_\theta + E_z(r,z)\,\mathbf{u}_z\right) e^{im\theta},$$

and similarly for $\mathbf{B}$. If $m = 0$, the waves preserve the axisymmetry of the underlying static plasma medium, as the wave vector satisfies $\mathbf{k} = (k_r, 0, k_z)$. We will restrict our attention to this simple special case, in which

$$\nabla \times \mathbf{E} = \frac{1}{r}\left[E_{z,\theta}\mathbf{u}_r + E_{r,z}(r\mathbf{u}_\theta) + (rE_\theta)_r\,\mathbf{u}_z\right]$$

$$- \frac{1}{r}\left[E_{z,r}(r\mathbf{u}_\theta) + rE_{\theta,z}\mathbf{u}_r + E_{r,\theta}\mathbf{u}_z\right]$$

$$= -E_{\theta,z}\mathbf{u}_r + (E_{r,z} - E_{z,r})\,\mathbf{u}_\theta + \frac{1}{r}(rE_\theta)_{,r}\,\mathbf{u}_z. \tag{2.62}$$

Thus (2.56) implies in particular that

$$-iE_{\theta,z} = B_r \tag{2.63}$$

and

$$\frac{(rE_\theta)_{,r}}{r} = -iB_z. \tag{2.64}$$

Just as Eqs. (2.63) and (2.64) imply, using (2.56), that B_r and B_z can each be expressed in terms of derivatives of E_θ, so it is possible to use (2.57) to show that the other cylindrical components of **E** and **B** can be expressed as appropriate derivatives of E_θ and B_θ. This will allow the angular components of **E** and **B** to play the role of potentials for the two fields.

Applying (2.56) to the middle term of the last identity in (2.62) yields

$$E_{r,z} - E_{z,r} = -iB_\theta. \tag{2.65}$$

Because **E** and **B** have exactly analogous forms and the left-hand and middle terms of (2.57) is exactly analogous to (2.56) with a change of sign, we can immediately write

$$D_r = iB_{\theta,r}$$

and

$$D_z = -i\frac{(rB_\theta)_{,z}}{r}.$$

Now the extreme right-hand side of (2.57) yields E_r and E_z (see Eqs. (22), (23) of Ref. 39) and one obtains

$$B_{r,z} - B_{z,r} = iD_\theta = i\left(K_{\theta r}E_r + K_{\theta\theta}E_\theta + K_{\theta z}E_z\right), \tag{2.66}$$

completing the system of equations for E_θ and B_θ.

2.5.2. *Variational interpretation*

Continuing to adopt the special hypotheses and special notation of Sec. 2.5.1, we continue to review the analysis in Ref. 39 of geometry-preserving plane waves in an axisymmetric plasma.

Equations (2.65), (2.66) can be associated to an energy functional:

$$\mathcal{E} = \int \{[\nabla(rE_\theta^*) \cdot \nabla(rE_\theta)]/r^2 + [\nabla(rB_\theta^*) \cdot \mathbf{K}\nabla(rB_\theta)]/r^2\Delta$$
$$+iE_\theta\left[(rB_\theta^*)_{,r}(K_{zr}K_{r\theta} - K_{z\theta}K_{rr})/r + B_{\theta,z}^*(K_{r\theta}K_{zz} - K_{rz}K_{z\theta})\right]/\Delta$$
$$-iE_\theta^*\left[(rB_\theta)_{,r}(K_{zr}K_{\theta r} - K_{\theta z}K_{rr})/r + B_{\theta,z}(K_{\theta r}K_{zz} - K_{zr}K_{\theta z})\right]/\Delta$$
$$-B_\theta^*B_\theta - E_\theta^*E_\theta\left[\det(\mathbf{K})\right]/\Delta\}r\,dr\,dz, \tag{2.67}$$

where

$$\nabla = \frac{\partial}{\partial r} r + \frac{\partial}{\partial z} z,$$

and

$$\Delta = K_{rr} K_{zz} - K_{rz} K_{zr}.$$

Provided $\mathbf{K}$ can be made self-adjoint, so can $\mathcal{E}$. Form a right-handed orthogonal set $(\mathbf{v}, \theta, \mathbf{u})$, where

$$\mathbf{u} = \sin \beta r + \cos \beta k$$

and

$$\mathbf{v} = \cos \beta r - \sin \beta k.$$

The basis is to be chosen so that $\mathbf{u}$ lies in the poloidal direction and $\mathbf{v}$ lies orthogonal to it; so we write the magnetic field vector in the form

$$\mathbf{B} = B_0 \left[\cos \alpha \theta + \sin \alpha \left(\sin \beta r + \cos \beta z \right) \right],$$

where α, β, and B_0 depend only on r and z. In this notation, the variational equations of $\mathcal{E}$ form a second-order system in which the differential operator for one of the equations is essentially the Laplacian $\mathcal{L}$. We ignore that equation, as standard analytic methods can be applied to it. The differential operator for the other equation looks like

$$\mathcal{L} + \left(\mathbf{u} \cdot \nabla \right)^2, \tag{2.68}$$

and that is the equation — in particular, the second of the two differential operators in that equation — that we will study in the remainder of this review. The term (2.68) in Eq. (2.67) can be written explicitly, in terms of the chosen basis, in the form

$$r\nabla \cdot \left[\left(\frac{\xi}{r^2 \Delta} \right) \nabla \left(rB_\theta \right) \right] - r\nabla \cdot \left[\left(\frac{\zeta \sin^2 \alpha}{r^2 \Delta} \right) \left(\mathbf{u} \cdot \nabla \right) \left(rB_\theta \right) \mathbf{u} \right]$$

$$-i\theta \cdot \nabla \left(rB_\theta \right) \times \nabla \left(\frac{\mu \cos \alpha}{r\Delta} \right) + B_\theta$$

$$= \left(r\Delta \right)^{-1} \left[\mu \left(\zeta - \xi \right) \sin \alpha \, \mathbf{u} \cdot \nabla \left(rE_\theta \right) + i \left(\mu^2 - \xi \zeta \right) \sin \alpha \cos \alpha \, \mathbf{v} \cdot \nabla \left(rE_\theta \right) \right], \tag{2.69}$$

where

$$\xi = 1 + \sum_{\nu=1}^{N} \frac{\Pi_\nu^2}{\Omega_\nu^2 - \omega^2},$$

$$\zeta = \xi + \sum_{\nu=1}^{N} \frac{\Pi_\nu^2}{\omega^2} - 1,$$

and

$$\mu = \sum_{\nu=1}^{N} \frac{\Pi_\nu^2 \Omega_\nu}{\omega \left(\Omega_\nu^2 - \omega^2 \right)}.$$

Equation (2.69) is only elliptic for negative values of ξ. Physically, this is the condition for so-called *lower-hybrid* frequencies, at which

$$1 + \frac{\Pi_e^2}{\Omega_e^2} < \frac{\Pi_i^2}{\omega^2},$$

c.f. (2.37). Noticing that ξ is a function of r and z, define a new variable $\eta\left(r, z\right)$ so the curves $\eta = $ constant are orthogonal to the curves $\xi = $ constant. Rewriting (2.69) in (ξ, η)-coordinates, the behavior of the solution depends on whether or not

$$\mathbf{u} \cdot \nabla \xi = 0.$$

This identity implies that flux surfaces coincide with resonance surfaces. In that case, Eq. (2.69) is analogous to Eq. (2.46) of Sec. 2.4.2 and is not of Tricomi type. The second-order terms of that equation can be written in the form

$$L(u) = f\left(\xi, \eta\right) \left[\xi u_{\xi\xi} + M\left(\xi, \eta\right) u_{\eta\eta}\right], \tag{2.70}$$

where $u = u\left(\xi, \eta\right)$ is a scalar function; f and M are given well behaved scalar functions near $\xi = 0$ and, in addition, M is positive.

The physical model allows two further alternatives: If the curve representing the flux surface in two dimensions is collinear with the resonance curve as in (2.70), then the plasma can be treated as a perpendicularly stratified medium, which is essentially the case considered in Sec. 2.4.1. If the resonance curve is tangent to the curve representing the flux surface, then we are in a more mathematically and physically interesting case. In this latter case, the simplest model for the operator L of (2.69) is an operator for which the highest-order terms have the form

$$\tilde{L}(u) = \left(x - y^2\right) u_{xx} + u_{yy}. \tag{2.71}$$

Note that this operator is closely related to the differential operator of Eq. (2.53). The two operators can be made virtually identical by replacing the coordinate transformation (2.51) by

$$x \to \tilde{x} = -x/a. \tag{2.72}$$

2.6. *A conjecture about the singular set*

Methods for deriving the smoothness of solutions to the Tricomi equation appear to fail for an operator of the form (2.71) whenever the domain includes the origin of coordinates. This suggests the existence of a singular point at the origin, a conjecture which is supported by an analysis of characteristic lines.

In order for a characteristic line to pass through the origin, the point (x, y) would need to satisfy the identity

$$x = \lambda y^2 \tag{2.73}$$

for some constant λ, and also the characteristic equation for (2.71). Substituting (2.73) into the characteristic equation

$$\left(x - y^2\right) dy^2 + dx^2 = 0, \tag{2.74}$$

one obtains the equation

$$\frac{dy^2}{(2\lambda y dy)^2} = \frac{1}{(1 - \lambda) y^2}, \tag{2.75}$$

or

$$4\lambda^2 + \lambda - 1 = 0.$$

This polynomial has two real solutions; considering that the characteristic equation (2.75) has two roots, one concludes[26] that four characteristic lines must pass through the origin — two more than pass through any other hyperbolic point. This motivates the suspicion that solutions of at least the equation $\tilde{L}u = 0$ will tend to be singular at the origin. It has been observed that an energy sink or plasma heating zone might be associated with such a singularity; see Refs. 16, 26, 33, 39, and 40 for details on this and other issues raised in this section.

3. Analysis of the model equation

Physical reasoning suggests that the *closed* Dirichlet problem, in which data are prescribed along the entire boundary of the domain, should be well-posed for the cold plasma model. However, the closed Dirichlet problem has been shown to be ill-posed, in the classical sense, for the equation

$$\left(x - y^2\right) u_{xx} + u_{yy} + \frac{1}{2} u_x = 0$$

on a typical domain.[26] This leads us to ask whether a well-posed problem with closed boundary data can be formulated in a suitably weak sense. In this section we address the "existence" part of that question.

Because the operator introduced in Eq. (2.71) is not of Tricomi type at the origin, where it satisfies a condition of the form (2.42), we expect weaker regularity than we have for operators which are uniformly of Tricomi type. In particular, although we can show the existence of very weak solutions in L^2, we do not expect H^1 regularity for the closed Dirichlet problem. This lack of optimism is supported by numerical experiments.[26] Moreover, our methods are insufficient to determine the uniqueness of a solution.

Denote by Ω a bounded, connected domain of $\mathbb{R}^2$ having piecewise smooth boundary $\partial\Omega$, oriented in a counterclockwise direction; the domain includes both an arc of the sonic curve and the origin of coordinates in $\mathbb{R}^2$. (This insures that our equation will be elliptic-hyperbolic but not equivalent to an equation of Tricomi type.)

Define,[21] for a given C^1 function $\mathcal{K}(x,y)$, the space $L^2(\Omega; |\mathcal{K}|)$ and its dual. These spaces consist, respectively, of functions u for which the norm

$$||u||_{L^2(\Omega;|\mathcal{K}|)} = \left(\int_\Omega |\mathcal{K}| u^2 dx dy \right)^{1/2}$$

is finite, and functions $u \in L^2(\Omega)$ for which the norm

$$||u||_{L^2(\Omega;|\mathcal{K}|^{-1})} = \left(\int_\Omega |\mathcal{K}|^{-1} u^2 dx dy \right)^{1/2}$$

is finite. Standard arguments allow us to define the space $H_0^1(\Omega; \mathcal{K})$ as the closure of $C_0^\infty(\Omega)$ with respect to the norm

$$||u||_{H^1(\Omega;\mathcal{K})} = \left[\int\int_\Omega \left(|\mathcal{K}| u_x^2 + u_y^2 + u^2 \right) dx dy \right]^{1/2}. \tag{3.1}$$

The $H_0^1(\Omega; \mathcal{K})$-norm has the explicit form

$$||u||_{H_0^1(\Omega;\mathcal{K})} = \left[\int\int_\Omega \left(|\mathcal{K}| u_x^2 + u_y^2 \right) dx dy \right]^{1/2}, \tag{3.2}$$

which can be derived from (3.1) via a weighted Poincaré inequality.

In the following we denote by C generic positive constants, the value of which may change from line to line.

3.1. *The closed Dirichlet problem for distribution solutions*

Consider the equation

$$Lu = f, \tag{3.3}$$

where f is a given, sufficiently smooth function of (x, y) and

$$L = \left(x - y^2 \right) \frac{\partial^2}{\partial x^2} + \frac{\partial^2}{\partial y^2} + \kappa \frac{\partial}{\partial x} \tag{3.4}$$

for a given constant κ. By a *distribution solution* of equations (3.3), (3.4) with the boundary condition

$$u(x, y) = 0 \, \forall (x, y) \in \partial\Omega \tag{3.5}$$

we mean a function $u \in L^2(\Omega)$ such that $\forall \xi \in H_0^1(\Omega; \mathcal{K})$ for which $L^*\xi \in L^2(\Omega)$, we have

$$(u, L^*\xi) = \langle f, \xi \rangle. \tag{3.6}$$

Here L^* is the adjoint operator; $(\,,\,)$ denotes the L^2 inner product on Ω; $\langle\,,\,\rangle$ is the *duality bracket* associated to the H^{-1} norm

$$\|w\|_{H^{-1}(\Omega;\mathcal{K})} = \sup_{0 \neq \xi \in C_0^\infty(\Omega)} \frac{|\langle w, \xi\rangle|}{\|\xi\|_{H_0^1(\Omega;\mathcal{K})}}.$$

Such solutions have also been called *weak*; *c.f.* Eq. (2.13) of Ref. 4, Sec. II.2. In fact they are a little smoother than generic distribution solutions, as they lie in a classical function space.

The existence of solutions to boundary value problems can be shown to follow from *energy inequalities* having the general form

$$\|v\|_V \leq C\|L^*v\|_U,$$

where U and V are suitable function spaces. We will combine such an inequality with the Riesz Representation Theorem to prove the existence of distribution solutions; see Ref. 4 for a general reference.

Lemma 3.1. (Ref. 28). *The inequality*

$$\|u\|_{H_0^1(\Omega;\mathcal{K})} \leq C\|Lu\|_{L^2(\Omega)},$$

is satisfied for $u \in C_0^2(\Omega)$, where the positive constant C depends on Ω and $\mathcal{K}$; L is defined by (3.4) with $\kappa \in [0,2]$; $\mathcal{K} = x - y^2$.

Proof.   We outline the proof; for details, see Ref. 28, Sec. 2. Initially, let $1 \leq \kappa \leq 2$, and let δ be a small, positive constant. Define an operator M by the identity

$$Mu = au + bu_x + cu_y$$

for $a = -1$, $c = 2(2\delta - 1)y$, and

$$b = \begin{cases} \exp(2\delta\mathcal{K}/Q_1) \text{ if } (x,y) \in \Omega^+ \\ \exp(6\delta\mathcal{K}/Q_2) \text{ if } (x,y) \in \Omega^- \end{cases},$$

where $\Omega^+ = \{(x,y) \in \Omega \,|\, \mathcal{K} > 0\}$ and $\Omega^- = \Omega\backslash\Omega^+$. Choose $Q_1 = \exp(2\delta\mu_1)$, where $\mu_1 = \max_{(x,y)\in\overline{\Omega^+}} \mathcal{K}$. Define the negative number $\mu_2 = \min_{(x,y)\in\overline{\Omega^-}} \mathcal{K}$ and let $Q_2 = \exp(\mu_2)$. Notice that $b \leq Q_1$ on Ω^+ and $b > Q_2$ on Ω^-.

We will estimate the quantity (Mu, Lu) from above and below. As in the Tricomi case,[21] one of the coefficients in Mu fails to be continuously differentiable on all of Ω. When integrating this quantity, a cut should be introduced along the line $\mathcal{K} = 0$. The boundary integrals involving a, b, and c on either side of this line will cancel.

The boundary terms vanish by the compact support of u. Integration by parts yields the identity

$$(Mu, Lu) = \int\!\!\int_{\Omega^+\cup\Omega^-} \alpha u_x^2 + 2\beta u_x u_y + \gamma u_y^2 \, dxdy,$$

where

$$\alpha = \left(\frac{c_y}{2} - a - \frac{b_x}{2}\right)\mathcal{K} + \left(\kappa - \frac{1}{2}\right)b - cy,$$

for

$$\alpha_{|\Omega^+} = \left(2 - \frac{b}{Q_1}\right)\delta\mathcal{K} + 2\left(1 - 2\delta\right)y^2 + \left(\kappa - \frac{1}{2}\right)b$$

and

$$\alpha_{|\Omega^-} = \left(3\frac{b}{Q_2} - 2\right)\delta|\mathcal{K}| + 2\left(1 - 2\delta\right)y^2 + \left(\kappa - \frac{1}{2}\right)b;$$

$$\beta = \frac{1}{2}\left[c\left(\kappa - 1\right) - b_y\right] = \begin{cases} y\left[2\delta\left(b/Q_1\right) + \left(\kappa - 1\right)\left(2\delta - 1\right)\right] \le |y| \ \text{in} \ \Omega^+ \\ y\left[6\delta\left(b/Q_2\right) + \left(\kappa - 1\right)\left(2\delta - 1\right)\right] \le \kappa|y| \ \text{in} \ \Omega^- \end{cases};$$

$$\gamma = \frac{1}{2}\left(b_x - c_y\right) - a = \begin{cases} 2\left(1 - \delta\right) + \delta\left(b/Q_1\right) \ \text{in} \ \Omega^+ \\ 2\left(1 - \delta\right) + 3\delta\left(b/Q_2\right) \ \text{in} \ \Omega^- \end{cases}.$$

On Ω^+, for any scalars ξ and η, we have

$$\alpha\xi^2 + 2\beta\xi\eta + \gamma\eta^2 \ge \alpha\xi^2 - \left(y^2\xi^2 + \eta^2\right) + \gamma\eta^2$$

$$= \left[\left(2 - \frac{b}{Q_1}\right)\delta\mathcal{K} + \left(1 - 4\delta\right)y^2 + \left(\kappa - \frac{1}{2}\right)b\right]\xi^2 + \left[\left(1 - 2\delta\right) + \frac{6b}{Q_1}\right]\eta^2$$

$$\ge \delta\left(\mathcal{K}\xi^2 + \eta^2\right),$$

provided δ is sufficiently small. On Ω^-,

$$\alpha\xi^2 + 2\beta\xi\eta + \gamma\eta^2 \ge \alpha^2\xi^2 - 2\left(y^2\xi^2 + \eta^2\right) + \gamma\eta^2$$

$$= \left[\left(3\frac{b}{Q_2} - 2\right)\delta|\mathcal{K}| - 4\delta y^2 + \left(\kappa - \frac{1}{2}\right)b\right]\xi^2 + \delta\left(3\frac{b}{Q_2} - 2\right)\eta^2$$

$$\ge \delta\left(|\mathcal{K}|\xi^2 + \eta^2\right).$$

Arguing in this way on each subdomain (and taking $\xi = u_x$, $\eta = u_y$), we obtain

$$(Mu, Lu) \ge \delta\|u\|_{H_0^1(\Omega;\mathcal{K})}^2. \tag{3.7}$$

For the upper estimate, we have[21]

$$(Mu, Lu) \le \|Mu\|_{L^2}\|Lu\|_{L^2} \le C\left(K, \Omega\right)\|u\|_{H_0^1(\Omega;\mathcal{K})}\|Lu\|_{L^2(\Omega)}. \tag{3.8}$$

Combining (3.7) and (3.8), and dividing through by the weighted H_0^1-norm of u, completes the proof for the case $\kappa \in [1, 2]$.

Now let $0 \leq \kappa < 1$. Again subdivide the domain into Ω^+ and Ω^- by introducing a cut along the curve $K = 0$. Integrate by parts, choosing $a = -1$;

$$b = \begin{cases} -N\mathcal{K} \text{ in } \Omega^+ \\ N\mathcal{K} \text{ in } \Omega^- \end{cases},$$

where N is a constant satisfying

$$\frac{1 + \tilde{\delta}}{3 - \kappa} < N < \frac{1 - \tilde{\delta}}{\kappa + 1} \tag{3.9}$$

for a sufficiently small positive constant $\tilde{\delta}$; $c = -4Ny$. The boundary integrals involving a and c on either side of the cut will cancel and the boundary integrals involving b will be zero along the cut. Inequality (3.7) can be derived by an argument broadly analogous to the case $\kappa \in [1, 2]$. Applying (3.8) completes the proof. $\square$

Theorem 3.1 (Ref. 28). *The Dirichlet problem (3.3), (3.4), (3.5) with $\kappa \in [0, 2]$ possesses a distribution solution $u \in L^2(\Omega)$ for every $f \in H^{-1}(\Omega; \mathcal{K})$.*

Proof. Again, we only outline the proof (*c.f.* Ref. 21, Theorem 2.2). Define for $\xi \in C_0^\infty$ a linear functional

$$J_f(L\xi) = \langle f, \xi \rangle.$$

This functional is bounded on a subspace of L^2 by the inequality

$$|\langle f, \xi \rangle| \leq ||f||_{H^{-1}(\Omega; \mathcal{K})} ||\xi||_{H_0^1(\Omega; \mathcal{K})} \tag{3.10}$$

and by applying Lemma 3.1 to the second term on the right. Precisely, J_f is a bounded linear functional on the subspace of $L^2(\Omega)$ consisting of elements having the form $L\xi$ with $\xi \in C_0^\infty(\Omega)$. Extending J_f to the closure of this subspace by Hahn-Banach arguments, we obtain a functional defined on all of L^2. The Riesz Representation Theorem then guarantees the existence of a distribution solution in the self-adjoint case. If $\kappa \neq 1$, then L is not self-adjoint, and

$$L^* = (x - y^2) \frac{\partial^2}{\partial x^2} + \frac{\partial^2}{\partial y^2} + (2 - \kappa) \frac{\partial}{\partial x}. \tag{3.11}$$

Estimating L for κ in $[0, 2]$ will also yield estimates for the adjoint L^*. Applying the preceding argument to the adjoint operator completes the proof of Theorem 3.1.$\square$

3.2. *Mixed boundary value problems with closed boundary data*

It is also possible to form *mixed Dirichlet-Neumann* problems for operators of the form (3.4). Mixed boundary value problems arise in various contexts in plasma physics (*e.g.*, Ref. 7) and in related topics from electromagnetic theory (*e.g.*, Ref. 19, which is related to the model of Sec. 2.4.1). However, the results of this section also imply — by taking the set of boundary points on which the Dirichlet conditions are imposed to be empty — the existence of weak solutions to a class of Neumann problems.

Denote by $\mathbf{u} = (u_1, u_2)$ and $\mathbf{w} = (w_1, w_2)$ measurable vector-valued functions on Ω. Define $\mathcal{H}_{\mathfrak{K}}$ to be the Hilbert space of measurable functions on Ω for which the norm induced in the obvious way by the weighted L^2 inner product

$$(\mathbf{u}, \mathbf{w})_{\mathfrak{K}} = \int \int_{\Omega} (|\mathcal{K}| u_1 w_1 + u_2 w_2) \, dx dy$$

is finite. In the notation for these spaces, $\mathfrak{K}$ denotes a diagonal matrix having entries $|\mathcal{K}|$ and 1.

By a *weak solution* of a mixed boundary-value problem in this context we mean an element $\mathbf{u} \in \mathcal{H}_{\mathfrak{K}}(\Omega)$ such that

$$-(\mathbf{u}, \mathcal{L}^* \mathbf{w})_{L^2(\Omega; \mathbb{R}^2)} = (\mathbf{f}, \mathbf{w})_{L^2(\Omega; \mathbb{R}^2)} \tag{3.12}$$

for every function $\mathbf{w} \in C^1(\overline{\Omega}; \mathbb{R}^2)$ for which $\mathfrak{K}^{-1} \mathcal{L}^* \mathbf{w} \in L^2(\Omega; \mathbb{R}^2)$ and for which

$$w_1 = 0 \, \forall \, (x, y) \in G \tag{3.13}$$

and

$$w_2 = 0 \, \forall \, (x, y) \in \partial\Omega \backslash G, \tag{3.14}$$

where $G \subset \partial\Omega$. Choose the differential operator $\mathcal{L}$ to have the form

$$\begin{pmatrix} \mathcal{K} \partial_x & \partial_y \\ \partial_y & -\partial_x \end{pmatrix} + \begin{pmatrix} \kappa & 0 \\ 0 & 0 \end{pmatrix}, \tag{3.15}$$

where κ is a number in $[0, 1]$.

Theorem 3.2 (Ref. 28). *Let G be a subset of $\partial\Omega$ and let $\mathcal{K} = x - y^2$. Define the functions $b(x, y) = m\mathcal{K} + s$ and $c(y) = \mu y - t$, where μ is a positive constant,*

$$m = \begin{cases} (\mu + \delta)/2 & \text{in } \Omega^+ \\ (\mu - \delta)/2 & \text{in } \Omega^- \end{cases}$$

for a small positive number δ, and t is a positive constant such that $\mu y - t < 0 \, \forall y \in \Omega$. Let s be a sufficiently large positive constant. In particular, choose s to be so large that the quantities $m\mathcal{K} + s$, $2cy + s$, and $b^2 + \mathcal{K}c^2$ are all positive. Let

$$bdy - cdx \leq 0 \tag{3.16}$$

on G and

$$\mathcal{K}(bdy - cdx) \geq 0 \tag{3.17}$$

on $\partial\Omega \backslash G$. Then there exists for every $\mathbf{f}$ such that $\mathfrak{K}^{-1} \mathcal{M}^T \mathbf{f} \in L^2(\Omega)$ a weak solution to the mixed boundary-value problem (3.12)–(3.14) for $\mathcal{L}$ given by Eq. (3.15) with $\kappa = 0$, where the superscripted T denotes matrix transpose.

Proof. We give the idea of the proof.[28] One shows that there exists a positive
constant C such that

$$(\mathbf{\Psi}, \mathcal{L}^*\mathcal{M}\mathbf{\Psi}) \geq C \int \int_\Omega \left(|\mathcal{K}|\Psi_1^2 + \Psi_2^2 \right) dxdy$$

for any sufficiently smooth 2-vector $\mathbf{\Psi}$, provided conditions (3.13), (3.14) are satisfied on the boundary for $\mathbf{w} = \mathcal{M}\mathbf{\Psi}$, where $\mathcal{L}^*$ is given by (3.15) with $\kappa = 1$
and

$$\mathcal{M} = \begin{pmatrix} b & c \\ -\mathcal{K}c & b \end{pmatrix}.$$

This inequality leads to an application of the Riesz Representation Theorem by
arguments which are roughly analogous to those used to prove Theorem 3.1. $\square$

Despite the technical nature of the hypotheses in Theorem 3.2, simple domains
which satisfy them are very easy to construct — *e.g.*, a box in the first quadrant
having a vertex at the origin of coordinates, or a narrow lens about the sonic curve
in the first quadrant. Note that by taking G to be the empty set, we obtain a
solution to the closed conormal problem (*c.f.* Ref. 32). But in order for Theorem
3.2 to guarantee a solution to the closed Dirichlet problem, we would need to find a
domain on which G could be taken to be the entire boundary; it is not obvious how
to construct such a domain. And, as was also the case in Theorem 3.1, the proof of
Theorem 3.2 does not establish the uniqueness of solutions.

In addition to its intrinsic mathematical and physical interest, the formulation of
boundary value problems illuminates other topics in the analysis of the cold plasma
model. For example, it is shown in Sec. 2.4.3, by a tedious analytic argument,
that away from the origin the governing equation for the model is of Tricomi type,
whereas in the neighborhood of the origin it is of Keldysh type. This distinction is
also suggested, without reference to such terminology, by other analytic arguments
in Ref. 33 and in Sec. 4 of Ref. 40. If we try to form a standard elliptic-hyperbolic
boundary value problem in which the hyperbolic region is composed of intersecting
characteristics, we might choose both these characteristics to originate at points
on the arc of the resonance curve $x = y^2$ that lies in the first quadrant, or both
of them to lie in the fourth quadrant. We then obtain a standard problem for a
vertical-ice-cream-cone-shaped region (in the former case, the ice-cream cone is held
upside down), similar to those formulated for the Tricomi equation (Eq. (2.44) with
$\mathcal{K}(\xi, \eta) = \xi$). The domain geometry is exactly analogous to, for example, Fig. 2
of Ref. 25, with the line AB in that figure replaced by an arc of the curve $x = y^2$,
lying either completely above or completely below the x-axis. But the origin will not
be included, as that is a singular point of the characteristic equation (2.74). If we
include the origin, we are led to a hyperbolic region bounded by characteristics in the
second and third quadrants, a horizontal-ice-cream-cone-shaped region similar to
those formulated for the Cinquini-Cibrario equation (Eq. (2.46) with $\mathcal{K}(\xi, \eta) = \xi$).

In this case typical domain geometry is analogous to Fig. 2 of Ref. 10, with the line MN in that figure replaced by an arc of the curve $x = y^2$ which is symmetric about the x-axis; see also Remark *i)* following Corollary 11 of Ref. 28. Thus the defining analytic character of the equation is clearly apparent in the geometry of the natural boundary value problems.

References

1. W. P. Allis, S. J. Buchsbaum, and A. Bers, *Waves in Anisotropic Plasmas.* (MIT Press, Cambridge, 1963).
2. W. P. Allis, Waves in a plasma, *Mass. Inst. Technol. Research Lab. Electronics Quart. Progr. Rep.* **54** (5) (1959).
3. E. O. Astrom, Waves in an ionized gas, *Arkiv. Fysik* **2**, 443 (1950).
4. Ju. M. Berezanskii, *Expansions in Eigenfunctions of Selfadjoint Operators.* (American Mathematical Society, Providence, 1968).
5. A. V. Bitsadze, *Equations of the Mixed Type*, translated from the Russian by P. Zador. (Pergammon, New York, 1964).
6. P. M. Bellan, *Fundamentals of Plasma Physics.* (Cambridge University Press, Cambridge, 2006).
7. M. S. Bobrovnikov and V. V. Fisanov, Plane wave diffraction by a wedge in a magnetoactive plasma under mixed boundary conditions and a thermodynamic paradox, *Russ. Phys. J.* **28**, 185–189 (1985).
8. H. G. Booker, *Cold Plasma Waves.* (Springer-Verlag, Berlin, 2004).
9. M. Cibrario, Sulla riduzione a forma canonica delle equazioni lineari alle derivate parziali di secondo ordine di tipo misto, *Rendiconti del R. Insituto Lombardo* **65** (1932).
10. M. Cibrario, Intorno ad una equazione lineare alle derivate parziali del secondo ordine di tipe misto iperbolico-ellittica, *Ann. Sc. Norm. Sup. Pisa, Cl. Sci., Ser. 2*, **3**, Nos. 3, 4, 255–285 (1934).
11. A. Czechowski and S. Grzedzielski, A cold plasma layer at the heliopause, *Adv. Space Res.* **16** (9), 321–325 (1995).
12. R. Fitzpatrick, *An Introduction to Plasma Physics: a graduate course*, e-notes (2006).
13. V. L. Ginzburg, *Propagation of Electromagnetic Waves in Plasma*, translated from the Russian by J. B. Sykes and R. J. Tayler. (Pergamon, New York, 1970).
14. T. V. Gramchev, An application of the analytic microlocal analysis to a class of differential operators of mixed type, *Math. Nachr.* **121**, 41–51 (1985).
15. R. J. P. Groothuizen, *Mixed Elliptic-Hyperbolic Partial Differential Operators: A Case Study in Fourier Integral Operators.* (CWI Tract, Vol. 16, Centrum voor Wiskunde en Informatica, Amsterdam, 1985).
16. W. Grossman and H. Weitzner, A reformulation of lower-hybrid wave propagation and absorption, *Phys. Fluids* **27**, 1699–1703 (1984).
17. T. C. Killian, T. Pattard, T. Pohl, J.M. Rost, Ultracold neutral plasmas, *Phys. Rep.* **449**, 77–130 (2007).
18. W. S. Kurth, Waves in space plasmas, e-note (n.d.)
19. O. Laporte and R. G. Fowler, Weber's mixed boundary-value problem in electrodynamics, *J. Math. Phys.* **8** (3), 518–522 (1967).
20. E. Lazzaro and C. Maroli, Lower hybrid resonance in an inhomogeneous cold and collisionless plasma slab, *Nuovo Cim.* **16B** (1), 44–54 (1973).
21. D. Lupo, C. S. Morawetz, and K. R. Payne, On closed boundary value problems for

equations of mixed elliptic-hyperbolic type, *Commun. Pure Appl. Math.* **60**, 1319–1348 (2007).

22. D. Lupo, C. S. Morawetz, and K. R. Payne, Erratum: "On closed boundary value problems for equations of mixed elliptic-hyperbolic type," [*Commun. Pure Appl. Math.* **60** 1319–1348 (2007)] *Commun. Pure Appl. Math.* **61**, 594 (2008).

23. K. T. McDonald, An electrostatic wave, arXiv:physics/0312025v1 [physics.plasm-ph] (2003).

24. C. S. Morawetz, A weak solution for a system of equations of elliptic-hyperbolic type, *Commun. Pure Appl. Math.* **11**, 315–331 (1958).

25. C. S. Morawetz, Mixed equations and transonic flow, *Rend. Mat.* **25**, 1–28 (1966).

26. C. S. Morawetz, D. C. Stevens, and H. Weitzner, A numerical experiment on a second-order partial differential equation of mixed type, *Commun. Pure Appl. Math.* **44**, 1091–1106 (1991).

27. T. H. Otway, A boundary-value problem for cold plasma dynamics, *J. Appl. Math.* **3**, 17–33 (2003).

28. T. H. Otway, Energy inequalities for a model of wave propagation in cold plasma, *Publ. Mat.* **52**, 195–234 (2008).

29. T. H. Otway, Variational equations on mixed Riemannian-Lorentzian metrics, *J. Geom. Phys.* **58**, 1043–1061 (2008).

30. K. R. Payne, Interior regularity of the Dirichlet problem for the Tricomi equation, *J. Mat. Anal. Appl.* **199**, 271–292 (1996).

31. K. R. Payne, Solvability theorems for linear equations of Tricomi type, *J. Mat. Anal. Appl.* **215**, 262–273 (1997).

32. M. Pilant, The Neumann problem for an equation of Lavrent'ev-Bitsadze type, *J. Math. Anal. Appl.* **106**, 321–359 (1985).

33. A. D. Piliya and V. I. Fedorov, Singularities of the field of an electromagnetic wave in a cold anisotropic plasma with two-dimensional inhomogeneity, *Sov. Phys. JETP* **33**, 210–215 (1971).

34. K. S. Riedel, Geometric optics at lower hybrid frequencies, *Phys. Fluids* **29**, 3643–3647 (1986).

35. A. G. Sitenko and K. N. Stepanov, On the oscillations of an electron plasma in a magnetic field, *Z. Eksp. Teoret. Fiz.* [in Russian] **31**, 642 (1956) [*Sov. Phys. JETP* **4**, 512 (1957).].

36. T. H. Stix, *The Theory of Plasma Waves*. (McGraw-Hill, New York, 1962).

37. D. G. Swanson, *Plasma Waves*. (Institute of Physics, Bristol, 2003).

38. L. Tonks and I. Langmuir, Oscillations of ionized gases, *Phys. Rev.* **33**, 195–210 (1929).

39. H. Weitzner, "Wave propagation in a plasma based on the cold plasma model," Courant Inst. Math. Sci. Magneto-Fluid Dynamics Div. Report MF-103, August, 1984.

40. H. Weitzner, Lower hybrid waves in the cold plasma model, *Commun. Pure Appl. Math.* **38**, 919–932 (1985).

41. Y. Yamamoto, "Existence and uniqueness of a generalized solution for a system of equations of mixed type," Ph.D. Dissertation, Polytechnic University of New York, 1994.

Perspectives in Mathematical Sciences
Interdisciplinary Mathematical Sciences, Volume 9, 2009
pp. 211–239

Chapter 10

Gravitating Yang–Mills Fields in All Dimensions

Eugen Radu[†] and D. H. Tchrakian[‡*]

*†Institut für Physik, Universität Oldenburg,
Postfach 2503 D-26111 Oldenburg, Germany*
*‡School of Theoretical Physics, Dublin Institute for Advanced Study,
10 Burlington Road, Dublin 4, Ireland*
*⋆Department of Computer Science,
National University of Ireland Maynooth, Maynooth, Ireland*

A classification of gravitating Yang–Mills systems in all dimensions is presented. These systems are set up so that they support finite energy solutions. Both regular and black hole solutions are considered, the former being the limit of the latter for vanishing event horizon radius. Special attention is paid to systems necessarily involving higher order Yang–Mills curvature terms, along with the option of incorporating higher order terms in the Riemann curvature. The scope here is restricted to Einstein systems, with or without cosmological constant, and the Yang–Mills(–Higgs) systems.

1. Introduction

By gravitating Yang–Mills (YM) fields we understand YM fields on curved backgrounds whose dynamics includes the *backreaction* of the gravitational field on it. These are the particle-like and black hole solutions to the Einstein-Yang–Mills (EYM) systems.

Initially, EYM solutions[1,2] in $3 + 1$ dimensional space with Minkowskian signature were primarily of interest because they presented configurations with *non-Abelian hair*[a]. More recently however, gravitating non-Abelian solutions have found extensive application in string inspired theories, *e.g.*, in various supergravity and $D-$brane models. These results point to the physical relevance of classical gravitating non-Abelian solutions. Moreover in this context, it is mainly such classical solutions in dimensions higher that $3+1$ that find applications. It is our aim here to review gravitating non-Abelian YM solutions in higher dimensions[b], together with their four dimensional counterparts.

---

[a]These were preceded historically by gravitating Skyrmions[3] which exhibited *Skyrmion hair*, but the latter have not been studied as intensively since.
[b]It is conceivable that gravitating Skyrmions may also be of relevance to field theories in higher dimensions, but to date there has been no work on them reported in the literature.

Soon after the original particle-like (regular) solutions of[1] were constructed, the corresponding black hole solutions were found.[4-6] Not long after the discovery of the EYM solutions, both regular and black hole gravitating monopole solutions of the EYM–Higgs (EYMH) systems were constructed in.[7-9] EYM solutions in the presence of a cosmological constant Λ were constructed later, with negative Λ in,[10-12] and for $\Lambda > 0$ in.[13] These results indicate that the solutions of the Einstein equations coupled to non-Abelian matter fields possess a much richer structure than in the better known U(1) case. They also show that our intuition based on solutions with linear field sources may fail in more general situations.

In this review, our description of gravitating YM systems will in addition to the EYM fields also include the Higgs fields. The study of EYM-Higgs (EYMH) systems enables a more extensive description of physical phenomena as a result of the symmetry breaking mechanism which (gauged) Higgs models describe. This is a result of the special nature of Higgs fields here as dimensional descendents of gauge fields, as will be explained below. Therefore we shall restrict our considerations to such Higgs multiplets that afford topological stability to the solutions.

EYM solutions in (higher) $d > 4$ dimensions were considered relatively recently. Here two main possibilities have been included so far in the literature, which are distinguished by the different asymptotic structure of the spacetime. In the first case, of main interest here, the metric is asymptotically Minkowski (or (anti-)de Sitter, if a cosmological constant is considered in the action). Such solutions provide the natural counterparts of the $d = 3 + 1$ non-Abelian configurations mentioned above. In the second case, a number of p codimensions are included. Such configurations are important if one posits the existence of extra-dimensions in the universe, which are likely to be compact and described by a Kaluza-Klein (KK) theory. All known KK non-Abelian solutions have no dependence on the extra-dimensions, *i.e.* these are frozen, as in the case of the $z-$coordinate of the Abrikosov-Nielsen-Olesen vortex.

A number of results in the literature show that, in the absence of codimensions, the mass of gravitating non-Abelian solutions which asymptote to a Minkowski spacetime, is infinite[14] (the same results holds for asymptotically (anti-)de Sitter solutions[15]). This is not surprising because the usual EYM system in $d \geq 5$ dimensions does not have the requisite scaling properties for there to exist finite energy solutions.

However, the scaling properties of the usual EYM system can be altered by the addition of higher order terms of the Yang–Mills curvatures. Such terms can occur in the low energy effective action of string theory.[16,17] The hierarchies of both YM and of gravitational (*i.e.*, Einstein) systems will be defined below. Employing suitably defined EYM systems featuring higher order YM curvature [c] terms such that the equations of motion remain second order, finite mass/energy static spherically

---

[c]Higher order Riemann curvature terms cannot be employed for this purpose for when they are of sufficiently higher order to satisfy the scaling requirement, they invariably vanish or are *total divergence* terms.

symmetric non-Abelian solutions in $d > 4$ dimensions were constructed. The existence of this type of configurations is a nonperturbative effect, since they cannot be predicted in a perturbative approach around the solutions of the usual EYM system.

With zero cosmological constant, $d > 4$ regular solutions of the extended EYM system were found and analysed in,[18,19] and black hole solutions in,[20].[21] Both regular and black hole with negative cosmological constant Λ were presented in.[21] With positive Λ likewise, such EYM solutions were given in.[22] Thus there is a comprehensive sample of finite mass, static, spherically symmetric EYM field configurations in $d = D + 1$ dimensional spacetimes. These are the higher dimensional counterparts of the EYM solutions in four dimensions, and in general their non-Abelian matter content is composed of several YM terms with various (appropriate) scaling properties. Such models proliferate with increasing dimension and exhibit additional features, absent in the usual $3 + 1$ dimensional EYM case[d].

Concerning EYMH solutions describing gravitating monopoles in higher dimensions, recently a very particular hierarchy of YMH models in $d = 4p$ dimensions has been studied where regular and black hole solutions are constructed.[24] The reason for restricting to a particular family of models is the ubiquity of Higgs models in dimensions higher than $3 + 1$, as will be described below.

All the above noted EYM and EYMH solutions are static and spherically symmetric. But in $3 + 1$ dimensions there are axially symmetric solutions[25–27] with very interesting properties, so it is natural to seek their higher dimensional counterparts. To the best of our knowledge, the only result in this direction is that in,[28] for the simplest EYM system in $d = 4 + 1$ dimensions. The symmetry imposed in that case was bi-azimuthal symmetry in the 4 spacelike dimensions.

Concerning the case of non-Abelian solutions with codimensions, the situation is less explored, the case $d = 5$ being the only one discussed in a systematic way in the literature. Both non-Abelian vortices and black strings with non-Abelian hair have been studied in the literature, starting with the pioneering work of.[14]

Finally, we note that all the work referred to above pertains to gravitating YM in $D + 1$ dimensional spacetime with Minkowskian signature. This is because to date all of the work on Euclidean EYM in all dimensions is effectively restricted to the study of YM fields on fixed (gravitational) backgrounds. This, even though the earliest EYM solutions were constructed in 4 Euclidean dimensions (see[29,30] and[31,32]). As such, these results are not central to the present review which is mainly concerned with fully backreacting matter-gravity solutions.

The outline of this paper is as follows. In the next section, **2**, we introduce the hierarchies of gravitational and Yang–Mills (and Higgs) systems, in all dimensions. Then in the following section, **3**, we review known results in gravitating YM and

---

[d]The exception is a very particular hierarchy of gravitating YM models in $d = 4p$ dimensions, consisting of a single higher order YM curvature term which scales appropriately. In this case[23] all the qualitative features of the EYM solutions in $d = 4$ are preserved.

YMH solutions in higher dimensions, including a brief description of EYM solutions in Euclidean space. There we pay special attention to the existence and the generic properties of the non-Abelian solutions and not to their physical applications. In section 4 we give a summary and outlook.

2. Layout

The gravitating Yang–Mills solutions covered in this review are both in spaces with Minkowskian and with Euclidean signature. In the Minkowskian case, which is that of greater interest, these are static spherically symmetric solutions, both regular and black hole. In the Euclidean case, these are black hole solutions with periodic dependence on the Euclidean time which are spherically symmetric in the space dimensions, and also black hole solutions that are spherically symmetric in the whole of the space time. They would describe spherically symmetric black hole solutions of horizon radius r_H, which include the regular particle like configuration in the limit of $r_H = 0$. As noted in section 1, the non-Abelian matter sector will consist both of YM, and YM-Higgs (YMH) fields. These systems are defined on Euclidean spaces since in the Minkowskian case the solutions in question are static so that the fields depend on the spacelike coordinates only.

Even before gravitating the YMH field, the existence of finite energy/action solutions is contingent on the requisite scaling requirement being satisfied. In the case of flat (Euclidean) space, this amounts to satisfying the familiar virial relation, as long as the (static) Hamiltonian/Lagrangian is positive definite. In the case of a gravitating system the property of positive definiteness is absent, since all Einstein systems are defined such that they are not bounded from below. One can proceed nonetheless heuristically seeking to satisfy the virial relation. A systematic analysis of this for higher dimensional EYM systems was presented in,[21] and we will return to this question in section 3 when the static systems are gravitated and will elaborate on the scaling arguments. For now, we restrict our attention to the YMH systems on a flat background. It is the virial constraint that necessitates the introduction of higher order curvature terms in the (static) Hamiltonian/Lagrangian. These are the members of the Einstein and YM hierarchies to be defined below.

There is a marked difference in the status of the Einstein hierarchy and the YM hierarchy in the present context. The higher order YM terms are necessary for rendering the scaling properties of the EYM system in question appropriate for supporting finite energy solutions, while the corresponding terms in the gravitational hierarchy do not play this role. All the higher order gravitational terms (*e.g.*, Gauss–Bonnet) with the required scaling property either vanish or are *total divergence*.

Anticipating the definition of the Einstein hierarchy in the next subsection, we note that the p–Einstein system $e R_{(p)}$ as defined by the relations (2.1), or (2.3) bellow, scales as L^{-2p}. (Note that the usual Einstein-Hilbert term, (or 1–Einstein system in the terminology of this work), scales as L^{-2} and the usual YM, or 1−YM system, scales as L^{-4}.) According to the heuristic scaling argument, adding the p–Einstein term to the usual EYM system would result in the correct scaling if $d < 2p+1$. In the limiting case when $d = 2p$, $e\,R_{(p)}$ is a *total divergence* and beyond that it vanishes, as will be seen from the definitions below.

On the other, hand higher order members of the gravitational hierarchy play a quantitatively interesting role in highlighting certain qualitative features of the solutions, that repeat in dimensions *modulo 4p*. They are also of intrinsic interest in some considerations for the case of Euclidean signature. For those reasons, they are included in this review.

In the next subsection, **2.1**, we present the definition of gravitational systems in all dimensions, which in this review we refer to as the hierarchy of Einstein systems. In general, higher dimensional gravitational systems are composed of the superposition of individual members of the Einstein hierarchy. In subsection **2.2** we define the Yang–Mills (YM) systems in all even (spacelike) dimensions, referred to as the hierarchy of YM systems here. YM systems in all dimensions, both odd and even, can then be constructed from the superposition of individual members of the YM hierarchy. Then in subsection **2.3** we introduce that subclass of (gauged) Higgs models that are relevant to the presentation here, namely those YM–Higgs (YMH) models that have been gravitated to date, along with an example of the next most natural candidate. Finally in subsection **2.4** we state the expressions for the reduced Lagrangians of the systems introduced, subject to the appropriate symmetries. This will be mainly the case of most interest involving the static fields which are spherically symmetric in the spacelike dimensions.

2.1. *Gravitational systems in all dimensions: Einstein hierarchy*

The gravitational systems which we shall refer to as the Einstein hierarchy in $d = 2p + k$ spacetime dimensions are defined as the product of the determinant of the *Vielbeine* e^n_ν, $e = \det e^m_\mu$, with Ricci scalars $R_{(p)}$, defined by

$$\mathcal{L}_{p-\mathrm{E}} = e\,R_{(p)} = \varepsilon^{\mu_1\mu_2...\mu_{2p}\nu_1\nu_2...\nu_k}\,e^{n_1}_{\nu_1}e^{n_{2q}}_{\nu_2}\ldots e^{n_k}_{\nu_k}\,\varepsilon_{m_1m_2...m_{2p}n_1n_2...n_k}\,R^{m_1m_2...m_{2p}}_{\mu_1\mu_2...\mu_{2p}},$$
$$\tag{2.1}$$

$R^{m_1m_2...m_{2p}}_{\mu_1\mu_2...\mu_{2p}} = R(2p)$ being the p-fold totally antisymmetrised product of the Riemann curvature, in component notation

$$R(2) = R^{m_1m_2}_{\mu_1\mu_2} = \partial_{[\mu_1}\,\omega^{m_1m_2}_{\mu_2]} + \omega^{m_1n}_{[\mu_1}\,\omega^{nm_2}_{\mu_2]},$$

ω^{mn}_μ being the Levi-Civita spin connection. It is clear from the definition (2.1) that the spacetime dimensionality $d = 2p + k$ sets an upper limit on the highest order nontrivial member of this hierarchy, the term $R_{(p)(k=0)}$ being the (total divergence) Euler-Hirzebruch density. We shall refer to $e\,R_{(p)}$ in (2.1) as p–Einstein systems (*i.e.* $p = 1$ is the usual Einstein-Hilbert Lagrangian, $p = 2$ is the Gauss-Bonnet term etc.).

Subjecting (2.1) to the variational principle with respect to the arbitrary variation of the *Vielbeine* one arrives at what we refer to as the p-Einstein equation

$$G_{(p)\,\mu}{}^{m} = R_{(p)\,\mu}{}^{m} - \frac{1}{2p} R_{(p)} e_{\mu}^{m}, \qquad (2.2)$$

in terms of the p-Einstein tensor $G_{(p)\,\mu}{}^{m}$, with $R_{(p)}$ and $R_{(p)\,\mu}{}^{m}$ being the p-th order Ricci scalar and the p-th order Ricci tensor defined respectively by

$$R_{(p)} = R^{m_1 m_2 \ldots m_{2p}}_{\mu_1 \mu_2 \ldots \mu_{2p}} e^{\mu_1}_{m_1} e^{\mu_1}_{m_2} \ldots e^{\mu_{2p}}_{m_{2p}} \qquad (2.3)$$

$$R_{(p)\,\mu}{}^{m} = R^{m\, m_2 m_3 \ldots m_{2p}}_{\mu \mu_2 \mu_3 \ldots \mu_{2p}} e^{\mu_2}_{m_2} e^{\mu_3}_{m_3} \ldots e^{\mu_{2p}}_{m_{2p}}. \qquad (2.4)$$

The $p-$Einstein hierarchy of gravitational systems is defined by e times the $p-$Ricci scalar (2.3), and the most general Einstein system is given by the maximal number of nonvanishing superpositions of all $p-$Ricci scalars.

An interesting property of the $p-$Riemann and $q-$Riemann curvatures in even dimensions is the double-selfduality constraint

$$R^{m_1 m_2 \ldots m_{2p}}_{\mu_1 \mu_2 \ldots \mu_{2p}} = \pm \kappa^{2(p-q)} \frac{e}{[(2q)!]^2} \varepsilon_{\mu_1 \mu_2 \ldots \mu_{2p} \nu_1 \nu_2 \ldots \nu_{2q}} R^{\nu_1 \nu_2 \ldots \nu_{2q}}_{n_1 n_2 \ldots n_{2q}} \varepsilon^{m_1 m_2 \ldots m_{2p} n_1 n_2 \ldots n_{2q}}, \qquad (2.5)$$

where the Hodge dual of the $2q-$form curvature is equated to the $2p-$form curvature, with the dimensionful constant κ compensating for the difference in the respective dimensions. (2.5) can be stated for both Euclidean and Minkowskian signatures, with the $\pm$ sign respectively. For Euclidean signature, contracting the constraint (2.5) with the appropriate number of *vielbeine* one arrives at the vacuum Einstein equations for the $(p,q)-$Einstein system

$$\mathcal{L}_{(p,q)-\mathrm{E}} = e \left(R_{(p)} + \kappa^{2(p-q)} R_{(q)} + \Lambda \right), \qquad (2.6)$$

where $R_{(p)}$ and $R_{(q)}$ in (2.6) are defined by (2.3) and Λ is a cosmological constant whose value is related to the constant κ, as shown in.[33] This is not valid in the case of Minkowskian signature (see Appendix of[34] for details).

2.2. The Yang–Mills hierarchy

Since we seek finite energy/action solutions of the gravitating YM systems, the relevant members of the YM hierarchy to be defined, are those which support finite action solutions in the spacelike (Euclidean) subspace D of the spacetime $d = D + 1$. The flat space solutions in question will have topological stability when D is even. To avail of topological stability for these finite energy/action solutions in **odd** dimensions D, one has to consider the Higgs models derived from the dimensional descent of the YM systems in **even** $D + N$ with compact **odd** codimension N. The YM hierarchy is presented in the section **2.2.1** and the ensuing YMH models in section **2.2.2**, below.

Using the notation $F(2) = F_{\mu\nu}$ for the 2-form Yang–Mills (YM) curvature, the $2p$-form YM tensor

$$F(2p) = F(2) \wedge F(2) \wedge \ldots \wedge F(2), \quad p\text{-times} \qquad (2.7)$$

is a p fold totally antisymmetrised product of the 2-form curvature.

The p−YM system of the YM hierarchy is defined, on $\mathbb{R}^{4p}$, by the Lagrangian density

$$\mathcal{L}_{p-\text{YM}} = \text{Tr}F(2p)^2 \, . \tag{2.8}$$

In $2n$ dimensions, partitioning n as $n = p + q$, the Hodge dual of the $2q$-form field $F(2q)$, namely $(^\star F(2q))(2p)$, is a $2p$-form.

Starting from the inequality

$$\text{Tr}[F(2p) - \kappa \ ^\star F(2q)]^2 \geq 0 \, , \tag{2.9}$$

it follows that

$$\text{Tr}[F(2p)^2 + \kappa^2 \ F(2q)^2] \geq 2\kappa \ \mathcal{C}_n \, , \tag{2.10}$$

where $\mathcal{C}_n$ is the n-th Chern-Pontryagin density. In (2.9) and (2.10), the constant κ has the dimension of length to the power of $(p - q)$.

The element of the YM systems labeled by (p, q) in (even) $2(p + q)$ dimensions are defined by Lagrangians defined by the densities on the *left hand side* of (2.10). When in particular $p = q$, then these systems are conformally invariant and we refer to them as the p−YM members of the YM hierarchy.

The inequality (2.10) presents a topological lower bound which guarantees that finite action solutions to the Euler–Lagrange equations exist. Of particular interest are solutions to first order self-duality equations which solve the second order Euler–Lagrange equations, when (2.10) can be saturated.

For $\mathbf{M}^{2n} = \mathbb{R}^{2n}$, the self–duality equations support nontrivial solutions only if $q = p$,

$$F(2p) \ = \ ^\star F(2p) \, . \tag{2.11}$$

For $p = 1$, i.e. in four Euclidean dimensions, (2.11) is the usual YM selfduality equation supporting instanton solutions. Of these, the spherically symmetric[35,36] and axially symmetric[37–39] instantons on $\mathbb{R}^{4p}$ are the known. For $p \geq 2$, i.e. in dimensions eight and higher, only spherically symmetric[36] and axially symmetric[38,39] solutions can be constructed, because in these dimensions (2.11) are over-determined.[40]

In the $r \gg 1$ region, all these 'instanton' fields on $\mathbf{R}^{2n}$, whether self–dual or not, asymptotically behave as pure–gauge

$$A \rightarrow g d g^{-1}$$

For $\mathbf{M}^{2n} = G/H$, namely on compact coset spaces, the self–duality equations support nontrivial solutions for all p and q,

$$F(2p) \ = \ \kappa \ ^\star F(2q) \tag{2.12}$$

where the constant κ is some power of the 'radius' of the (compact) space. The simplest examples are $\mathbf{M}^{2n} = S^{2n}$, the $2n$-spheres,[41] and $\mathbf{M}^{2n} = \mathbf{CP}^n$, the complex projective spaces.[42]

The above definitions of the YM systems can be formally extended to all dimensions, including all odd dimensions. The only difference this makes is that all topological lower bounds enabling the construction of instantons are then lost, but this is immaterial from the viewpoint in the present review.

2.3. *Higgs models on* $\mathbf{R}^D$

Higgs fields have the same dimensions as gauge connections and appear as the extra components of the latter under dimensional reduction, when the extra dimension is a compact symmetric space. Dimensional reduction of gauge fields over a compact codimension is implemented by the imposition of the symmetry of the compact coset space on the coordinates of the codimensions. In this respect, the calculus of dimensional reduction does not differ from that of imposition of symmetries generally, which is the relevant formalism used in this, **2.3**, and the next subsection **2.3**.

The calculus of imposition of symmetry on gauge fields that has been used in the works being reviewed here is that of Schwarz.[43–45] This formalism was adapted to the dimensional reduction over arbitrary codimensions in.[33,46,47]

In general one can employ a linear combination of inequalities (2.10), for all $p \leq D/4$ and $q \leq D/4$. Restricting, for simplicity to the $4p$ dimensional conformal invariant systems in (2.10), i.e. to $p = q = D/4$, the descent over the compact space K^{4p-d} is described by

$$\int_{\mathbf{R}^D \times K^{4p-D}} \mathcal{F}(2p)^2 \geq \int_{\mathbf{R}^D \times K^{4p-D}} \mathcal{C}_{2p} , \qquad (2.13)$$

where $\mathcal{F}(2p)$ is the $2p$−form curvature of the 1−form connection $\mathcal{A}$ on the higher dimensional space $\mathbf{R}^D \times K^{4p-D}$. Imposing the symmetry appropriate to K^{4p-D} on the gauge fields results in the breaking of the original gauge group to, say, the residual gauge group g for the fields on $\mathbf{R}^D$. Performing then the integration over the compact space K^{4p-D} leads to the Lagrangian $\mathcal{L}[A, \phi]$, of the residual Higgs model on $\mathbf{R}^D$. A here is the connection taking values in the algebra of g and ϕ is the Higgs multiplet whose structure under g depends on the detailed choice of K^{4p-D}, implying the following gauge transformations

$$A \rightarrow gAg^{-1} + gd\,g^{-1}$$

and depending on the choice of K^{4p-D},

$$\phi \rightarrow g\phi\,g^{-1}, \quad \text{or}, \quad \phi \rightarrow g\phi, \quad \text{etc.}$$

The inequality (2.10) leads, after this dimensional descent, to

$$\int_{\mathbf{R}^D} \mathcal{L}[A, \phi] \geq \int_{\mathbf{R}^D} \nabla \cdot \mathbf{\Omega}[A, \phi] = \int_{\mathbf{\Sigma}^{D-1}} \mathbf{\Omega}[A, \phi] , \qquad (2.14)$$

where $\mathcal{L}[A, \phi] = \mathcal{L}[F, D\phi, |\phi|^2, \eta^2]$ is the residual Lagrangian in terms of the residual gauge connection A and its curvature F, the Higgs fields ϕ and its covariant derivative $D\phi$ and the inverse of the compactification 'radius' η. The latter is simply the

VEV of the Higgs field, seen clearly from the typical form of the components of the curvature F on the extra (compact) space K^{4p-D}

$$F|_{K^{4p-D}} \sim (\eta^2 - |\phi|^2) \otimes \Sigma \quad \Rightarrow \quad \lim_{r \to \infty} |\phi|^2 = \eta^2 \qquad (2.15)$$

where Σ are, symbolically, spin-matrices/Clebsch-Gordan coefficients.

It should be noted at this stage that subjecting the selfduality equations (2.11) to this dimensional descent results in Bogomol'nyi equations on $\mathbb{R}^D$, which for $p \geq 2$ in $\mathbb{R}^D \times K^{4p-D}$ (cf. (2.13)) turn out to be over-determined[40] with few exceptions.

There arise a plethora of Higgs models, depending on the mode of dimensional descent, namely on the particular choice of the compact codimension K^{4p-d}. We will not dwell on various modes of descent and the detailed properties of the descended YMH models here, and will limit our attention to those models that have been gravitated to date.

Perhaps the most interesting, or useful, family of YMH models on $\mathbb{R}^D$ arrived at *via* this descent mechanism are those in which the residual gauge group is $SO(D)$, and the Higgs field multiplet is an isovector of $SO(D)$[e]. It turns out that only when $D = 2$ and when $D = 4p - 1$ in the descent over $\mathbb{R}^D \times K^{4p-D}$, the resulting Bogomol'nyi equations are **not** over-determined.[40] The $D = 2$ case is uninteresting from our present perspective since the Abelian Higgs systems in that case live in $2 + 1$ spacetime dimensions and gravitating them is unproductive. This leaves the family of $SO(D)$ Higgs models that live in $D = 4p - 1$ space, or $d = 4p$ spacetime dimensions, for which the flat space Bogomol'nyi equations can be saturated. These are the only Higgs models to date, that are gravitated.[24] (The flat space solutions of this system was studied in,[48] which are direct generalisations of the usual BPS monopoles with $p = 1$.)

The YM field in the YMH models on R^D discussed thus far, is purely magnetic supporting a 'magnetic' monopole. But when it comes to YMH models, as stated earlier, the presence of the Higgs field enables the support of a dyon in $d = D + 1$ dimensional spacetime. In the usual[49] sense as the dyon in $3 + 1$ dimensions, the Higgs field partners the newly introduced 'electric' YM potential A_0. Thus we can describe $SO(d)$ dyons[f] in d−dimensional spacetime.

When the dimension of the spacetime d is even, then the chiral representations of the algebra of $SO(d)$ can be employed, namely replacing the Dirac representation matrices $\Gamma_{\mu\nu} = (\Gamma_{ij}, \Gamma_{i,d})$ with $\Sigma_{\mu\nu}^{(\pm)} = (\Sigma_{ij}^{(\pm)}, \Sigma_{i,d}^{(\pm)})$, the precise definition of these matrices to be given explicitly in the next subsection.

---

[e]Employing Dirac matrix representations for the algebra of $SO(D)$ in terms of Γ_{ij}, $i, j = 1, 2, \ldots,$, the Higgs field takes its values in the matrix basis $\Gamma_{i,D+1}$. with $(\Gamma_{ij}, \Gamma_{i,D+1})$ representing the algebra of $SO(d) = SO(D+1)$

[f]While both the Higgs field and A_0 take their values in the Dirac matrix basis $\Gamma_{i,D+1}$, in the in the $3 + 1$ dimensional case the 'enlarged' algebra $SO(4)$ splits in the two $SU(2)$ subalgebras, whence the dyon and the monopole are both described by $SU(2)$ matrices. In all higher dimensions, this is not the case and the full algebra employed is that is $SO(D+1)$, where $d = D+1$ is the dimension of the spacetime.

When by contrast the dimension of the spacetime d is odd, then the Dirac represenation matrices are the appropriate ones to be used, *e.g.*, in $d = 4 + 1$ spacetime[50] when $D = 4$. (No such higher dimensional monopoles are gravitated to date.)

The family of YMH in $d = 4p$ dimensional spacetime that are gravitated result from the simplest mode of descent over $\mathbb{R}^D \times K^1$, with $D = 4p - 1$, *i.e.*, over one codimension. The Lagrangian densities can be expressed for arbitrary p in flat space as[48]

$$\mathcal{L}_{p-\text{YMH}} = \text{Tr}\ \left[F(2p)^2 + 2p\ (F(2p-2) \wedge D\Phi)^2 \right]$$
$$= \text{Tr}\ \left[(F_{m_1 m_2 \dots m_{2p}})^2 + 2p\ (F_{[m_1 m_2 \dots m_{2p-2}} D_{i_{2p-1}]}\Phi)^2 \right], \quad (2.16)$$

in an obvious notation.

2.4. *Static spherically symmetric fields*

Since almost all the work reviewed in this article involves static spherically symmetric only, we will subject the above introduced systems to this symmetry only.

The usual metric Ansatz with spherical symmetry in $d-1$ dimensional subspace is

$$ds^2 = \mp N(r)\sigma^2(r)d\tau^2\ +\ N(r)^{-1}dr^2\ +\ r^2 d\Omega_{(d-2)}^2\,, \quad (2.17)$$

the $\mp$ sign pertaining to Lorentzian and Euclidean signatures, and with $d\Omega_{(d-2)}^2$ being the metric on S^{d-2}.

Subject to the Ansatz (2.17), the reduced one dimensional Lagrangian of the $p-$Einstein system (2.1) (or (2.3)), in $d-$dimensional spacetime is calculated. After neglecting the appropriate surface terms, this can be expressed compactly as

$$L_{(\text{grav})}^{(p,d)} = \frac{\kappa_p}{2^{2p-1}} \frac{(d-2)!}{(d-2p-1)!}\ \sigma\ \frac{d}{dr} \left[r^{d-2p-1}(1-N)^p \right]\,. \quad (2.18)$$

We next state the static spherically symmetric Ansatz for the $p-$YM system (2.8) and the $p-$YMH system,

$$\Phi = \eta\, h(r)\, \hat{x}_j\, \Gamma_{j,d}\,, \quad A_0 = u(r)\, \hat{x}_j\, \Gamma_{j,d}\,, \quad A_i = \left(\frac{1-w(r)}{r} \right) \Gamma_{ij}\hat{x}_j\,. \quad (2.19)$$

for odd dimensional spacetime d, and, for even d

$$\Phi = \eta\, h(r)\, \hat{x}_j\, \Sigma_{j,d}^{(\pm)}\,, \quad A_0 = u(r)\, \hat{x}_j\, \Sigma_{j,d}^{(\pm)}\,, \quad A_i = \left(\frac{1-w(r)}{r} \right) \Sigma_{ij}^{(\pm)}\hat{x}_j\,, \quad (2.20)$$

where

$$\Sigma_{ij}^{(\pm)} = -\frac{1}{4} \left(\frac{1 \pm \Gamma_{d+1}}{2} \right) [\Gamma_i, \Gamma_j]\,,$$

are the chiral representations of $SO(d)$. The dimensionful constant η in (2.19)-(2.20) is the Higgs VEV.

It should be stated here that the choice of gauge group here is made with the purpose of enabling the construction of nontrivial finite energy solutions, and that this is the minimal size of gauge group in each case. Larger gauge groups, with the appropriate representations containing these, can be employed in case of necessity, *e.g.*, when a Chern–Simons term is to be introduced in the Lagrangian.

Imposing this symmetry, *i.e.*, substituting (2.19) or (2.20) into the appropriate Lagrange density, results in exactly the same one dimensional reduced Lagrangians in both cases (except for an overall factor of 2). This is because the algebraic manipulations involved in both cases are identical. The situation changes if Fermions are introduced, but we do not do that here.

The resulting reduced one dimensional Lagrangian of the $p-$YM system (2.8), augmented by the Higgs kinetic term in (2.16), in $d-$spacetime dimensions is

$$
\begin{aligned}
L_{\text{YMH}}^{(p,d)} = \frac{\tau_p\, r^{d-2}}{2\cdot(2p)!} &\left\{ \frac{(d-2)!}{(d-[2p+1])!}\, \sigma \left(\frac{1-w^2}{r^2}\right)^{2(p-1)} \right. \\
&\times \left[(2p)N\left(\frac{w'}{r}\right)^2 + (d-[2p+1])\left(\frac{1-w^2}{r^2}\right)^2 \right] \\
\mp \frac{(d-2)!}{(d-2p)!}\frac{(2p-1)}{\sigma} &\left[([(1-w^2)^{p-1}u]')^2 \right. \\
&\left. + (d-2p)\frac{(2p-1)}{N}[(1-w^2)^{p-1}u]^2\left(\frac{w}{r}\right)^2 \right] \\
\mp \frac{(d-2)!}{(d-2p)!}\frac{(2p-1)}{\sigma}\, \eta^2 &\left[([(1-w^2)^{p-1}h]')^2 \right. \\
&\left. \left. + (d-2p)\frac{(2p-1)}{N}[(1-w^2)^{p-1}h]^2\left(\frac{w}{r}\right)^2 \right] \right\}
\end{aligned}
\tag{2.21}
$$

This is the most general matter part consisting of the YMH system for this *particular* family of Higgs models.

Before proceeding to describe the various types of gravitating YM and YMH solutions in the next section, let us make some remarks concerning the particular choices of the various models employed.

- To recover the formula used for the $p-$YM system (2.8), one simply replaces $h = 0$ in the third line of (2.21), thus eliminating the Higgs field.
- To recover the matter Lagrangian used for the gravitating monopoles in $d = 4p$ dimensions presented in,[24] one replaces $u \to 0$ in the second line of (2.21), and, sets $d = 4p$ since the gravitating monopoles there are constructed only in those spacetime dimensions. Of course, it would be possible also to gravitate dyons in higher dimensions just as in $d = 4$ (see for example[51] and references therein), but this has not been done to date.

The choice of model for the monopoles gravitated in[24] was made firstly such that the YM term and the Higgs kinetic term have the same dimensions. This is precisely with the criterion that there should not be a mismatch of dimensions giving rise to a *conical fixed point* singularity, so as not to cloud an otherwise more complicated system. This family of models is dimensionally descended from the $p-$YM system and hence the Bogomol'nyi equations can be saturated, which is unimportant since the backreaction of gravity prevents this saturation anyway. Secondly, we opted for gravitating the $p-$Higgs models with $p-$gravity for the more or less *aesthetic* reasons of that choice in the EYM case of.[23]

- In the $d \geq 5$ EYM case, various matter systems consisting of the superpositions of $p-$YM systems (2.8) are gravitated with the $1-$Einstein gravity (usual Einstein-Hilbert Lagrangian). This immediately introduces a mismatch between the dimensions of the constituent terms in the Lagrangian. It is found that one result of this is the absence of the radial excitations (higher node solutions) observed in the $p = 1$, $d = 4$ Bartnik-McKinnon[1] case. Another result is that in addition to the Reissner-Nordström fixed point, there arises a new singularity[20] in $d = 4+1$ dimensions. This singularity was found[19] to repeat in $d = 4p+1$ dimensions, *modulo $4p$*. The fixed point corresponding to it was called a *conical fixed point*.

- The electric potential necessarily vanishes for asymptotically flat finite energy solutions of the EYM system, *i.e.* $u(r) = 0$ if $h(r) = 0$. The proof here is similar to that found in,[52,53] for $d = 3+1$ dimensions. One starts with the equation for the electric potential $u(r)$, which, for a generic model with a number of P terms in the YM hierarchy can be rewritten as

$$\sum_{p=1}^{P} \tau_p \frac{1}{2} \left(\frac{r^{d-2}}{\sigma} (W^2 u^2)' \right)'$$
$$= \sum_{p=1}^{P} \tau_p \frac{r^{d-2}}{\sigma} \left(((Wu)')^2 + \frac{2(d-2p)(2p-1)}{N} W^2 \frac{u^2 w^2}{r^2} \right), \qquad (2.22)$$

(here, to simplify the relations, we denote $W = (1 - w^2)^{p-1}$). One can easily see that the r.h.s. of (2.22) is a strictly positive quantity. Thus the integral of the l.h.s. should also be positive,

$$\sum_{p=1}^{P} \tau_p \frac{1}{2} \left(\frac{r^{d-2}}{\sigma} (W^2 u^2)' \right) \Bigg|_{r_0}^{\infty} > 0 , \qquad (2.23)$$

(where $r_0 = 0, r_H$ for particle-like and black hole solutions, respectively). However, the regularity of the solutions together with finite energy requirements impose that, in the above relation, both the contributions at $r = r_0$ and at infinity vanish. As a result, $u(r)$ should vanish for any reasonable solution. The same proof generalises for anti-de Sitter solutions, the only exception being the systems featuring exclusively the p-th terms of the hierarchy, in $d = 4p$ dimensions.

- Departing from these generic models, there is a family of models for which the mismatch of the dimensionality of the constituent terms is removed. Like in the usual EYM system consisting of the $1-$Einstein and $1-$YM systems in $d = 3 + 1$, this family of models[23] consists exclusively of the $p-$Einstein and $p-$YM systems in $d = 4p$. The result is that all qualitative features of the EYM solutions of[1] are preserved. Indeed, if $p-$Einstein is replaced by $q-$Einstein, $q \neq p$, the salient features persist but are quantitatively somewhat deformed.

3. Non-Abelian solutions in $d-$dimensions

3.1. *Solutions with Lorentzian signature*

3.1.1. *Einstein–Yang-Mills solutions in four dimensions*

The closed form solutions in Chakrabarti *et al*,[31][32] are probably the first examples of black holes with non-Abelian hair, albeit with no backreaction between gravity and the non-Abelian matter. Fully selfgravitating EYM solutions were constructed somewhat after those on fixed backgrounds, by Bartnik and McKinnon.[1] These were regular particle-like solutions, and were soon followed by their black hole counterparts in.[4–6]

Subsequently, a large literature has developed on this subject, extending to systems with a cosmological constant, and separately, to systems whose Lagrangian contains also a Higgs field, supporting gravitating monopoles. Extensions of the EYM system to include other fields which enter various stringy models have been considered as well, in particular for a Gauss-Bonnet quadratic curvature term coupled with a dilaton.[54] However, these solutions are beyond the scope of the present review. (A detailed review of the various $d = 4$ gravitating solutions with non-Abelian fields was presented a decade ago in.[2] The case of $d = 4$ asymptotically anti-de Sitter (AdS) solutions which was not covered in,[2] was the subject of the recent review.[55])

These were all static spherically symmetric solutions, some of whose salient properties will be contrasted in their higher dimensional counterparts to be reported in the next subsection. Restricting to solutions with a gauge group $SU(2)$, their basic properties are:

In the EYM case, the asymptotically flat solutions

- were sphalerons, i.e. that they were unstable,[56][57] since there was no topological charge to supply the energy with a lower bound, and,
- they present radial excitations characterised by a number k of nodes of the function magnetic gauge $w(r)$ in (2.20);

- the non-Abelian electric potential necessarily vanishes for both globally regular and black hole solutions with finite energy.[52,53]

When a negative cosmological constant is added to the EYM Lagrangian,[10,11] the asymptotically AdS solutions exhibit new and interesting features, namely that now

- the asymptotic value of the function $w(r)$ in (2.20) is not fixed *a priori*, which leads to finite mass solutions with a nonvanishing non-Abelian magnetic charge, even without a Higgs field;
- stable solutions have been shown to exist[10,58] (this corresponds basically to the case where the profile of the function $w(r)$ presents no nodes);
- black holes with non-Abelian hair and a nonspherical topology of the event horizon have been found for $\Lambda < 0$ in;[59,60]
- most importantly in this case, it becomes possible to construct finite energy solutions with a nonvanishing A_0,[11] *i.e.* non-Abelian dyons;
- moreover, finite mass solutions with AdS asymptotics exist for any $\Lambda < 0$.

In the case of a positive cosmological constant,[13,61–63] by contrast,

- the EYM solutions with de Sitter (dS) asymptotics exist for sufficiently small values of Λ only;
- all solutions have been shown to be unstable, since $w(r)$ necessarily presents nodes (although the asymptotic value of the magnetic gauge potential is not fixed *a priori*);
- the electric potential A_0 necessarily vanishes for all dS solutions.

In the EYMH (gravitating monopole) case, which differs from the EYM in that a dimensionful constant (the Higgs vacuum expectation value) appears in the Lagrangian, the solutions with $\Lambda \leq 0$

- are topologically stable in the YMH sector, stabilised by the monopole charge;
- they present radial excitations characterised by multinode profiles in the function $w(r)$ in (2.20) as in the YMH case, and in addition,
- due to the presence of the dimensionful constant in the Lagrangian, they exhibit a Reissner-Nordström fixed point, which results in the absence of solutions for a range of the gravitational coupling constant.

The picture is more complicated for asymptotically dS gravitating monopole solutions. Refs.[62,64] presented arguments that

- although the total mass within the cosmological horizon of the monopoles is finite, their mass evaluated at timelike infinity generically diverges;
- no solutions exist in the absence of a Higgs potential.

The $d = 4$ asymptotically Minkowski (or AdS) EYM solutions discussed above have axially symmetric generalisations. The first work in this direction was,[25] which presented a generalization of the Bartnik-McKinnon solutions characterized by a pair of integers (k, n), where n is an integer – the winding number and k is the node number of the amplitude $w(r)$. The black hole counterparts of these configurations were discussed in,[26] which shows that Israel's theorem[65] does not generalise to the non-Abelian case (*i.e.* a static black hole is not necessarily spherically symmetric). These asymptotically flat solutions were extended afterwards in various directions, see *e.g.*[66–69] They present also generalizations with a negative cosmological constant, which were discussed in.[70–72] The case of axially symmetric non-Abelian solutions with dS asymptotics was not considered yet in the literature.

Interestingly, although non-Abelian generalizations of the Kerr-Newmann black hole were shown to exist,[73,74] it turns out that the Bartnik-McKinnon globally regular solutions admit no asymptotically flat rotating generalizations[73,75] (however, note that they were predicted in a perturbative approach[76]). Not completely unexpected, spinning EYM solitons were found to exist for AdS asymptotics.[72,77]

Finally, let us remark that both the EYM and EYMH systems present nontrivial solutions with a NUT charge.[78,79] These solutions approach asymptotically the Taub-NUT spacetime[80] and provide the non-Abelian counterparts of the U(1) Brill solution.[81] The nonexistence results in[52,53] are circumvented by these asymptotically locally flat solutions, which necessarily present a nonzero electric part A_0 of the non-Abelian potential.

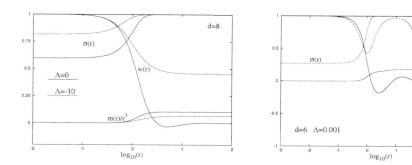

Fig. 10.1. Typical particle-like higher dimensional EYM-$F(2)^2$ solutions in asymptotically flat and anti-de Sitter spacetimes (left) and de Sitter spacetime (right). The function $m(r)$ corresponds to the local mass-energy density.

3.1.2. *Einstein–Yang-Mills solutions in higher dimensions*

Gravitating non-Abelian fields in higher dimensions have been considered for the first time in[14] for $d = 5$ and a YM model containing the usual $F(2)^2$ term only. For spherically symmetric regular solutions asymptotic to the Minkowski background, it was found that their energy is infinite. Then in Ref.[15] it was proven that the energy of the black hole is also infinite. Moreover,[15] extended these results to asymptotically AdS solutions. When employing the Einstein-Hilbert gravity only, one usually defines

$$N(r) = 1 - \frac{m(r)}{r^{d-3}} - \frac{2\Lambda r^2}{(d-2)(d-1)}, \tag{3.1}$$

the function $m(r)$ being related to the local mass-energy density up to some d−dependent factor. The results in[14,15] prove that, in five dimensions, $m(r) \to \log r$, as $r \to \infty$.

As discussed in,[21] this is a generic feature of all higher dimensional EYM solutions with a F^2 term only (*i.e.* $L_{YM} = \tau_1 F^a_{\mu\nu} F^{a\mu\nu}$). Although these configurations are still asymptotically Minkowski, their mass function generically diverges as r^{d-5} (or as $\log r$ for $d = 5$). A similar conclusion is reached when considering[21] solutions of a EYM-Λ model containing the usual F^2 term only and approaching asymptotically an AdS (or dS) background[g]. This can most easily be seen by considering the simplest $w(r) = 0$ solution of the EYM equations. This corresponds to the gravitating Dirac–Yang monopoles,[86,87] which are non-Abelian configurations (except in $d = 4$ where one has the Abelian Dirac monopole). These fields are singular at the origin and hence an event horizon should be present.

The result is a black hole solution, which for $d > 5$ has a line element

$$ds^2 = \frac{dr^2}{1 - \frac{\mu^2}{r^2} - \frac{M_0}{r^{d-3}} - \frac{2\Lambda r^2}{(d-2)(d-1)}}$$
$$+ r^2 d\Omega^2_{(d-2)} - \left(1 - \frac{\mu^2}{r^2} - \frac{M_0}{r^{d-3}} - \frac{2\Lambda r^2}{(d-2)(d-1)}\right) dt^2, \tag{3.2}$$

where μ is a constant fixed by the YM coupling parameter,[21] $M_0 > 0$ is an arbitrary constant and Λ the cosmological constant. This infinite mass configuration generalises to higher dimensions the $d = 4$ magnetic Reissner-Nordström black hole and has a number of interesting properties which are discussed in.[88] The generic solutions of the F^2 EYM model have a more complicated pattern (including particle-like configuration with a regular origin), but always approach asymptotically the line element (3.2) (see Figure 10.1).

The nonexistence result on finite mass solutions is circumvented by adding the appropriate p−YM term(s) to the matter Lagrangian. As a result, the EYM system presents (at least) one more coupling constant $\alpha^2 = \sqrt{\tau_1^3/\kappa_1^2\tau_2}$, which usually implies a rich structure of the solutions. Various examples were studied:

- in[18] the particle-like solutions of the system consisting of 1− and 2−Einstein-terms (*i.e.*, the Einstein-Gauss-Bonnet system), 1−YM and

---

[g]Asymptotically AdS solutions with diverging mass have been considered by some authors, mainly for a scalar field in the bulk (see e.g.[82]). In this case it might be possible to relax the standard asymptotic conditions without loosing the original symmetries, but modifying the charges in order to take into account the presence of matter fields. A similar approach has been used in ref.[21] to assign a finite mass to $d > 4$ EYM solutions in a F^2 theory with a negative cosmological constant.

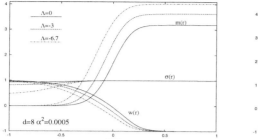 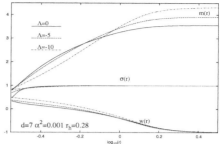

Fig. 10.2. Typical higher dimensional particle-like (left) and black hole (right) finite mass solutions of the $p = 1, 2$ EYM theory in asymptotically flat and anti-de Sitter spacetimes.

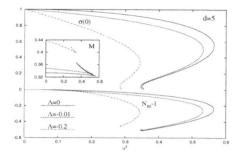 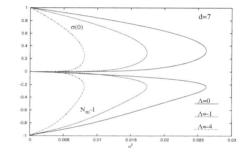

Fig. 10.3. The value N_m of the minimum of the metric function $N(r)$, the mass parameter M as well as the value of the metric function σ at the origin, $\sigma(0)$, are shown for $d = 5$, $d = 7$ asymptotically flat and anti-de Sitter particle solutions of the $p = 1, 2$ EYM theory, as functions of the coupling parameter $\alpha^2 = \sqrt{\tau_1^3/\kappa_1^2 \tau_2}$ and several values of the cosmological constant Λ.

2−YM terms in spacetime dimensions $d = 6, 7, 8$, thus exhibiting a dimensionful constant. Although a pure gauge configuration is approached in the far field, these solutions however are not quite direct analogues of the Bartnik-McKinnon solutions because of the presence of the dimensionful τ_2 constant in the Lagrangian. This is analogous with the gravitating monopole,[7] where a dimensionful constant is also involved. Unlike the latter, however, there were no radial excitations in this case. Moreover, the Gauss-Bonnet in the gravity action does not lead to any new qualitative features of the solutions,

- in[20] for particle like solutions of the system consisting of 1−Einstein, 1−YM and 2−YM subsystems as above, but in spacetime dimensions $d = 5$. In addition, in,[20] asymptotically flat black hole solutions are constructed. The fixed point properties in $d = 5$ solutions however differ substantially from those of the $d = 6, 7, 8$ solutions for the same model (see Figure 10.3).

- The fixed point analysis for this model is carried out in,[19] where it is found that in addition to the Reissner-Nordström fixed point, a new type of fixed point appears. While the Reissner-Nordström fixed point is typified by the value of the function $w(r) = 0$, the new type of fixed point is typified by the value of the function $w(r) = 1$ and is referred to as a *conical fixed point* in.[19] It is further shown in,[19] by extending the model judiciously for higher values of d (always keeping only the $1-$Einstein terms) by higher p YM terms, that this conical singularity appears *modulo* every $4p$ dimensions.

- EYM systems with negative cosmological constant in higher dimensions are also studied in.[21] The finite energy solutions in these models exhibit all the properties seen in[18-20] for $\Lambda = 0$. Different from the $d = 4$ case, the higher dimensional AdS solutions necessarily have $w(r) \rightarrow -1$ as $r \rightarrow \infty$, *i.e.* a pure gauge configuration is approached in the far field (see Figure 10.2). As a consequence, asymptotically AdS solutions with negative cosmological constant cannot support nonvanishing A_0 solutions, a *per* the argument given in the last but one item in section **2.4**.

- Higher dimensional EYM systems with positive cosmological constant are studied in.[22] The presence of a cosmological horizon leads to a more complicated pattern, where again a conical fixed point appears for $d = 5$.

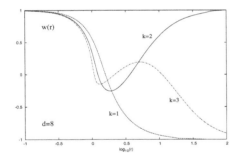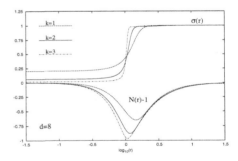

Fig. 10.4. The profiles of the metric functions $N(r)$, $\sigma(r)$ and gauge function $w(r)$ are presented for $k-$node globally regular solutions of the $p = 2$ gravity-Yang-Mills model in $d = 8$ dimensions.

All the above listed EYM solutions pertain to models motivated entirely by the criterion of satisfying the scaling requirement for finite energy. In a further work[23] in $d = 4p$, the models were chosen according to the criterion that only $p-$Einstein and $p-$YM terms appear in the Lagrangian[h]. In these cases no dimensionful constant appears in the Lagrangian and the properties of the solutions are entirely similar to those of the Bartnik-McKinnon solution (in particular the existence of radial excitations), except that qualitative features are appreciably magnified with increasing p (see Figure 10.4).

---

[h]The ref.[23] presents also an exact solution for the p-th Einstein-Yang-Mills system in $d = 2p + 1$ dimensions.

There is also the question of non–spherically symmetric EYM solutions in higher dimensions. Since all EYM solutions are constructed numerically, the problem here is to relax this symmetry such that the numerical process remains tractable. The most straightforward step would be the imposition of *axial* symmetry in the $D-$spacelike dimensions, *i.e.*, by imposing spherical symmetry in the $(D-1)-$dimensional subspace, thus reducing the problem to a $2-$dimensional PDE. This can be done readily for arbitrary $d = D + 1$, but unfortunately the implementation of the numerical integration becomes problematic when removing the gauge arbitrariness. Instead, for $d = 5$, a system of $2-$dimensional PDE's can be obtained when an azimuthal symmetry is imposed in each of the two planes of the $4-$dimensional $t = const.$ spacelike subspace. In principle, this can be generalized for any odd, $d = 2n + 1$, dimensional spacetime, and then the reduced problem will be that of a $n-$dimensional PDE's. This limits one to the bi-azimuthal regime in $d = 4 + 1$, for practical reasons.

Other than this static result, there are two other indirect $d > 4$ results in the literature which are not spherically symmetric. One is the case where there is a rotation in the two spacelike sub-planes in the $4+1$ dimensional case, and the other concerns a rather different topology of the spacetime.

- Ref.[28] discussed static solutions in $d = 4 + 1$ dimensions of a EYM system with bi-azimuthal symmetry in four spacelike dimensions. They generalise the configurations in,[18,20] both particle-like and black hole solutions being found to exist. It is interesting that the fixed point structures discovered in the spherically symmetric cases in[18–22] in $d = 5$, manifest themselves for these bi-azimuthally symmetric solutions, although a rigorous fixed point analysis like in[19] is not analytically accessible in this case.

- $d = 5$ rotating EYM black hole solutions in the usual EYM model (*i.e.*, with a F^2 term only) with negative cosmological constant were constructed in.[84] The rotation in question was that of two equal angular momenta in the two spacelike 2-planes, which is known to reduce the $2-$dimensional PDE problem to a $1-$dimensional ODE one (*i.e.* the angular dependence is factorised in the ansatz). In this respect, it is a system defined by a single radial variable, but does not not describe a spherically symmetric field configuration. Different from the static case, no spinning regular solution is found for a vanishing event horizon radius. As expected, the mass of these solutions as defined in the usual way, diverges. However, a finite mass can be assigned by using a suitable version of the boundary counterterm regularization method.[85] (We would of course expect finite energy solutions in this case too, had the F^4 YM term been included.)

- In AdS spacetime the topology of the horizon of a black hole solution is no longer restricted to be *spherical*. It can be *planar*, or, *hyperbolic* instead. A surprising result reported in[83] is that, for $\Lambda < 0$, there are $d > 4$

asymptotically AdS, finite mass black hole solutions with a *planar* topology of the event horizon, even in a theory without higher derivative terms in the YM curvature. This contrasts with the corresponding black holes with a *spherical* topology of the event horizon, which have infinite mass.[21] The case of a *hyperbolic* topology of the horizon has not been considered yet in the literature for $d > 4$. As in the previous example, the (consistent) Ansatz used for the YM field does reduce the PDE's to a system of one dimensional ODE's in terms of a radial variable, but likewise does not describe a spherically symmetric field configuration.

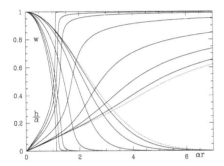 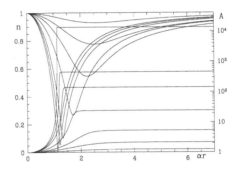

Fig. 10.5. Fundamental solutions corresponding to several values of the coupling constant $\alpha = \eta(\tau_p/\kappa_p)^{1/2p}$ are shown for $p = 4$ higher dimensional gravitating YMH monopoles in;[24] the dotted curves are for the monopole in flat space.

In addition to the higher dimensional gravitating YM fields described above, there has been some work also studying gravitating monopoles in higher dimensions. Recently a very particular hierarchy of selfgravitating YMH models in $4p$ dimensions was studied in detail in.[24] This family of models, whose flat space monopoles were constructed in,[48] is the most direct generalisation of the $d = 3 + 1$ Georgi-Glashow model (in the BPS limit), and the only one for which the Bogomol'nyi inequalities can be saturated. Both regular and black hole solutions have been constructed in,[24] which exhibit all the generic properties of the well known $d = 4$ gravitating monopoles (the profiles of typical solutions are exhibited in Figure 10.5). In higher dimensions, there are many other types of monopoles, *e.g.*, that in $d = 4 + 1$ in,[50] and that in $d = 3 + 1$ in.[89] The selfgravitating versions of these are not studied to date.

Finally, let us mention the case of $d > 4$ non-Abelian solutions with codimensions[i]. As mentioned already, the situation of $d = 5$ with one codimension is the only case discussed in a systematic way in the literature, mainly for a metric ansatz

---

[i]A detailed review of these solutions is presented in.[90,91]

which is spherically symmetric in a four dimensional perspective

$$ds^2 = e^{-2\phi(r)/\sqrt{3}} \left(\frac{dr^2}{N(r)} + r^2 d\Omega_{(2)}^2 - N(r)\sigma^2(r)dt^2 \right) + e^{4\phi(r)/\sqrt{3}}(dx^5)^2, \quad (3.3)$$

x^5 being the extra-direction and $\phi(r)$ corresponding to a dilaton field.

- The KK theory possesses in this case a variety of interesting non-Abelian configurations, including axially symmetric generalizations.[14,15,92,93] After performing a KK reduction, they correspond to $d = 4$ particle like and black hole solutions in a Einstein-Yang-Mills-Higgs-U(1)-dilaton theory.[94]
- EYM black strings and vortices with a cosmological constant were discussed in ref.[95]
- The inclusion of higher order terms of the YM curvature is optional for $d = 5$ black strings and vortices, since they possess a finite mass per unit length of the extra dimension already in the usual F^2 theory. In this case, the higher derivative terms do not affect the basic properties of the solutions.

3.2. *EYM solutions with Euclidean signature*

While this review concerns primarily fully gravitating EYM and EYMH solutions in Lorentzian signature, it is reasonable to allude to EYM solutions in Euclidean signature, especially since these were the first such solutions that appeared in the literature. Such solutions, including those on (Euclidean) Schwarzschild and de Sitter[29-31] and Taub-NUT[96] backgrounds have been studied long ago. More recently there have been further investigations[97-99] of YM fields on fixed backgrounds.

All of the known EYM fields with $A_0 \neq 0$ and Euclidean signature in the literature are given on fixed gravitational backgrounds. None with gravity backreacting on the YM fields is known, which puts these solutions on a different footing to those with Lorenzian signature discussed above.

Starting with backgrounds which are analytic continuations of relevant solutions with Lorenzian signatures, we note that

- The exact solutions of Charap and Duff,[29,30] as also their higher dimensional counterparts,[34] are by construction given on gravitational backgrounds for which the $2p-$form Riemann curvature is double–self-dual. Such metrics satisfy the hierarchy of vacuum Einstein equations (with or without cosmological constant) so that by construction, these are EYM solutions on backgrounds of fixed curvature.
- In an effort to go away from fixed backgrounds, a direct numerical method was employed in,[100] using a (Euclidean) Schwarzschild metric and static YM fields in $d = 4$. But being a static field configuration, the 'electric' YM potential A_0 assumed the role of a Higgs field and the resulting solutions turned out to be self-dual 'deformed Prasad-Sommerfield monopoles', again on a fixed background. These solutions are different from those in,[29,30]

as shown *e.g.* by a computation of their action. The higher dimensional analogues of this type of solutions in $d = 4p$ were constructed in.[34]

- In a further development beyond,[100] a number of other static spherically symmetric $d = 4$ metric backgrounds were employed in[101] to construct Euclidean non-Abelian solutions. All these resulted in selfdual YM solutions, on the basis of which it was conjectured that for any $d = 4$ (Euclidean) static spherically symmetric metric, the solutions satisfy the $d = 4$ Yang-Mills self-duality equations. An analytic proof of this conjecture has been given in,[34] where the $4p$−dimensional analogues of these were also constructed, satisfying the self-duality equations (2.11).

  The solutions described in the above three items generalise both the Charap-Duff[29,30] and the solutions in[101] in $d = 4$, to $d = 4p$, the properties of the four dimensional case being generic.

- Another property of the known gravitating instantons which is related to the fact they are given on fixed backgrounds is, that they are always (Euclidean) time independent. Even when an explicit time dependence is built into the YM Ansatz, it turns out that the solutions are either time independent,[102] or in the presence of a cosmological constant, that the Pontryagin charge of the instanton is noninteger.[103] We believe that this is a result of having used a static metric. Relaxing this last property may be interesting but promises to lead to a nontrivial numerical problem.

The metric Ansatz (2.17) in all above described Euclidean EYM fields, makes a distinction between the (Euclidean) time and the space coordinates. A different type of solutions were found in another setting, the metric Ansatz being spherically symmetric in d dimensions[34,104]

$$ds^2 = d\rho^2 + f^2(\rho)d\Omega^2_{(d-1)}, \tag{3.4}$$

where $f(\rho)$ is a function fixed by the gravity-matter field equations, ρ being the radial coordinate, $\rho = \sqrt{|x_\mu|^2}$ and $\hat{x}_\mu = x_\mu/\rho$ is the unit radius vector. The YM ansatz compatible with the symmetries of the above line element is

$$A_\mu = \left(\frac{1-w(\rho)}{\rho}\right)\Sigma^{(\pm)}_{\mu\nu}\hat{x}_\nu, \tag{3.5}$$

where the spin matrices are precisely those used in (2.19), (2.20).

The resulting reduced one dimensional YM Lagrangian for the p-th term in the YM hierarchy read

$$L^{(p,d)}_{\mathrm{YM}} = \frac{\tau_p}{2 \cdot (2p)!} \frac{(d-1)!}{(d-2p)!} f^{d-4p+1}(w^2-1)^{2p-2}\left(w'^2 + \frac{d-2p}{2p}\frac{(w^2-1)^2}{f^2}\right) \tag{3.6}$$

For any choice of the metric function $f(\rho)$, the solution of the YM self-duality equation (2.11) in $d = 4p$ dimensions reads

$$w(\rho) = \frac{1 - c_0 e^{\mp 2\int \frac{d\rho}{f(\rho)}}}{1 + c_0 e^{\mp 2\int \frac{d\rho}{f(\rho)}}}, \tag{3.7}$$

where c_0 is an arbitrary positive constant. As discussed in,[34] the action of these non-Abelian solutions is finite for any value of p. For $f(\rho) = \rho$ one recovers the $d = 4p$ generalisation of the BPST instanton first found in,[36] with $w = (\rho^2 - c)/(\rho^2 + c)$. An AdS background $f(\rho) = \rho_0 \sinh \rho/\rho_0$ leads to a $d = 4p$ generalisation of the $d = 4$ AdS selfdual instantons in,[104] with $w = (\tanh^2(\rho/2\rho_0) - c)/(\tanh^2(\rho/2\rho_0) + c)$. The $d = 4p$ selfdual instantons on a sphere (Euclideanised dS space) are found by taking $\rho_0 \to i\rho_0$ in the corresponding AdS relations.

As selfdual solutions, these are also fixed background YM fields. Moreover, on curved backgrounds the selfduality equation saturating the inequality (2.9) can be solved. Thus in Ref.[34] systems consisting of the superposition of two members of the YM hierarchy, say those labeled by p and q, with $d = 2(p + q)$ were considered as well. Solutions of such selfduality equations were also discussed in.[105,106]

4. Summary and outlook

We have reviewed a number of results on Einstein–Yang-Mills (EYM) solutions, with special emphasis on the new features that arise for spacetime dimensions $d > 4$, mainly in Lorentzian signature. The EYM solutions that we have considered are those for fully backreacting gravity with matter. It turns out that these solutions are all constructed numerically and no relevant closed form solutions are known. As such, the question of imposition of symmetries with the aim of reducing the dimensionality of the Euler–Lagrange equations becomes a very important feature of these investigations, at least as important as in the case of solutions that can be expressed in closed form.

Higher dimensional EYM fields with Euclidean signature are also mentioned, but only in passing since they are not on the same footing as their Minkowskian counterparts. They are exclusively YM fields on fixed backgrounds, in all dimensions.

A salient feature of higher dimensional non-Abelian solutions is that in all $d = D + 1$ dimensions, for $D \geq 4$ the usual EYM system cannot support asymptotically flat, finite energy solutions. This is because of the inappropriate scaling properties of that system and is remedied by the addition of higher order curvature terms. The higher order terms in question are exclusively ones that are constructed with higher order YM curvature forms, which in our nomenclature are the p−YM members of the YM hierarchy, the 1−YM being the usual YM system. The resulting YM equations contain no higher derivatives of the gauge potential than second. The higher order gravitational curvature systems (*e.g.*, Gauss-Bonnet and higher) are not possible to exploit for this purpose. Higher order gravities have nonetheless been employed in some contexts, not out of necessity, but for emphasising qualitative features of certain EYM and EYM-Higgs (EYMH) solutions, which get magnified if dimensions of the constituent terms in the system are suitably matched (see e.g.[23]).

The necessity of employing higher curvature members of the YM hierarchy to enable finite energy holds both for asymptotically flat EYM, as well as in the presence of a cosmological constant. There is however an exception to this rule, namely in the case of asymptotically AdS EYM black hole solutions with a Ricci flat horizon geometry. In that case, which is of particular interest for applications to AdS/CFT, it turns out that the usual EYM system can support finite energy solutions in all dimensions.[83] There, the nonexistence proof for solutions with nonvanishing 'electric' YM potential does not hold.

Concerning the physical justification of employing higher order YM and Riemann curvature terms, one notes that these occur in the low energy effective action of string theory.[16,17] Indeed there is some controversy on the precise structure of such terms, especially in the YM case,[107–109] but we do not take account of these considerations here. From the point of view of applications for higher dimensional EYM, these can be used to extend the results of[110–112] to higher dimensions. Some of these authors, and,[113,114] employ only solutions in closed form, but further developments would necessitate the fully gravitating solutions which are evaluated only numerically. In any case, our focus here was exclusively on the existence and the generic properties of EYM solutions in higher dimensions, rather than their physical applications.

Nearly all gravitating non-Abelian solutions in $d = D + 1$ dimensions with $D \geq 4$ reviewed in this work are static and spherically symmetric. This contrasts with the situation in $3 + 1$ spacetime dimensions where axial symmetry in 3 space dimensions is really *azimuthal* symmetry and the non-Abelian field configurations are encoded with a vortex (winding) number. The latter turns out to be an essential tool in the numerical constructions.[25,26] In higher than four spacetime dimensions however, axial symmetry implies spherical symmetry in one dimension lower than D and there is no winding number associated with the axially symmetric fields. However, the removal of the gauge arbitrariness turns out to be a much harder problem in this case, presenting a technical obstacle. Thus, the only non-spherically symmetric EYM solutions studied to date are ones in $4 + 1$ spacetime with bi-azimuthal symmetry, when there are two vortex numbers encoding the symmetries of the field configurations.[28]

Looking further ahead we should note that in recent years it has become clear that as the dimension d increases, the phase structure of the (non-spherically symmetric) solutions of the Einstein equations becomes increasingly intricate and diverse, already in the vacuum case (see *e.g.* the recent work[115]). It is very likely that, given the interplay in this case between internal group symmetries and the spacetime symmetries, the extension of known such vacuum solutions to a non-Abelian matter content would lead to a variety of new unexpected configurations. This is a promising but technically difficult direction for the future.

Finally, we mention the higher dimensional Euclidean EYM fields, which is of marginal interest here since none of the known such solutions are genuinely selfgravitating, but rather YM fields on fixed backgrounds. In all dimensions, including four, these turn out to be selfdual YM fields, and as such are restricted to even dimensions only. Another aspect of this restriction turns out to be that these YM fields appear to be (Euclidean-) time independent and are not YM instantons at all. It would be interesting to construct non-selfdual Euclidean EYM fields and inquire whether these, if they exist, describe genuine time dependent instantons.

5. Acknowledgments

This material was presented at the Heraeus School in Bremen, September 2008. We thank the organisers, Jutta Kunz and Claus Laemmerzahl for giving us this opportunity. We have greatly benefited from discussions and collaboration with Peter Breitenlohner, Yves Brihaye, Amithabha Chakrabarti, Betti Hartmann, Jutta Kunz, Burkhard Kleihaus, Dieter Maison, Yasha Shnir, Michael Volkov and Yisong Yang. This work is carried out in the framework of Science Foundation Ireland (SFI) project RFP07-330PHY. The work of ER was supported by a fellowship from the Alexander von Humboldt Foundation.

References

1. R. Bartnik and J. Mckinnon, Phys. Rev. Lett. **61** (1988) 141.
2. M. S. Volkov and D. V. Gal'tsov, Phys. Rept. **319** (1999) 1 [arXiv:hep-th/9810070].
3. H. Luckock, *Black hole skyrmions*, in H.J. De Vega and N. Sanches (editors), *"String Theory, Quantum Cosmology and Quantum Gravity, Integrable and Conformal Integrable Theories"*, page 455. World Scientific, 1987.
4. M. S. Volkov and D. V. Galtsov, JETP Lett. **50** (1989) 346 [Pisma Zh. Eksp. Teor. Fiz. **50** (1989) 312].
5. H. P. Kuenzle and A. K. M. Masood- ul- Alam, J. Math. Phys. **31** (1990) 928.
6. P. Bizon, Phys. Rev. Lett. **64** (1990) 2844.
7. P. Breitenlohner, P. Forgacs and D. Maison, Nucl. Phys. B **383** (1992) 357.
8. P. Breitenlohner, P. Forgacs and D. Maison, Nucl. Phys. B **442** (1995) 126 [arXiv:gr-qc/9412039].
9. K. M. Lee, V. P. Nair and E. J. Weinberg, Phys. Rev. D **45** (1992) 2751 [arXiv:hep-th/9112008].
10. E. Winstanley, Class. Quant. Grav. **16** (1999) 1963 [arXiv:gr-qc/9812064].
11. J. Bjoraker and Y. Hosotani, Phys. Rev. Lett. **84** (2000) 1853 [arXiv:gr-qc/9906091].
12. J. Bjoraker and Y. Hosotani, Phys. Rev. D **62** (2000) 043513 [arXiv:hep-th/0002098].
13. M. S. Volkov, N. Straumann, G. V. Lavrelashvili, M. Heusler and O. Brodbeck, Phys. Rev. D **54** (1996) 7243 [arXiv:hep-th/9605089].
14. M. S. Volkov, Phys. Lett. B **524** (2002) 369 [arXiv:hep-th/0103038].
15. N. Okuyama and K. i. Maeda, Phys. Rev. D **67** (2003) 104012 [arXiv:gr-qc/0212022].
16. M.B. Green, J.H. Schwarz and E. Witten, *Superstring Theory*, Cambridge University Press, Cambridge, 1987.
17. J. Polchinski, *TASI lectures on D-branes*, hep-th/9611050.
18. Y. Brihaye, A. Chakrabarti and D. H. Tchrakian, Class. Quant. Grav. **20** (2003) 2765 [arXiv:hep-th/0202141].
19. P. Breitenlohner, D. Maison and D. H. Tchrakian, Class. Quant. Grav. **22** (2005) 5201 [arXiv:gr-qc/0508027].

20. Y. Brihaye, A. Chakrabarti, B. Hartmann and D. H. Tchrakian, Phys. Lett. B **561** (2003) 161

21. E. Radu and D. H. Tchrakian, Phys. Rev. D **73** (2006) 024006 [arXiv:gr-qc/0508033].

22. Y. Brihaye, E. Radu and D. H. Tchrakian, Phys. Rev. D **75** (2007) 024022 [arXiv:gr-qc/0610087].

23. E. Radu, C. Stelea and D. H. Tchrakian, Phys. Rev. D **73** (2006) 084015 [arXiv:gr-qc/0601098].

24. P. Breitenlohner and D. H. Tchrakian, "Gravitating BPS Monopoles in all d=4p Spacetime Dimensions," Class. Quant. Grav. **26** (2009) 145008 [arXiv:0903.3505 [gr-qc]].

25. B. Kleihaus and J. Kunz, Phys. Rev. Lett. **78** (1997) 2527 [arXiv:hep-th/9612101].

26. B. Kleihaus and J. Kunz, Phys. Rev. Lett. **79** (1997) 1595 [arXiv:gr-qc/9704060].

27. B. Hartmann, B. Kleihaus and J. Kunz, Phys. Rev. Lett. **86** (2001) 1422 [arXiv:hep-th/0009195].

28. E. Radu, Y. Shnir and D. H. Tchrakian, Phys. Lett. B **657** (2007) 246 [arXiv:0705.3608 [hep-th]].

29. J. M. Charap and M. J. Duff, Phys. Lett. B **71** (1977) 219.

30. J. M. Charap and M. J. Duff, Phys. Lett. B **69** (1977) 445.

31. H. Boutaleb-Joutei, A. Chakrabarti and A. Comtet, Phys. Rev. D **20** (1979) 1884.

32. A. Chakrabarti, Fortsch. Phys. **35** (1987) 1.

33. G. M. O'Brien and D. H. Tchrakian, J. Math. Phys. **29** (1988) 1212.

34. E. Radu, D. H. Tchrakian and Y. Yang, Phys. Rev. D **77** (2008) 044017 [arXiv:0707.1270 [hep-th]].

35. A. A. Belavin, A. M. Polyakov, A. S. Shvarts and Y. S. Tyupkin, Phys. Lett. B **59** (1975) 85.

36. D. H. Tchrakian, Phys. Lett. B **150** (1985) 360.

37. E. Witten, Phys. Rev. Lett. **38** (1977) 121.

38. A. Chakrabarti, T. N. Sherry and D. H. Tchrakian, Phys. Lett. B **162** (1985) 340.

39. J. Spruck, D. H. Tchrakian and Y. Yang, Commun. Math. Phys. **188** (1997) 737.

40. D. H. Tchrakian and A. Chakrabarti, J. Math. Phys. **32** (1991) 2532.

41. D. O'Se and D. H. Tchrakian, Lett. Math. Phys. **13** (1987) 211.

42. Z. Ma and D. H. Tchrakian, J. Math. Phys. **31** (1990) 1506.

43. A. S. Schwarz, Commun. Math. Phys. **56** (1977) 79.

44. V. N. Romanov, A. S. Schwarz and Yu. S. Tyupkin, Nucl. Phys. B **130** (1977) 209.

45. A. S. Schwarz and Yu. S. Tyupkin, Nucl. Phys. B **187** (1981) 321.

46. Z. Q. Ma, G. M. O'Brien and D. H. Tchrakian, Phys. Rev. D **33** (1986) 1177.

47. Z. Q. Ma and D. H. Tchrakian, Phys. Rev. D **38** (1988) 3827.

48. E. Radu and D. H. Tchrakian, Phys. Rev. D **71** (2005) 125013 [arXiv:hep-th/0502025].

49. B. Julia and A. Zee, Phys. Rev. D **11** (1975) 2227.

50. G. M. O'Brien and D. H. Tchrakian, Mod. Phys. Lett. A **4** (1989) 1389.

51. R. Ibadov, B. Kleihaus, J. Kunz and U. Neemann, Phys. Lett. B **659** (2008) 421 [arXiv:0709.0285 [gr-qc]].

52. D. V. Galtsov and A. A. Ershov, Phys. Lett. A **138** (1989) 160.

53. P. Bizon and O. T. Popp, Class. Quant. Grav. **9** (1992) 193.

54. P. Kanti, N. E. Mavromatos, J. Rizos, K. Tamvakis and E. Winstanley, Phys. Rev. D **54** (1996) 5049 [arXiv:hep-th/9511071].

55. E. Winstanley, Lect. Notes Phys. **769** (2009) 49 [arXiv:0801.0527 [gr-qc]].

56. N. Straumann and Z. H. Zhou, Phys. Lett. B **243** (1990) 33.

57. P. Bizon, Phys. Lett. B **259** (1991) 53.

58. P. Breitenlohner, D. Maison and G. Lavrelashvili, Class. Quant. Grav. **21** (2004) 1667 [arXiv:gr-qc/0307029].

59. J. J. Van der Bij and E. Radu, Phys. Lett. B **536** (2002) 107 [arXiv:gr-qc/0107065].

60. S. S. Gubser, Phys. Rev. Lett. **101** (2008) 191601 [arXiv:0803.3483 [hep-th]].

61. P. Breitenlohner, P. Forgacs and D. Maison, Commun. Math. Phys. **261** (2006) 569 [arXiv:gr-qc/0412067].

62. Y. Brihaye, B. Hartmann, E. Radu and C. Stelea, Nucl. Phys. B **763** (2007) 115 [arXiv:gr-qc/0607078].

63. T. Torii, K. i. Maeda and T. Tachizawa, Phys. Rev. D **52** (1995) 4272 [arXiv:gr-qc/9506018].

64. Y. Brihaye, B. Hartmann and E. Radu, Phys. Rev. Lett. **96** (2006) 071101 [arXiv:hep-th/0508247].

65. W. Israel, Commun. Math. Phys. **8** (1968) 245.

66. R. Ibadov, B. Kleihaus, J. Kunz and Y. Shnir, Phys. Lett. B **609** (2005) 150 [arXiv:gr-qc/0410091].

67. B. Kleihaus, J. Kunz and U. Neemann, Phys. Lett. B **623** (2005) 171 [arXiv:gr-qc/0507047].

68. R. Ibadov, B. Kleihaus, J. Kunz and M. Wirschins, Phys. Lett. B **627** (2005) 180 [arXiv:gr-qc/0507110].

69. B. Kleihaus, J. Kunz, F. Navarro-Lerida and U. Neemann, Gen. Rel. Grav. **40** (2008) 1279 [arXiv:0705.1511 [gr-qc]].

70. E. Radu, Phys. Rev. D **65** (2002) 044005 [arXiv:gr-qc/0109015].

71. E. Radu and E. Winstanley, Phys. Rev. D **70** (2004) 084023 [arXiv:hep-th/0407248].

72. R. B. Mann, E. Radu and D. H. Tchrakian, Phys. Rev. D **74** (2006) 064015 [arXiv:hep-th/0606004].

73. B. Kleihaus and J. Kunz, Phys. Rev. Lett. **86** (2001) 3704 [arXiv:gr-qc/0012081].

74. B. Kleihaus, J. Kunz and F. Navarro-Lerida, Phys. Rev. D **66** (2002) 104001 [arXiv:gr-qc/0207042].

75. J. J. Van der Bij and E. Radu, Int. J. Mod. Phys. A **17** (2002) 1477 [arXiv:gr-qc/0111046].

76. O. Brodbeck, M. Heusler, N. Straumann and M. S. Volkov, Phys. Rev. Lett. **79** (1997) 4310 [arXiv:gr-qc/9707057].

77. E. Radu, Phys. Lett. B **548** (2002) 224 [arXiv:gr-qc/0210074].

78. E. Radu, Phys. Rev. D **67** (2003) 084030 [arXiv:hep-th/0211120].

79. Y. Brihaye and E. Radu, Phys. Lett. B **615** (2005) 1 [arXiv:gr-qc/0502053].

80. E. T. Newman, L.Tamburino and T. Unti, J. Math. Phys. **4** (1963) 915;
C. W. Misner, J. Math. Phys. **4** (1963) 924;
C. W. Misner and A. H. Taub, Sov. Phys. JETP **28** (1969) 122.

81. D. R. Brill, Phys. Rev. **133** (1964) B845.

82. T. Hertog and K. Maeda, JHEP **0407** (2004) 051 [arXiv:hep-th/0404261];
T. Hertog and K. Maeda, Phys. Rev. D **71** (2005) 024001 [arXiv:hep-th/0409314];
M. Henneaux, C. Martinez, R. Troncoso and J. Zanelli, Phys. Rev. D **70** (2004) 044034 [arXiv:hep-th/0404236];
E. Radu and D. H. Tchrakian, Class. Quant. Grav. **22** (2005) 879 [arXiv:hep-th/0410154];
J. T. Liu and W. A. Sabra, Phys. Rev. D **72** (2005) 064021 [arXiv:hep-th/0405171].

83. R. Manvelyan, E. Radu and D. H. Tchrakian, arXiv:0812.3531 [hep-th].

84. Y. Brihaye, E. Radu and D. H. Tchrakian, Phys. Rev. D **76** (2007) 105005 [arXiv:0707.0552 [hep-th]].

85. V. Balasubramanian and P. Kraus, Commun. Math. Phys. **208** (1999) 413 [arXiv:hep-th/9902121].

86. C. N. Yang, J. Math. Phys. **19** (1978) 320.

87. T. Tchrakian, Phys. Atom. Nucl. **71** (2008) 1116.

88. G. W. Gibbons and P. K. Townsend, Class. Quant. Grav. **23** (2006) 4873 [arXiv:hep-th/0604024];
P. Diaz and A. Segui, Phys. Rev. D **76** (2007) 064033 [arXiv:0704.0366 [gr-qc]];
A. Belhaj, P. Diaz and A. Segui, arXiv:0906.0489 [hep-th];
S. H. Mazharimousavi and M. Halilsoy, Phys. Lett. B **659** (2008) 471 [arXiv:0801.1554 [gr-qc]].

89. B. Kleihaus, D. O'Keeffe and D. H. Tchrakian, Nucl. Phys. B **536** (1998) 381 [arXiv:hep-th/9806088].

90. B. Hartmann, arXiv:0811.4076 [gr-qc]. "Non-abelian black holes and black strings in higher dimensions," AIP Conf. Proc. **1122** (2009) 137 [arXiv:0811.4076 [gr-qc]].

91. M. S. Volkov, *Gravitating non-Abelian solitons and hairy black holes in higher dimensions,* in *"Berlin 2006 Marcel Grossmann Meeting on General Relativity"*, 1379-1396 [arXiv:hep-th/0612219]. THE ELEVENTH MARCEL GROSSMANN MEETING On Recent Developments in Theoretical and Experimental General Relativity, Gravitation and Relativistic Field Theories (Proceedings of the MG11 Meeting on General Relativity Berlin, Germany, 23 - 29 July 2006; edited by Hagen Kleinert and Robert T Jantzen), World Scientific, Singapore (2007).

92. Y. Brihaye and E. Radu, Phys. Lett. B **605** (2005) 190 [arXiv:hep-th/0409065].

93. B. Hartmann, Phys. Lett. B **602** (2004) 231 [arXiv:hep-th/0409006];
Y. Brihaye, B. Hartmann and E. Radu, Phys. Rev. D **72** (2005) 104008 [arXiv:hep-th/0508028].

94. Y. Brihaye, B. Hartmann and E. Radu, Phys. Rev. D **71** (2005) 085002 [arXiv:hep-th/0502131].

95. Y. Brihaye and E. Radu, Phys. Lett. B **658** (2008) 164 [arXiv:0706.4378 [hep-th]].

96. C. N. Pope and A. L. Yuille, Phys. Lett. B **78** (1978) 424.

97. M. Bianchi, F. Fucito, G. Rossi and M. Martellini, Nucl. Phys. B **473** (1996) 367 [arXiv:hep-th/9601162].

98. G. Etesi and T. Hausel, Commun. Math. Phys. **235** (2003) 275 [arXiv:hep-th/0207196].

99. S. A. Cherkis, arXiv:0902.4724 [hep-th].

100. Y. Brihaye and E. Radu, Phys. Lett. B **636** (2006) 212 [arXiv:gr-qc/0602069].

101. Y. Brihaye and E. Radu, Europhys. Lett. **75** (2006) 730 [arXiv:hep-th/0605111].

102. B. Tekin, Phys. Rev. D **65** (2002) 084035 [arXiv:hep-th/0201050].

103. O. Sarioglu and B. Tekin, Phys. Rev. D **79** (2009) 104024 [arXiv:0903.3803 [hep-th]].

104. J. M. Maldacena and L. Maoz, JHEP **0402** (2004) 053 [arXiv:hep-th/0401024].

105. H. Kihara and M. Nitta, Phys. Rev. D **77** (2008) 047702 [arXiv:hep-th/0703166].

106. H. Kihara and M. Nitta, Phys. Rev. D **76** (2007) 085001 [arXiv:0704.0505 [hep-th]].

107. A. A. Tseytlin, *Born–Infeld action, suersymmetry and string theory,* in *"Yuri Golfand memorial volume"*, ed. M. Shifman, World Scientific, 2000.

108. E. Bergshoeff, M. de Roo and A. Sevrin, **49** (2001) 433-440; *ibid.* (2001) 50-55.

109. M. Cederwall, B. Nilsson and D. Tsimpis, JHEP 0106 (2001) 034.

110. J. P. Gauntlett, J. A. Harvey and J. T. Liu, Nucl. Phys. B **409** (1993) 363 [arXiv:hep-th/9211056].

111. E. A. Bergshoeff, G. W. Gibbons and P. K. Townsend, Phys. Rev. Lett. **97** (2006) 231601 [arXiv:hep-th/0607193].

112. G. W. Gibbons, D. Kastor, L. A. J. London, P. K. Townsend and J. H. Traschen, Nucl. Phys. B **416** (1994) 850 [arXiv:hep-th/9310118].
113. R. Minasian, S. L. Shatashvili and P. Vanhove, Nucl. Phys. B **613** (2001) 87 [arXiv:hep-th/0106096].
114. J. Polchinski, JHEP **0609** (2006) 082 [arXiv:hep-th/0510033].
115. R. Emparan, T. Harmark, V. Niarchos and N. A. Obers, Phys. Rev. Lett. **102** (2009) 191301 [arXiv:0902.0427 [hep-th]].

Perspectives in Mathematical Sciences
Interdisciplinary Mathematical Sciences, Volume 9, 2009
pp. 241–252

Chapter 11

Hamiltonian Constraint and Mandelstam Identities over Extended Knot Families $\{\psi_i\}_2^2$, $\{\psi_i\}_3^4$ and $\{\psi_i\}_4^6$ in Extended Loop Gravity

Dan Shao [1], Liang Shao [2] and Changgui Shao [3]

[1] *Institute of Light and Electronic Information Jianghan University, Wuhan 430056 China,*

[2] *Department of Applied physics, Wuhan University of Science and Technology, Wuhan 430081 China,*

[3] *Institute of Theoretical Physics, Hubei University of Education, Wuhan 430205 China*

The actions of the Hamiltonian constraint onto the members of the extended knot families $\{\psi_i\}_2^2$, $\{\psi_i\}_3^4$ and $\{\psi_i\}_4^6$, and the check of their invariances under the Mandelstam identities are given in the extended loop representation of loop quantum gravity.

We know that, different from general relativistic quantum field,[1] extended loops offer a new avenue in the investigation of Ashtekar's new variable reformulation of general relativity.[2] The extended loop representation can be viewed as an alternative approach to develop the calculation of canonical quantization of general relativity with the extended knot invariants.[3,4] In recent research on quantum general relativity, a systematic method to obtain the extended knot families as the solutions of diffeomorphism constraints has been developed.[5] This method has particular significance for the search for gravitational quantum states. The situation of the extended knot families in view of satisfying the Hamiltonian constraint and the Mandelstam identities are important for considering suitable candidates of the quantum states. In this paper the calculation concerning the actions of the Hamiltonian constraint and the Mandelstam identities over the members of the extended knot families $\{\psi_i\}_2^2$, $\{\psi_i\}_3^4$ and $\{\psi_i\}_4^6$ is given.

1. Extended knot families

Extended loops are generalization of ordinary loops to allow various new insights.[6] The elements of the extended loop group are given by infinite strings of multivector density fields of the form $X = (X, X^{\mu_1}, \cdots, X^{\mu_1 \cdots \mu_R}, \cdots)$, where X is a real number, and a Greek index represents a paired vector index a_i and space point x_i, $\mu_i \equiv a_i x_i$. The number of paired indices defines the rank of the multivector field. The combination

$$R^{\mu_1\cdots\mu_r} := \frac{1}{2}[X^{\mu_1\cdots\mu_r} + (-1)^r X^{\mu_r\cdots\mu_1}]$$

satisfies the following symmetry property under the inversion of the indices:

$$R^{\mu_1\cdots\mu_r} = (-1)^r R^{\mu_r\cdots\mu_1}. \tag{1.1}$$

The one-point R and the two-point R are given as follows

$$[R^{(ax)}]^{\mu_1\cdots\mu_r} \equiv R^{(ax)\mu_1\cdots\mu_r} := R^{(ax\mu_1\cdots\mu_r)_c},$$

$$[R^{(ax,bx)}]^{\mu_1\cdots\mu_r} \equiv R^{(ax,bx)\mu_1\cdots\mu_r} := \sum_{r=0}^{n} (-1)^{r-k} R^{(ax\mu_1\cdots\mu_k bx\mu_n\cdots\mu_{k+1})_c}.$$

The extended loop wavefunctions are linear in the multivector fields and they are written in general as

$$\psi(X) \equiv \psi(R) = \psi_\mu^i R^\mu = \sum_{r=0}^{\infty} \psi_{\mu_1\cdots\mu_r} R^{\mu_1\cdots\mu_r}, \tag{1.2}$$

where the propagators $\psi_{\mu_1\cdots\mu_r}$ should satisfy a set of symmetry properties, that is, the Mandelstam identities.

The extended knot families are given by sets $\{\psi_i\}_n^N$ of linear extended loop wavefunctions with the same maximum rank N and minimum rank n. We shall evaluate the operations of the Hamiltonian constraint and the Mandelstam identities over the families $\{\psi_i\}_2^2 = \{\varphi_G\}$, $\{\psi_i\}_3^4 = \{J_2, \psi_0\}$, and $\{\psi_i\}_4^6 = \{\psi_1, \psi_2, \psi_3, \psi_4\}$; where

$$\psi_0 = \frac{1}{2}(*\varphi_G)^2 - J_2, \quad \psi_1 = J_3 - J_2 * \varphi_G + \frac{1}{3!}(*\varphi_G)^3,$$

$$\psi_2 = \varphi_G * J_2 - J_3, \quad \psi_3 = J_3, \quad \psi_4 = \frac{1}{3!}(*\varphi_G)^3 - J_3,$$

and φ_G is the extended Gauss invariant, J_2 and J_3 coincide with the second and the third coefficients of a certain expansion of the Jones polynomial, the "product *" is the *-product of diffeomorphism invariants.

With definitions and some basic calculation the analytic expressions of φ_G, J_2 and ψ_0 are given as follows:

$$\varphi_G = g_{\mu_1\mu_2} R^{\mu_1\mu_2}, \tag{1.3}$$

$$J_2 = g_{\mu_1\mu_3} g_{\mu_2\mu_4} R^{\mu_1\cdots\mu_4} + h_{\mu_1\mu_2\mu_3} R^{\mu_1\mu_2\mu_3},$$

$$\psi_0 = C_{\mu_1\cdots\mu_4} R^{\mu_1\cdots\mu_4} - h_{\mu_1\mu_2\mu_3} R^{\mu_1\mu_2\mu_3}, C^1_{\mu_1\cdots\mu_4} = g_{\mu_1\mu_2} g_{\mu_3\mu_4} + g_{\mu_1\mu_4} g_{\mu_2\mu_3},$$

where $g_{..}$ and $h_{...}$ are respectively two and three point propagators of the Chern-Simons theory, that is,

$$g_{\mu_1\mu_2} := \varepsilon_{a_1a_2c}\phi^{cx_1}_{x_2},$$

$$h_{\mu_1\mu_2\mu_3} = \varepsilon^{\alpha_1\alpha_2\alpha_3} g_{\mu_1\alpha_1} g_{\mu_2\alpha_2} g_{\mu_3\alpha_3}, \tag{1.4}$$

with

$$\phi^{cx_1}_{x_2} = -\frac{1}{4\pi}\frac{(x_1 - x_2)^c}{|x_1 - x_2|^3} = -\frac{\partial^c}{\nabla^2}\delta(x_1 - x_2),$$

$$\varepsilon^{\alpha_1\alpha_2\alpha_3} = \varepsilon^{c_1c_2c_3}\delta(x_1 - x_2)\delta(x_1 - x_3).$$

For the diffeomorphism invariants $\psi_1, \cdots, \psi_4$, one has following analytic expressions:[5]

$$\psi_1 = (C^2_{\mu_1\cdots\mu_6} + C^3_{\mu_1\cdots\mu_6})R^{\mu_1\cdots\mu_6} - C^1_{\mu_1\cdots\mu_5}R^{\mu_1\cdots\mu_5} + C^2_{\mu_1\cdots\mu_4}R^{\mu_1\cdots\mu_4}, \tag{1.5}$$

$$\psi_2 = (C^1_{\mu_1\cdots\mu_6} + C^4_{\mu_1\cdots\mu_6} + C^5_{\mu_1\cdots\mu_6})R^{\mu_1\cdots\mu_6} + C^1_{\mu_1\cdots\mu_5}R^{\mu_1\cdots\mu_5} - C^2_{\mu_1\cdots\mu_4}R^{\mu_1\cdots\mu_4}, \tag{1.6}$$

$$\psi_3 = J_3 = (2C^1_{\mu_1\cdots\mu_6} + C^4_{\mu_1\cdots\mu_6})R^{\mu_1\cdots\mu_6} + C^2_{\mu_1\cdots\mu_5}R^{\mu_1\cdots\mu_5} + C^2_{\mu_1\cdots\mu_4}R^{\mu_1\cdots\mu_4}, \tag{1.7}$$

$$\psi_4 = (-C^1_{\mu_1\cdots\mu_6} + C^2_{\mu_1\cdots\mu_6} + C^3_{\mu_1\cdots\mu_6} + C^5_{\mu_1\cdots\mu_6})R^{\mu_1\cdots\mu_6}$$

$$-C^2_{\mu_1\cdots\mu_5}R^{\mu_1\cdots\mu_5} - C^2_{\mu_1\cdots\mu_4}R^{\mu_1\cdots\mu_4}, \tag{1.8}$$

here

$$C^2_{\mu_1\cdots\mu_4} = h_{\mu_1\mu_2\alpha}g^{\alpha\beta}h_{\beta\mu_3\mu_4} - h_{\mu_1\mu_4\alpha}g^{\alpha\beta}h_{\beta\mu_2\mu_3}, \tag{1.9}$$

$$C^1_{\mu_1\cdots\mu_5} = g_{(\mu_1\mu_2}h_{\mu_3\mu_4\mu_5)_c}, \tag{1.10}$$

$$C^2_{\mu_1\cdots\mu_5} = g_{(\mu_1\mu_3}h_{\mu_2\mu_4\mu_5)_c},$$

$$C^1_{\mu_1\cdots\mu_6} = g_{\mu_1\mu_4}g_{\mu_2\mu_5}g_{\mu_3\mu_6},$$

$$C^2_{\mu_1\cdots\mu_6} = g_{\mu_1\mu_2}g_{\mu_3\mu_4}g_{\mu_5\mu_6} + g_{\mu_1\mu_6}g_{\mu_2\mu_3}g_{\mu_4\mu_5},$$

$$C^3_{\mu_1\cdots\mu_6} = g_{\mu_1\mu_2}g_{\mu_3\mu_6}g_{\mu_4\mu_5} + g_{\mu_1\mu_4}g_{\mu_2\mu_3}g_{\mu_5\mu_6} + g_{\mu_1\mu_6}g_{\mu_2\mu_5}g_{\mu_3\mu_4},$$

$$C^4_{\mu_1\cdots\mu_6} = g_{\mu_1\mu_3}g_{\mu_2\mu_5}g_{\mu_4\mu_6} + g_{\mu_1\mu_4}g_{\mu_2\mu_6}g_{\mu_3\mu_5} + g_{\mu_1\mu_5}g_{\mu_2\mu_4}g_{\mu_3\mu_6},$$

$$C^5_{\mu_1\cdots\mu_6} = g_{\mu_1\mu_2}g_{\mu_3\mu_5}g_{\mu_3\mu_6} + g_{\mu_1\mu_3}(g_{\mu_2\mu_6}g_{\mu_4\mu_5} + g_{\mu_2\mu_4}g_{\mu_5\mu_6})$$

$$+g_{\mu_1\mu_6}g_{\mu_2\mu_4}g_{\mu_3\mu_5} + g_{\mu_1\mu_5}(g_{\mu_2\mu_3}g_{\mu_4\mu_6} + g_{\mu_2\mu_6}g_{\mu_3\mu_4}).$$

2. Actions of Hamiltonian H_0 over families $\{\psi_i\}_2^2$ and $\{\psi_i\}_3^4$

The action of the vacuum Hamiltonian constraint H_0 onto extended loop wavefunctions $\psi(R)$is given by the following expression:

$$\frac{1}{2}H_0(x)\psi(R) = \sum_{r=0}^{\infty} \psi_{\mu_1\cdots\mu_r}[F_{ab}^{\mu_1}(x)R^{(ax,bx)\mu_2\cdots\mu_r} + F_{ab}^{\mu_1\mu_2}(x)R^{(ax,bx)\mu_3\cdots\mu_r}], \quad (2.1)$$

where

$$F_{ab}^{\mu_1}(x) = \delta_{ab}^{a_1 d}\partial_d(x_1 - x),$$

$$F_{ab}^{\mu_1\mu_2}(x) = \delta_{ab}^{a_1 a_2}\delta(x_1 - x)\delta(x_2 - x).$$

The following expressions are useful to the evaluation of the Hamiltonian constraint:

$$F_{ab}^{\overset{*}{\mu_1}}(x)g_{\mu_1\mu_2} = -\overset{*}{\varepsilon}_{abc}\delta_{T\mu_2}^{cx} = -\varepsilon_{abc}[\delta_{a_2}^c\delta(x - x_2) - \phi_{x_2,a_2}^{cx}], \quad (2.2)$$

$$F_{ab}^{\mu_1}(x)h_{\mu_1\mu_2\mu_3} = -\varepsilon_{abc}\delta_{Td_k}^{cx} \in^{\alpha_k\alpha_l\alpha_m} g_{\mu_2\alpha_l}g_{\mu_3\alpha_m}$$

$$= -g_{\mu_2[ax}g_{bx]\mu_3} + \varepsilon_{abc}\phi_z^{cx}\delta_{T[\mu_2}^{dz}g_{\mu_3]}dz, \quad (2.3)$$

$$F_{ab}^{\mu_1\mu_2}(x)g_{\mu_1\mu_2} = 0,$$

$$F_{ab}^{\mu_1\mu_2}(x)h_{\mu_1\mu_2\mu_3} = 2h_{axbx\mu_3}$$

$$F_{ab}^{\mu_1\mu_2}(x)g_{\mu_1\mu_3}g_{\mu_2\mu_4} = g_{\mu_3[ax}g_{bx]\mu_4}.$$

We shall calculate the actions of the vacuum Hamiltonian H_0 over the extended knot families $\{\psi_i\}_2^2$ and $\{\psi_i\}_3^4$ respectively.

2.1. *Family* $\{\psi_i\}_2^2$

Introducing the expression (1.3) into expression (2.1), we get that the action of the vacuum Hamiltonian H_0 onto the extended Gauss invariant φ_G is given as follows:

$$\frac{1}{2}H_0\varphi_G = g_{\mu_1\mu_2}[F_{ab}^{\mu_1}(x)R^{(ax,bx)\mu_2} + F_{ab}^{\mu_1\mu_2}(x)R^{(ax,bx)}].$$

Because the second term in the above expression vanishes, so one has

$$\frac{1}{2}H_0\varphi_G = F_{ab}^{\mu_1}(x)g_{\mu_1\mu_2}R^{(ax,bx)\mu_2}, \quad (2.4)$$

putting the expression of the "rank-one part" of curvature $F_{ab}^{\mu_1}(x)$into expression (2.4), one has

$$\frac{1}{2}H_0\varphi_G = \varepsilon_{abc}(-R^{(ax,bx)cx} + \phi_{x_2,a_2}^{cx}R^{(ax,bx)\mu_2})$$

$$= -\varepsilon_{abc}(R^{(ax,bx)cx} + \phi_{x_2}^{cx}\partial_{\mu_2}R^{(ax,bx)\mu_2}) = -\varepsilon_{abc}R^{(ax,bx)cx}.$$

This result means that the extended Gauss invariant φ_G is not a Hamiltonian invariant, that is, the family $\{\psi_i\}_2^2 = \{\varphi_G\}$ shall not vanish under the action of the vacuum Hamiltonian constraint.

2.2. Family $\{\psi_i\}_3^4$

The family $\{\psi_i\}_3^4$ has two members J_2 and ψ_0, so we shall compute them respectively. J_2 is a Hamiltonian invariant state, The action of Hamiltonian H_0 onto J_2 is given by[4]

$$\frac{1}{2}H_0 J_2 = h_{\mu_1\mu_2\mu_3}[F_{ab}^{\mu_1}(x)R^{(ax,bx)\mu_2\mu_3} + F_{ab}^{\mu_1\mu_2}(x)R^{(ax,bx)\mu_3}] + g_{\mu_1\mu_3}g_{\mu_2\mu_4}$$

$$[F_{ab}^{\mu_1}(x)R^{(ax,bx)\mu_2\mu_3\mu_4} + F_{ab}^{\mu_1\mu_2}(x)R^{(ax,bx)\mu_3\mu_4}].$$

One can in above expression compute the actions of $F_{ab}^{\mu_1}(x)$ and $F_{ab}^{\mu_1\mu_2}(x)$ over the propagators $g_{..}$ and $h_{...}$, then obtain

$$\frac{1}{2}H_0 J_2 = -\varepsilon_{abc}g_{\mu_1\mu_2}R^{(ax,bx)\mu_1 cx\mu_2} + (2h_{axbx\mu_1} - \varepsilon^{def}g_{axbx}g_{\mu_1 dz}g_{exfz})R^{(ax,bx)\mu_1}.$$

In the above expression, the quantity in brackets vanishes identically. The first term is also zero, because the contribution of the rank five term vanishes due to symmetry consideration. Thus one has

$$\frac{1}{2}H_0 J_2 = 0.$$

For the diffeomorphism invariant ψ_0, we have the action of H_0 on it as

$$\frac{1}{2}H_0 \psi_0 = C_{\mu_1\cdots\mu_4}^1 [F_{ab}^{\mu_1}(x)R^{(ax,bx)\mu_2\mu_3\mu_4} + F_{ab}^{\mu_1\mu_2}(x)R^{(ax,bx)\mu_3\mu_4}]$$

$$-h_{\mu_1\mu_2\mu_3}[F_{ab}^{\mu_1}(x)R^{(ax,bx)\mu_2\mu_3} + F_{ab}^{\mu_1\mu_2}(x)R^{(ax,bx)\mu_3}]. \tag{2.5}$$

The computation of the first tern in the expression (2.5) is as follows:

$$F_{ab}^{\mu_1}(x)(g_{\mu_1\mu_2}g_{\mu_3\mu_4} + g_{\mu_1\mu_4}g_{\mu_2\mu_3})R^{(ax,bx)\mu_2\mu_3\mu_4}$$

$$= -\varepsilon_{abc}[g_{\mu_3\mu_4}(X^{(ax,bx)cx\mu_3\mu_4} + \phi_{x_2}^{cx}\partial_{\mu_2}R^{(ax,bx)\mu_2\mu_3\mu_4})$$

$$+g_{\mu_2\mu_3}(R^{(ax,bx)\mu_2\mu_3 cx} + \phi_{x_3}^{cx}\partial_{\mu_4}R^{(ax,bx)\mu_2\mu_3\mu_4})] = -2\varepsilon_{abc}g_{\mu_1\mu_2}R^{(ax,bx)cx\mu_1\mu_2}. \tag{2.6}$$

The second term in the expression (2.5) becomes

$$F_{ab}^{\mu_1\mu_2}(x)(g_{\mu_1\mu_2}g_{\mu_3\mu_4} + g_{\mu_1\mu_4}g_{\mu_2\mu_3})R^{(ax,bx)\mu_3\mu_4}$$

$$= F_{ab}^{\mu_1\mu_2}(x)g_{\mu_1\mu_4}g_{\mu_2\mu_3}R^{(ax,bx)\mu_3\mu_4} = g_{\mu_4[ax}g_{bx]\mu_3}R^{(ax,bx)\mu_3\mu_4}. \tag{2.7}$$

The third term has the following form:

$$F_{ab}^{\mu_1}(x)h_{\mu_1\mu_2\mu_3}R^{(ax,bx)\mu_2\mu_3} = g_{\mu_2[ax}g_{bx]\mu_3}R^{(ax,bx)\mu_2\mu_3} - \varepsilon_{abc}\phi_z^{cx}\{g_{[\mu_3 dz]}R^{(ax,bx)dz\mu_3}$$

$$+[\phi_{x_2}^{dz}g_{\mu_3 dz}(\delta(x_2 - x) - \delta(x_2 - x_3)R^{(ax,bx)\mu_3} + 3 \leftrightarrow 2]\}$$

$$= g_{\mu_2[ax}g_{bx]\mu_3}R^{(ax,bx)\mu_2\mu_3} - \varepsilon_{abc}g_{\mu_1 dz}\phi_z^{cx}\phi_x^{dz}R^{(ax,bx)\mu_1}. \tag{2.8}$$

The fourth one in (2.5) becomes the following form:

$$F_{ab}^{\mu_1}(x)h_{\mu_1\mu_2\mu_3}R^{(ax,bx)\mu_2\mu_3} = -h_{axbx\mu_3}R^{(ax,bx)\mu_1}. \tag{2.9}$$

Taking summation of the expressions (2.6)-(2.9), one has

$$\frac{1}{2}H_0\psi_0 = -2\varepsilon_{abc}g_{\mu_1\mu_2}R^{(ax,bx)cx\mu_1\mu_2}. \tag{2.10}$$

Using the expression

$$R^{(ax,bx)cx\mu_1\mu_2} = 3R^{(axcxbx\mu_2\mu_1)_c} - R^{(axcx\mu_1bx\mu_2)_c},$$

the expression (2.10) becomes

$$\frac{1}{2}H_0\psi_0 = -2\varepsilon_{abc}g_{\mu_1\mu_2}(3R^{(axcxbx\mu_2\mu_1)_c} - R^{(axcx\mu_1bx\mu_2)_c}). \tag{2.11}$$

Thus we conclude that the member J_2 of the family $\{\psi_i\}_3^4$ is annihilated by the vacuum Hamiltonian constraint, however the member ψ_0 is not. The result (2.11) is same as that obtained by computing $\frac{1}{2}H_0\{*\varphi_G\}^2$ and $-H_0J_2$ Separately.

3. Action of Hamiltonian H_0 over family $\{\psi_i\}_4^6$

The four diffeomorphism invariants $\psi_1, \cdots, \psi_4$ given by the expressions (1.5)-(1.8) are the members of the family $\{\psi_i\}_4^6$, we shall evaluate the actions of the Hamiltonian constraint H_0 over them through the calculations of different rank contributions.

3.1. *Contribution of the rank four term $C_{\mu_1\cdots\mu_4}^2 R^{(ax,bx)\mu_1\cdots\mu_4}$*

Using the expression (2.1) and the analytic expression (1.9), the action of the Hamiltonian H_0 on the term of rank four of ψ_1 is[7]

$$\frac{1}{2}H_0(C_{\mu_1\cdots\mu_4}^2 R^{(ax,bx)\mu_1\cdots\mu_4}) = 2[g_{\mu_1[bx}h_{ax]\mu_2\mu_3} + \varepsilon_{abc}\phi_{x_1}^{cx}h_{\mu_1\mu_2\mu_3} + \varepsilon_{abc}\phi_z^{cx}$$

$$(\phi_{x_2}^{dz} - \phi_{x_3}^{dz})g_{\mu_1 dz}g_{\mu_2\mu_3}]R^{(ax,bx)\mu_1\mu_2\mu_3}. \tag{3.1}$$

3.2. *Contribution of the rank five term $C_{\mu_1\cdots\mu_5}^1 R^{\mu_1\cdots\mu_5}$*

Introducing (1.10) into the expression (2.1), we have

$$\frac{1}{2}H_0(C_{\mu_1\cdots\mu_4}^1 R^{\mu_1\cdots\mu_4}) = C_{\mu_1\cdots\mu_5}^1[F_{ab}^{\mu_1}(x)R^{(ax,bx)\mu_2\cdots\mu_5} + F_{ab}^{\mu_1\mu_2}(x)R^{(ax,bx)\mu_3\mu_4\mu_5}]. \tag{3.2}$$

Using the symmetry property (1.1) of the two-point R, and the expressions (2.2) and (2.3), the first term in expression (3.2) becomes

$$C_{\mu_1\cdots\mu_5}^1 F_{ab}^{\mu_1}(x)R^{(ax,bx)\mu_2\cdots\mu_5} = F_{ab}^{\mu_1}(x)(g_{(\mu_1\mu_2}h_{\mu_3\mu_4\mu_5)_c})R^{(ax,bx)\mu_2\cdots\mu_5}$$

$$= F_{ab}^{\mu_1}(x)(2g_{\mu_1\mu_2}h_{\mu_3\mu_4\mu_5} + 2g_{\mu_2\mu_3}h_{\mu_1\mu_4\mu_5} + g_{\mu_3\mu_4}h_{\mu_1\mu_2\mu_5})R^{(ax,bx)\mu_2\cdots\mu_5}$$

$$= (-2\varepsilon_{abc}\delta^{cx}_{T\mu_2}h_{\mu_3\mu_4\mu_5} - 2g_{\mu_2\mu_3}g_{\mu_4[ax}g_{bx]\mu_5} + 2g_{\mu_2\mu_3}\varepsilon_{abc}$$

$$\phi^{cx}_z\delta^{dz}_{T[\mu_4}g_{\mu_5]dz} - g_{\mu_3\mu_4}g_{\mu_2[ax}g_{bx]\mu_5} + \varepsilon_{abc}\phi^{cx}_z g_{\mu_3\mu_4}\delta^{dz}_{T[\mu_2}g_{\mu_5]dz})R^{(ax,bx)\mu_2\cdots\mu_5}.$$

Expending operator δ_T and integrating by parts, above result becomes

$$2\varepsilon_{abc}[\phi^{cx}_{x_1}h_{\mu_1\mu_2\mu_3} + \phi^{cx}_z(\phi^{dz}_{x_2} - \phi^{dz}_{x_3})g_{\mu_2\mu_3}g_{\mu_1 dz} + 2\phi^{cx}_z\phi^{dz}_x g_{\mu_2\mu_3}g_{\mu_1 dz}]$$

$$R^{(ax,bx)\mu_1\mu_2\mu_3} + 2[g_{\mu_1\mu_2}g_{\mu_4[ax}g_{bx]\mu_3} + \varepsilon_{abc}(\phi^{cx}_{x_1} - \phi^{cx}_{x_2})g_{\mu_1\mu_2}g_{\mu_3\mu_4}$$

$$+g_{\mu_2\mu_3}g_{\mu_4[ax}g_{bx]\mu_1} + \varepsilon_{abc}\phi^{cx}_{x_1}g_{\mu_2\mu_3}g_{\mu_1\mu_4}]R^{(ax,bx)\mu_1\cdots\mu_4}$$

$$-2\varepsilon_{abc}h_{\mu_1\mu_2\mu_3}R^{(ax,bx)cx\mu_1\mu_2\mu_3}. \tag{3.3}$$

For the second term in the expression (3.2), using (1.4) one has

$$C^1_{\mu_1\cdots\mu_5}F^{\mu_1\mu_2}_{ab}(x)R^{(ax,bx)\mu_3\mu_4\mu_5}$$

$$= F^{\mu_1\mu_2}_{ab}(x)(g_{\mu_1\mu_2}h_{\mu_3\mu_4\mu_5} + g_{\mu_2\mu_3}h_{\mu_1\mu_4\mu_5} + g_{\mu_3\mu_4}h_{\mu_1\mu_2\mu_5}h_{\mu_5\mu_1}h_{\mu_2\mu_3\mu_4})R^{(ax,bx)\mu_3\mu_4\mu_5}$$

$$= 2(-g_{\mu_1[ax}h_{bx]\mu_2\mu_3} + 2g_{\mu_2\mu_3}h_{axbx\mu_1})R^{(ax,bx)\mu_1\mu_2\mu_3}. \tag{3.4}$$

Taking summation of the expressions (3.3) and (3.4), and using the equality

$$\varepsilon_{abc}\phi^{cx}_z\phi^{dz}_x g_{\mu_1 dz} + h_{axbx\mu_1} = 0,$$

the contribution of the term $C^1_{\mu_1\cdots\mu_5}R^{\mu_1\cdots\mu_5}$ is

$$\frac{1}{2}H_0(C^1_{\mu_1\cdots\mu_5}R^{\mu_1\cdots\mu_5})$$

$$= 2[\varepsilon_{abc}\phi^{cx}_{x_1}h_{\mu_1\mu_2\mu_3} - g_{\mu_1[ax}g_{bx]\mu_2\mu_3} + \varepsilon_{abc}\phi^{cx}_z(\phi^{dz}_{x_2} - \phi^{dz}_{x_3})g_{\mu_2\mu_3}g_{\mu_1 dz}]$$

$$R^{(ax,bx)\mu_1\mu_2\mu_3} + 2[g_{\mu_1\mu_2}g_{\mu_4[ax}g_{bx]\mu_3} + \frac{1}{2}g_{\mu_2\mu_3}g_{\mu_4[ax}g_{bx]\mu_1}$$

$$+\varepsilon_{abc}(\phi^{cx}_{xx_1} - \phi^{cx}_{x_2})g_{\mu_1\mu_2}g_{\mu_3\mu_4} + \varepsilon_{abc}\phi^{cx}_{x_1}g_{\mu_1\mu_4}g_{\mu_2\mu_3}]$$

$$R^{(ax,bx)\mu_1\cdots\mu_4} - 2\varepsilon_{abc}h_{\mu_1\mu_2\mu_3}R^{(ax,bx)cx\mu_1\mu_2\mu_3}. \tag{3.5}$$

3.3. Contribution of the rank six term $(C^2_{\mu_1\cdots\mu_6} + C^3_{\mu_1\cdots\mu_6})R^{\mu_1\cdots\mu_6}$

Putting the term $(C^2_{\mu_1\cdots\mu_6} + C^3_{\mu_1\cdots\mu_6})R^{\mu_1\cdots\mu_6}$ into (2.1), one has

$$\frac{1}{2}H_0(C^2_{\mu_1\cdots\mu_6} + C^3_{\mu_1\cdots\mu_6})R^{\mu_1\cdots\mu_6}$$

$$= (C^2_{\mu_1\cdots\mu_6} + C^3_{\mu_1\cdots\mu_6})(F^{\mu_1}_{ab}(x)R^{(ax,bx)\mu_2\cdots\mu_6} + F^{\mu_1\mu_2}_{ab}(x)R^{(ax,bx)\mu_3\cdots\mu_6}). \quad (3.6)$$

In above expression, owing to the symmetry property of the two-point R and integrating by parts, the action of the rank-one part of curvature $F^{\mu_1}_{ab}(x)$ becomes

$$(C^2_{\mu_1\cdots\mu_6} + C^3_{\mu_1\cdots\mu_6})F^{\mu_1}_{ab}(x)R^{(ax,bx)\mu_2\cdots\mu_6}$$

$$= F^{\mu_1}_{ab}(x)(2g_{\mu_1\mu_2}g_{\mu_3\mu_4}g_{\mu_5\mu_6} + 2g_{\mu_1\mu_2}g_{\mu_3\mu_6}g_{\mu_4\mu_5} + g_{\mu_1\mu_4}g_{\mu_2\mu_3}g_{\mu_5\mu_6})R^{(ax,bx)\mu_2\cdots\mu_6}$$

$$= -\varepsilon_{abc}[2\delta_T{}_{\mu_2}{}^{cx}(g_{\mu_3\mu_4}g_{\mu_5\mu_6} + g_{\mu_3\mu_6}g_{\mu_4\mu_5}) + \delta_T{}^{cx}_{\mu_4}g_{\mu_2\mu_3}g_{\mu_5\mu_6}]R^{(ax,bx)\mu_2\cdots\mu_6}$$

$$= 2\varepsilon_{abc}[g_{\mu_1\mu_2}g_{\mu_3\mu_4}(\phi^{cx}_{x_1} - \phi^{cx}_{x_2}) + \phi^{cx}_{x_1}g_{\mu_1\mu_4}g_{\mu_2\mu_3}]R^{(ax,bx)\mu_1\cdots\mu_4}$$

$$-\varepsilon_{abc}g_{\mu_1\mu_2}g_{\mu_3\mu_4}R^{(ax,bx)\mu_1\mu_2cx\mu_3\mu_4} - 2\varepsilon_{abc}(g_{\mu_1\mu_2}g_{\mu_3\mu_4}$$

$$+g_{\mu_1\mu_4}g_{\mu_2\mu_3})R^{(ax,bx)cx\mu_1\cdots\mu_4}. \quad (3.7)$$

In the expression (3.6), the action of the rank-two part of curvature $F^{\mu_1\mu_2}_{ab}(x)$ is

$$(C^2_{\mu_1\cdots\mu_6} + C^3_{\mu_1\cdots\mu_6})F^{\mu_1\mu_2}_{ab}(x)R^{(ax,bx)\mu_3\cdots\mu_6}$$

$$= F^{\mu_1\mu_2}_{ab}(x)(g_{\mu_1\mu_6}g_{\mu_2\mu_3}g_{\mu_4\mu_5} + g_{\mu_1\mu_4}g_{\mu_2\mu_3}g_{\mu_5\mu_6}$$

$$+g_{\mu_1\mu_6}g_{\mu_2\mu_5}g_{\mu_3\mu_{46}})R^{(ax,bx)\mu_3\cdots\mu_6}$$

$$= (g_{\mu_2\mu_3}g_{\mu_4[ax}g_{bx]\mu_1} + 2g_{\mu_1\mu_2}g_{\mu_4[ax}g_{bx]\mu_3})R^{(ax,bx)\mu_1\cdots\mu_4}. \quad (3.8)$$

Combining the results of the expressions (3.7) and (3.8), we get the following contribution:

$$\frac{1}{2}H_0(C^2_{\mu_1\cdots\mu_6} + C^3_{\mu_1\cdots\mu_6})R^{\mu_1\cdots\mu_6}$$

$$= [2g_{\mu_1\mu_2}g_{\mu_4[ax}g_{bx]\mu_3} + g_{\mu_2\mu_3}g_{\mu_4[ax}g_{bx]\mu_1} + 2\varepsilon_{abc}(\phi^{cx}_{x_1} - \phi^{cx}_{x_2})$$

$$g_{\mu_1\mu_2}g_{\mu_3\mu_4} + 2\varepsilon_{abc}\phi^{cx}_{x_1}g_{\mu_1\mu_4}g_{\mu_2\mu_3}]R^{(ax,bx)\mu_1\cdots\mu_4} - 2\varepsilon_{abc}$$

$$(g_{\mu_1\mu_2}g_{\mu_3\mu_4} + g_{\mu_1\mu_4}g_{\mu_2\mu_3})R^{(ax,bx)cx\mu_1\cdots\mu_4} - \varepsilon_{abc}$$

$$g_{\mu_1\mu_2}g_{\mu_3\mu_4}R^{(ax,bx)\mu_1\mu_2cx\mu_3\mu_4}. \quad (3.9)$$

3.4. Contribution of the rank six term $(C^1_{\mu_1\cdots\mu_6} + C^4_{\mu_1\cdots\mu_6} + C^5_{\mu_1\cdots\mu_6})R^{\mu_1\cdots\mu_6}$

Analogously, to this contribution we have

$$\frac{1}{2}H_0(C^1_{\mu_1\cdots\mu_6} + C^4_{\mu_1\cdots\mu_6} + C^5_{\mu_1\cdots\mu_6})R^{\mu_1\cdots\mu_6}$$

$$= (C^1_{\mu_1\cdots\mu_6} + C^4_{\mu_1\cdots\mu_6} + C^5_{\mu_1\cdots\mu_6})(F^{\mu_1}_{ab}(x)R^{(ax,bx)\mu_2\cdots\mu_6}F^{\mu_1\mu_2}_{ab}(x)R^{(ax,bx)\mu_3\cdots\mu_6}).$$

In the above expression, for the action of $F^{\mu_1}_{ab}(x)$, we have

$$(C^1_{\mu_1\cdots\mu_6} + C^4_{\mu_1\cdots\mu_6} + C^5_{\mu_1\cdots\mu_6})F^{\mu_1}_{ab}(x)R^{(ax,bx)\mu_2\cdots\mu_6}$$

$$= F^{\mu_1}_{ab}(x)(2g_{\mu_1\mu_3}g_{\mu_2\mu_5}g_{\mu_4\mu_6} + 2g_{\mu_1\mu_2}g_{\mu_3\mu_5}g_{\mu_4\mu_6} + 2g_{\mu_1\mu_3}g_{\mu_2\mu_6}g_{\mu_4\mu_5}$$

$$+2g_{\mu_1\mu_3}g_{\mu_2\mu_4}g_{\mu_5\mu_6} + g_{\mu_1\mu_4}g_{\mu_2\mu_5}g_{\mu_3\mu_6} + g_{\mu_1\mu_4}g_{\mu_2\mu_6}g_{\mu_3\mu_5})R^{(ax,bx)\mu_2\cdots\mu_6}$$

$$= -2\varepsilon_{abc}[g_{\mu_1\mu_4}g_{\mu_2\mu_3}\phi^{cx}_{x_1} + (\phi^{cx}_{x_2} - \phi^{cx}_{x_2})g_{\mu_1\mu_2}g_{\mu_3\mu_4}]R^{(ax,bx)\mu_1\cdots\mu_4}$$

$$-2\varepsilon_{abc}g_{\mu_1\mu_3}g_{\mu_2\mu_4}R^{(ax,bx)cx\mu_1\cdots\mu_4} - 2\varepsilon_{abc}(g_{\mu_1\mu_2}g_{\mu_3\mu_4} + g_{\mu_1\mu_3}g_{\mu_2\mu_4}$$

$$+g_{\mu_1\mu_4}g_{\mu_2\mu_3})R^{(ax,bx)\mu_1cx\mu_2\mu_3\mu_4} - \varepsilon_{abc}(g_{\mu_1\mu_3}g_{\mu_2\mu_4}$$

$$+g_{\mu_1\mu_4}g_{\mu_2\mu_3})R^{(ax,bx)\mu_1\mu_2cx\mu_3\mu_4} \tag{3.10}$$

for the action of $F^{\mu_1\mu_2}_{ab}(x)$, we have

$$(C^1_{\mu_1\cdots\mu_6} + C^4_{\mu_1\cdots\mu_6} + C^5_{\mu_1\cdots\mu_6})F^{\mu_1\mu_2}_{ab}(x)R^{(ax,bx)\mu_3\cdots\mu_6}$$

$$= (2g_{\mu_1\mu_2}g_{\mu_3[ax}g_{bx]\mu_4} + g_{\mu_2\mu_3}g_{\mu_1[ax}g_{bx]\mu_4})R^{(ax,bx)\mu_1\cdots\mu_4}. \tag{3.11}$$

From (3.10) and (3.11), we have

$$\frac{1}{2}H_0(C^1_{\mu_1\cdots\mu_6} + C^4_{\mu_1\cdots\mu_6} + C^5_{\mu_1\cdots\mu_6})R^{\mu_1\cdots\mu_6}$$

$$= 2[g_{\mu_1\mu_2}g_{\mu_3[ax}g_{bx]\mu_4} + \frac{1}{2}g_{\mu_2\mu_3}g_{\mu_1[ax}g_{bx]\mu_4} - \varepsilon_{abc}g_{\mu_1\mu_4}g_{\mu_2\mu_3}\phi^{cx}_{x_1}$$

$$-\varepsilon_{abc}g_{\mu_1\mu_2}g_{\mu_3\mu_4}(\phi^{cx}_{x_1} - \phi^{cx}_{x_2})]R^{(ax,bx)\mu_1\cdots\mu_4} - 2\varepsilon_{abc}g_{\mu_1\mu_3}g_{\mu_2\mu_4}R^{(ax,bx)cx\mu_1\cdots\mu_4}$$

$$-2\varepsilon_{abc}(g_{\mu_1\mu_2}g_{\mu_3\mu_4} + g_{\mu_1\mu_3}g_{\mu_2\mu_4} + g_{\mu_1\mu_4}g_{\mu_2\mu_3})R^{(ax,bx)\mu_1cx\mu_2\mu_3\mu_4}$$

$$-\varepsilon_{abc}(g_{\mu_1\mu_3}g_{\mu_2\mu_4} + g_{\mu_1\mu_4}g_{\mu_2\mu_3})R^{(ax,bx)\mu_1\mu_2cx\mu_3\mu_4}. \tag{3.12}$$

3.5. Action of H_0 over the family $\{\psi_i\}_4^6$

Collecting the partial results obtained above, we can evaluate the actions of Hamiltonian constraint H_0 on the members ψ, ψ_2, ψ_3 and ψ_4 of the family $\{\psi_i\}_4^6$. From the results of (3.1), (3.5) and (3.9), we get

$$\frac{1}{2}H_0\psi_1(R) = 2\varepsilon_{abc}h_{\mu_1\mu_2\mu_3}R^{(ax,bx)cx\mu_1\mu_2\mu_3} - 2\varepsilon_{abc}(g_{\mu_1\mu_2}g_{\mu_3\mu_4} + g_{\mu_1\mu_4}g_{\mu_2\mu_3})$$

$$R^{(ax,bx)cx\mu_1\cdots\mu_4} - \varepsilon_{abc}g_{\mu_1\mu_2}g_{\mu_3\mu_4}R^{(ax,bx)\mu_1\mu_2cx\mu_3\mu_4}$$

Combining the results of (3.5) and (3.12) we get

$$\frac{1}{2}H_0\psi_2(R) = -2\varepsilon_{abc}h_{\mu_1\mu_2\mu_3}R^{(ax,bx)cx\mu_1\mu_2\mu_3} - 2\varepsilon_{abc}g_{\mu_1\mu_3}g_{\mu_2\mu_4}R^{(ax,bx)cx\mu_1\cdots\mu_4}$$

$$-2\varepsilon_{abc}(g_{\mu_1\mu_2}g_{\mu_3\mu_4} + g_{\mu_1\mu_3}g_{\mu_2\mu_4} + g_{\mu_1\mu_4}g_{\mu_2\mu_3})R^{(ax,bx)\mu_1cx\mu_2\mu_3\mu_4}$$

$$-\varepsilon_{abc}(g_{\mu_1\mu_3}g_{\mu_2\mu_4} + g_{\mu_1\mu_4}g_{\mu_2\mu_3})R^{(ax,bx)\mu_1\mu_2cx\mu_3\mu_4}.$$

For the action of H_0 on the third coefficient of the Jones Polynomial J_3, we may use the result:[8]

$$\frac{1}{2}H_0J_3(R) = \frac{1}{2}H_0\psi_3(R)$$

$$= -2\varepsilon_{abc}h_{\mu_1\mu_2\mu_3}R^{(ax,bx)\mu_1cx\mu_2\mu_3} - 2\varepsilon_{abc}g_{\mu_1\mu_3}g_{\mu_2\mu_4}R^{(ax,bx)\mu_1cx\mu_2\mu_3\mu_4}$$

$$-\varepsilon_{abc}(2g_{\mu_1\mu_3}g_{\mu_2\mu_4} + g_{\mu_1\mu_4}g_{\mu_2\mu_3})R^{(ax,bx)\mu_1\mu_2cx\mu_3\mu_4}$$

Because in the family $\{\psi_i\}_4^6$ the diffeomorphism invariants $\psi_1, \cdots, \psi_4$ have a relation:

$$\psi_1 + \psi_2 = \psi_3 + \psi_4,$$

so the action of the Hamiltonian constraint H_0 on the member ψ_4 can obtain via the following way:

$$\frac{1}{2}H_0\psi_4(R) = \frac{1}{2}H_0\psi_1(R) + \frac{1}{2}H_0\psi_2(R) - \frac{1}{2}H_0\psi_3(R)$$

$$= 2\varepsilon_{abc}h_{\mu_1\mu_2\mu_3}R^{(ax,bx)\mu_1cx\mu_2\mu_3}$$

$$-2\varepsilon_{abc}(g_{\mu_1\mu_2}g_{\mu_3\mu_4} + g_{\mu_1\mu_3}g_{\mu_2\mu_4} + g_{\mu_1\mu_4}g_{\mu_2\mu_3})R^{(ax,bx)cx\mu_1\cdots\mu_4}$$

$$-2\varepsilon_{abc}(g_{\mu_1\mu_2}g_{\mu_3\mu_4} + g_{\mu_1\mu_4}g_{\mu_2\mu_3})R^{(ax,bx)\mu_1cx\mu_2\mu_3\mu_4}$$

$$-\varepsilon_{abc}(g_{\mu_1\mu_2}g_{\mu_3\mu_4} - g_{\mu_1\mu_3}g_{\mu_2\mu_4})R^{(ax,bx)\mu_1\mu_2cx\mu_3\mu_4}.$$

4. The Mandelstam identities

In the expression (1.2), the coefficients $\psi_{\mu_1\cdots\mu_r}$ contain all the information about $\psi(R)$ and should satisfy the following Mandelstam identities:

$$\psi_\mu = \psi_{(\mu)_c}, \tag{4.1}$$

$$\psi_\mu = \psi_{\bar\mu}, \tag{4.2}$$

$$\psi_{\alpha\beta\pi} + \psi_{\alpha\beta\bar\pi} = \psi_{\beta\alpha\pi} + \psi_{\beta\alpha\bar\pi}. \tag{4.3}$$

For the members of families $\{\psi_i\}_2^2$ and $\{\psi_i\}_3^4$, the check of satisfying the symmetry identities (4.1)-(4.3) can do easily.[5] And with respect to the members $\psi_1,\cdots,\psi_4$ of family $\{\psi_i\}_4^6$, the invariances under the cyclic symmetry and the inversion symmetry requirements (4.2) and (4.3) are also evident by computing straightforward. So below we shall give a brief demonstration only concerning the behaviors of the wavefunctions $\psi_1,\cdots,\psi_4$ under the property (4.3).

We may see that the propagators constructing the wavefunctions ψ_1 and ψ_2 shall satisfy the Mandelstam identity (4.3).

For the propagator $C^2_{\mu_1\cdots\mu_4}$, if π contains a single index ($\pi = \mu_4$) and α, β are arbitrary in choosing μ_1, μ_2 and μ_3, we have a identity:

$$C^2_{\alpha\beta\mu_4} + C^2_{\alpha\beta\bar\mu_4} = C^2_{\beta\alpha\mu_4} + C^2_{\beta\alpha\bar\mu_4} = 0;$$

if π contains two indices, that is, $\pi = \mu_3\mu_4$, $\alpha = \mu_1$, $\beta = \mu_2$, we also have a identity:

$$C^2_{\alpha\beta\pi} + C^2_{\alpha\beta\bar\pi} = C^2_{\beta\alpha\pi} + C^2_{\beta\alpha\bar\pi} = h_{\mu_1\mu_4\sigma}g^{\sigma\rho}h_{\rho\mu_3\mu_2} + h_{\mu_1\mu_3\sigma}g^{\sigma\rho}h_{\rho\mu_4\mu_2}.$$

About the propagator $C^1_{\mu_1\cdots\mu_5}$, we need to consider three cases: $\pi = \mu_5$, $\pi = \mu_4\mu_5$ and $\pi = \mu_3\mu_4\mu_5$. In the three cases we shall get following identities respectively:

$$C^1_{\alpha\beta\pi} + C^1_{\alpha\beta\bar\pi} = C^1_{\beta\alpha\pi} + C^1_{\beta\alpha\bar\pi} = 0,$$

$$C^1_{\alpha\beta\pi} + C^1_{\alpha\beta\bar\pi} = C^1_{\beta\alpha\pi} + C^1_{\beta\alpha\bar\pi} = g_{(\mu_1\mu_2}h_{\mu_3\mu_4\mu_5)_c} + g_{(\mu_1\mu_2}h_{\mu_3\mu_4\mu_5)_c},$$

and

$$C^1_{\alpha\beta\pi} + C^1_{\alpha\beta\bar\pi} = C^1_{\beta\alpha\pi} + C^1_{\beta\alpha\bar\pi} = g_{(\mu_1\mu_2}h_{\mu_3\mu_4\mu_5)_c} + g_{(\mu_1\mu_2}h_{\mu_5\mu_4\mu_3)_c},$$

in above three identities the α, β are arbitrary respectively in the choosing of other indices of $C^1_{\mu_1\cdots\mu_5}$.

The propagators $\psi^1_{\mu_1\cdots\mu_6} = C^2_{\mu_1\cdots\mu_6} + C^3_{\mu_1\cdots\mu_6}$ and $\psi^2_{\mu_1\cdots\mu_6} = C^1_{\mu_1\cdots\mu_6} + C^4_{\mu_1\cdots\mu_6} + C^5_{\mu_1\cdots\mu_6}$ are structured of some basic blocks of free g's, and have six rank, in their check we need to consider that the boldface index π have to runs μ_6, $\mu_5\mu_6$, $\mu_4\mu_5\mu_6$ and $\mu_3\mu_4\mu_5\mu_6$ respectively. The computation is similar to the computations of propagators $C^2_{\mu_1\cdots\mu_4}$ and $C^1_{\mu_1\cdots\mu_5}$, and the conclusion is same as them. Thus we can conclude the check that the wavefunctions ψ_1 and ψ_2 fulfill the requirement of identity (4.3).

Concerning the propagator $C^3_{\mu_1\cdots\mu_6} = 2C^1_{\mu_1\cdots\mu_6} + C^4_{\mu_1\cdots\mu_6}$ in the wavefunction ψ_3, we shall see that it does not fulfill the Mandelstam identity (4.3). To show this we may consider a case of $\alpha = \mu_1\mu_2$, $\beta = \mu_3$ and $\pi = \mu_4\mu_5\mu_6$ as follows:

$$\psi^3_{\alpha\beta\pi} + \psi^3_{\alpha\beta\bar{\pi}} = 2C^1_{\mu_1\cdots\mu_6} + C^4_{\mu_1\cdots\mu_6} - 2C^1_{\mu_1\mu_2\mu_3\mu_6\mu_5\mu_4} - C^4_{\mu_1\mu_2\mu_3\mu_6\mu_5\mu_4}$$

$$= g_{\mu_1\mu_4}(2g_{\mu_2\mu_5}g_{\mu_3\mu_6} + g_{\mu_2\mu_6}g_{\mu_3\mu_5}) + g_{\mu_1\mu_5}(g_{\mu_2\mu_4}g_{\mu_3\mu_6}$$

$$-g_{\mu_2\mu_6}g_{\mu_3\mu_4}) - g_{\mu_1\mu_6}(2g_{\mu_2\mu_5}g_{\mu_3\mu_4} + g_{\mu_2\mu_4}g_{\mu_3\mu_5}).$$

However

$$\psi^3_{\beta\alpha\pi} + \psi^3_{\beta\alpha\bar{\pi}} = 2C^1_{\mu_3\mu_1\mu_2\mu_4\mu_5\mu_6} + C^4_{\mu_3\mu_1\mu_2\mu_4\mu_5\mu_6} - 2C^1_{\mu_3\mu_1\mu_2\mu_6\mu_5\mu_4} - C^4_{\mu_3\mu_1\mu_2\mu_6\mu_5\mu_4}$$

$$= g_{\mu_3\mu_4}(2g_{\mu_1\mu_5}g_{\mu_2\mu_6} + g_{\mu_1\mu_6}g_{\mu_2\mu_5}) + g_{\mu_3\mu_5}(g_{\mu_1\mu_4}g_{\mu_2\mu_6} - g_{\mu_1\mu_6}g_{\mu_2\mu_4})$$

$$-g_{\mu_3\mu_6}(2g_{\mu_1\mu_5}g_{\mu_2\mu_4} + g_{\mu_1\mu_4}g_{\mu_2\mu_5}),$$

so we have

$$\psi^3_{\alpha\beta\pi} + \psi^3_{\alpha\beta\bar{\pi}} \neq \psi^3_{\beta\alpha\pi} + \psi^3_{\beta\alpha\bar{\pi}}.$$

The result is that the wavefunction ψ_3 is not invariant under the property (4.3).

Because the propagator $\psi^4_{\mu_1\cdots\mu_6} = C^1_{\mu_1\cdots\mu_6} + C^2_{\mu_1\cdots\mu_6} + C^3_{\mu_1\cdots\mu_6} + C^5_{\mu_1\cdots\mu_6}$ exists in the expression of ψ_4, and it being similar to $\psi^3_{\mu_1\cdots\mu_6}$ also does not fulfill the identity (4.3), so the wavefunction ψ_4 also is not invariant under the property (4.3).

References

1. Li Guoping, Guo Youzhong, The theory on General Relativistic Quantum Field, **I**, Hubei Science and Technology Press,1980.
2. Ashtekar A (Notes prepared in collaboration with Tate R), *Lectures on Nonperturbative Canonical Gravity*, World Scientific, Singapore 1991
3. Bartolo C.Di, Gambini R, Griego J, Commun, The extended loop group: An infinite dimensional manifold associated with the loop space Math. Phys, 1993, **158**: 217-240
4. Bartolo C. Di, Gambini R, Griego J, Extended loop representation of quantum gravity, Phys. Rev, 1995, **D51**: 502-516
5. Griego J, Extended knots and the space of states of quantum gravity, Nuc. Phys. 1996, **B 473**: 291-307
6. Bartolo C. Di, Gambini R, Griego J, Pullin J, Knot polynomial states of quantum gravity in terms of loops and extended loops: Some remarks, J. Math. Phys, 1995, **36**: 6510-6528
7. Griego J, The Kauffman bracket and the Jones polynomial in quantum gravity, Nuc. Phys, 1996, **B 467**: 332-352
8. Griego J, Is the third coefficient of the Jones knot polynomial a quantum state of gravity? Phys. Rev, 1996, **D53**: 6966 -6978

Perspectives in Mathematical Sciences
Interdisciplinary Mathematical Sciences, Volume 9, 2009
pp. 253–264

Chapter 12

Lattice Boltzmann Simulation of Nonlinear Schrödinger Equation with Variable Coefficients

Baochang Shi

School of Mathematics and Statistics
Huazhong University of Science and Technology
Wuhan 430074, China
shibc@hust.edu.cn

In this paper, the lattice Boltzmann model for diffusion equation with source term is applied directly to solve nonlinear Schrödinger equation with variable coefficients (NLSEvc), a very important mathematical-physical equation which describes different models in the optical fiber systems, by using complex-valued distribution function and relaxation time. Detailed simulations of NLSEvc are carried out by using the lattice Boltzmann model. Numerical results agree well with the analytical solutions, which show that the lattice Boltzmann model is an effective numerical solver for complex nonlinear systems.

1. Introduction

The lattice Boltzmann method (LBM) is an innovative computational fluid dynamics (CFD) approach for simulating fluid flows and modeling complex physics in fluids.[1] Compared with the conventional CFD approach, the LBM is easy for programming, intrinsically parallel, and it is also easy to incorporate complicated boundary conditions such as those in porous media. The LBM also shows potentials to simulate the nonlinear systems, including reaction-diffusion equation,[2–4] convection-diffusion equation,[5,6] Burgers equation[7] and wave equation,[3,8] etc. Recently, a generic LB model for advection and anisotropic dispersion equation was proposed.[9] However, almost all of the existing LB models are used for real nonlinear systems. Beginning in the middle of 1990s, based on quantum-computing ideas, several types of quantum lattice gases have been studied to model some real/complex mathematical-physical equations, such as Dirac equation, Schrödinger equation, Burgers equation and KdV equation.[10–17] Recently, Linhao Zhong, Shide Feng, Ping Dong, et al.[18] applied the LBM to solve one-dimensional nonlinear Schrödinger equation (NLSE) using the idea of quantum lattice-gas model[13,14] for treating the reaction term. Detailed simulation results in Ref.[18] have shown that the order of accuracy of the proposed LB schemes is higher than or equal to two. In Ref.,[19] motivated by the work in Ref.,[18] the LBM for n-dimensional (nD) convection-diffusion equation (CDE) with a source term was directly applied to some nonlinear complex equations, including the NLSE, coupled NLSEs, Klein-Gordon equation and coupled Klein-Gordon-Schrödinger equations, by adopting a complex-valued distribution function and relaxation time. In Ref.,[20] we presented a LB model for a general class of nD CDEs with nonlinear convection and isotropic-diffusion terms by selecting equilibrium distribution function properly. The model can be applied to both real and complex-valued nonlinear evolutionary equations. The studies in Refs.[18–20] show that the LBM may be an effective numerical solver for real and complex-valued nonlinear systems. Therefore, it is necessary to study the LBM for nonlinear complex equations further.

In this paper, using the idea of adopting complex-valued distribution function and relaxation time,[20] the LBM for n-dimensional (nD) diffusion equation (DE) with source term is applied directly to solve nonlinear Schrödinger equation with variable coefficients (NLSEvc). Detailed simulations of NLSEvc are carried out for accuracy test. Numerical results agree well with the analytical solutions.

2. Lattice Boltzmann Model

The NLSEvc considered in this paper is as follows

$$\mathrm{i}\psi_t + \frac{1}{2}\alpha(t)\psi_{xx} + \beta(t)|\psi|^2\psi = \mathrm{i}\gamma(t)\psi, (x,t) \in R^2. \tag{2.1}$$

This equation describes different models in the optical fiber systems. Since Eq.(2.1) can be taken as a diffusion equation (DE) with source term, we now give the LBM for the following nD DE

$$\partial_t\varphi = \nabla \cdot (\alpha\nabla\varphi) + F(\mathbf{x},t), \tag{2.2}$$

where ∇ is the gradient operator with respect to the spatial coordinate $\mathbf{x}$ in n dimensions. φ is a scalar function of time t and position $\mathbf{x}$. $\alpha = \alpha(\mathbf{x},t)$ is the diffusion coefficient. $F(\mathbf{x},t)$ is the source term, which is a known function of $(\mathbf{x},t)$ or $(\varphi,\mathbf{x},t)$. Several of DEs with source term form a reaction-diffusion system.

2.1. *LB model for DE*

The LB model for Eq.(2.2) is based on the DnQb lattice[1] with b velocity directions in nD space.

The evolution equation of the distribution function in the model reads

$$f_j(\mathbf{x} + \mathbf{c}_j\Delta t, t + \Delta t) - f_j(\mathbf{x},t) = -\frac{1}{\tau}(f_j(\mathbf{x},t) - f_j^{eq}(\mathbf{x},t))$$
$$+ \Delta t F_j(\mathbf{x},t) + \frac{\Delta t^2}{2}\partial_t F_j(\mathbf{x},t), j = 0,\ldots,b-1, \tag{2.3}$$

where $\{\mathbf{c}_j, j = 0,\ldots,b-1\}$ is the set of discrete velocity directions, Δx and Δt are the lattice spacing and the time step, respectively, $c = \Delta x/\Delta t$ is the particle speed,

τ is the dimensionless relaxation time, and $f_j^{eq}(\mathbf{x}, t)$ is the equilibrium distribution function which has a simple form

$$f_j^{eq}(\mathbf{x}, t) = \omega_j \varphi \tag{2.4}$$

such that

$$\sum_j f_j = \sum_j f_j^{eq} = \varphi, \sum_j \mathbf{c}_j f_j^{eq} = \mathbf{0}, \sum_j \mathbf{c}_j \mathbf{c}_j f_j^{eq} = c_s^2 \varphi \mathbf{I}, \tag{2.5}$$

where $\mathbf{I}$ is the unit tensor, ω_j are weights and c_s, so called sound speed in the LBM for fluids, is related to c and ω_j. They depend on the lattice model used.

For the D1Q3 model, $\{\mathbf{c}_0, \mathbf{c}_1, \mathbf{c}_2\} = \{0, c, -c\}$, $\omega_0 = 2/3$, $\omega_1 = \omega_2 = 1/6$, and for the D2Q9 one, $\{\mathbf{c}_j, j = 0, \ldots, 8\} = \{(0, 0), (\pm c, 0), (0, \pm c), (\pm c, \pm c)\}$, $\omega_0 = 4/9$, $\omega_{1 \sim 4} = 1/9$, $\omega_{5 \sim 8} = 1/36$, then $c_s^2 = c^2/3$ for both of them.

F_j in Eq.(2.3), corresponding to the source term in Eq.(2.2), is taken as

$$F_j = \omega_j F \tag{2.6}$$

such that $\sum_j F_j = F, \sum_j \mathbf{c}_j F_j = \mathbf{0}$. φ and α in Eq.(2.2) satisfy

$$\varphi = \sum_j f_j, \alpha = c_s^2(\tau - \tfrac{1}{2})\Delta t. \tag{2.7}$$

From Eq.(2.7) we can see that the relaxation time τ can be a function of $(\mathbf{x}, t)$.

The macroscopic equation (2.2) can be derived through the Chapman-Enskog expansion (See Appendix for details).

2.2. *Version of LB Model for Complex DE*

For the complex evolutionary equations, let us decompose the related complex functions and relaxation time into their real and imaginary parts by writing

$$f_j = g_j + \mathrm{i} h_j, f_j^{eq} = g_j^{eq} + \mathrm{i} h_j^{eq}, F_j = G_j + \mathrm{i} H_j, w = \frac{1}{\tau} = w_1 + \mathrm{i} w_2, \tag{2.8}$$

where $\mathrm{i}^2 = -1$.

Now we obtain the implemental version of Eq.(2.3) for complex DE (2.2)

$$g_j(\mathbf{x} + \mathbf{c}_j \Delta t, t + \Delta t) - g_j(\mathbf{x}, t) = - w_1(g_j(\mathbf{x}, t) - g_j^{eq}(\mathbf{x}, t))$$
$$+ w_2(h_j(\mathbf{x}, t) - h_j^{eq}(\mathbf{x}, t)) + \Delta t G_j(\mathbf{x}, t) + \frac{\Delta t^2}{2} \partial_t G_j(\mathbf{x}, t),$$
$$h_j(\mathbf{x} + \mathbf{c}_j \Delta t, t + \Delta t) - h_j(\mathbf{x}, t) = - w_2(g_j(\mathbf{x}, t) - g_j^{eq}(\mathbf{x}, t))$$
$$- w_1(h_j(\mathbf{x}, t) - h_j^{eq}(\mathbf{x}, t)) + \Delta t H_j(\mathbf{x}, t) + \frac{\Delta t^2}{2} \partial_t H_j(\mathbf{x}, t),$$
$$j = 0, \ldots, b - 1. \tag{2.9}$$

Let $\tau = \tau_1 + \mathrm{i}\tau_2, \alpha = \alpha_1 + \mathrm{i}\alpha_2$, then we have

$$\tau_1 = \frac{\alpha_1}{c_s^2 \Delta t} + \frac{1}{2}, \tau_2 = \frac{\alpha_2}{c_s^2 \Delta t}, w_1 = \frac{\tau_1}{\tau_1^2 + \tau_2^2}, w_2 = -\frac{\tau_2}{\tau_1^2 + \tau_2^2}. \tag{2.10}$$

3. Simulation Results

Now we apply the D1Q3 LB model presented above to simulate numerically the solitary waves for Eq.(2.1) to test mainly the numerical accuracy of the model. In simulations, the initial value of distribution function is taken as that of its equilibrium part at time $t = 0$, which is a commonly used strategy. If not specified, we use the nonequilibrium extrapolation scheme proposed by Guo *et al.*[21] to treat the boundary condition except for the periodic one, and the initial and boundary conditions of the test problems with analytical solutions are determined by their analytical solutions. The explicit difference scheme $\partial_t F_j(\mathbf{x}, t) = (F_j(\mathbf{x}, t) - F_j(\mathbf{x}, t - \Delta t))/\Delta t$, is used for computing $\partial_t F_j(\mathbf{x}, t)$. The following global relative error is used to measure the accuracy:

$$E = \frac{\sum_j |\psi(\mathbf{x}_j, t) - \psi^*(\mathbf{x}_j, t)|}{\sum_j |\psi^*(\mathbf{x}_j, t)|}, \tag{3.1}$$

where ψ and ψ^* are the numerical solution and analytical one, respectively, and the summation is taken over all grid points.

Example 3.1. We show an accuracy test for the 1D NLSEvc

$$i\psi_t + \alpha(t)\psi_{xx} + \theta(t)|\psi|^2\psi = 0 \tag{3.2}$$

with initial condition

$$\psi(x, 0) = \frac{1}{\sqrt{3}} \text{sech}(\frac{x}{3}) \exp(\frac{i(x^2 - 1)}{6}), \tag{3.3}$$

where

$$\alpha(t) = \frac{1}{2}\cos(t), \theta(t) = \frac{\cos(t)}{\sin(t) + 3}. \tag{3.4}$$

The problem has a periodically solitary wave solution[22]

$$\psi_p(x, t) = P_{1p}(x, t)P_{2p}(x, t)P_{3p}(x, t), \tag{3.5}$$

where

$$\begin{aligned}
P_{1p}(x, t) &= \frac{1}{(\sin(t)+3)^{\frac{1}{2}}}, \\
P_{2p}(x, t) &= \text{sech}(\frac{x}{\sin(t)+3}), \\
P_{3p}(x, t) &= \exp(\frac{i(x^2-1)}{2(\sin(t)+3)}).
\end{aligned} \tag{3.6}$$

Example 3.2. We show an accuracy test for the 1D NLSEvc (3.2) with initial condition

$$\psi(x, 0) = \frac{1}{\sqrt{5}} \text{sech}(\frac{x}{5}) \exp(\frac{i(x^2 - 1)}{10}), \tag{3.7}$$

where

$$\alpha(t) = \frac{1}{2}(\cos(t) + \sqrt{2}\cos(\sqrt{2}t)), \theta(t) = \frac{\cos(t) + \sqrt{2}\cos(\sqrt{2}t)}{\sin(t) + \sin(\sqrt{2}t) + 5}. \tag{3.8}$$

The problem has a quasi-periodically solitary wave solution[22,23]

$$\psi_{qp}(x,t) = P_{1qp}(x,t)P_{2qp}(x,t)P_{3qp}(x,t), \tag{3.9}$$

where

$$
\begin{aligned}
P_{1qp}(x,t) &= \frac{1}{(\sin(t)+\sin(\sqrt{2}t)+5)^{\frac{1}{2}}}, \\
P_{2qp}(x,t) &= \mathrm{sech}\!\left(\frac{x}{\sin(t)+\sin(\sqrt{2}t)+5}\right), \\
P_{3qp}(x,t) &= \exp\!\left(\frac{\mathbf{i}(x^2-1)}{2(\sin(t)+\sin(\sqrt{2}t)+5)}\right).
\end{aligned}
\tag{3.10}
$$

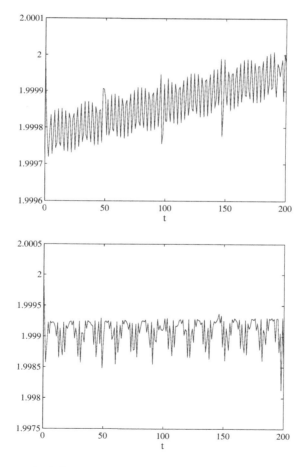

Fig. 12.1. *The numerical charge density evolves with time, the problem one (top) and the problem two (bottom).*

In simulations of Examples 3.1. and 3.2., we take $\Delta x = 0.01$, $\Delta t = 0.0001$.

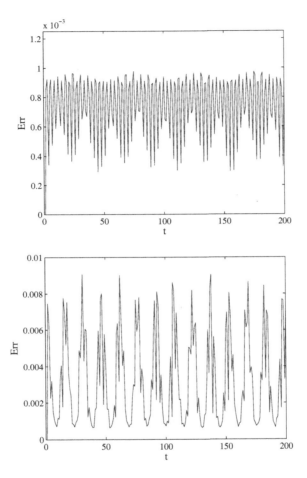

Fig. 12.2. *The global relative errors evolve with time, the problem one (top) and the problem two (bottom).*

It is well known that Eq.(3.2) has an important conversation law

$$E(\psi) = \int_R |\psi(x,t)|^2 dx = \text{constant} \tag{3.11}$$

under an appropriate boundary conditions, such as $\lim_{|x|\to\infty} \psi(x,t) = 0$. For the examples above $E(\psi) = 2$.

Fig. 12.1 shows the numerical conversation law for the two examples on $[-40, 40]$ for $t \in [0, 200]$. To test the LBGK model further, the global relative errors between their analytical solutions and numerical ones are computed on $[-40, 40]$ for $t \in [0, 200]$ and plotted in Fig. 12.2. From Fig. 12.1 and Fig. 12.2 we can find that the numerical results are in good agreement with the analytical ones. The real part and the imaginary part of the numerical solution evolving with time on $[-60, 60]$

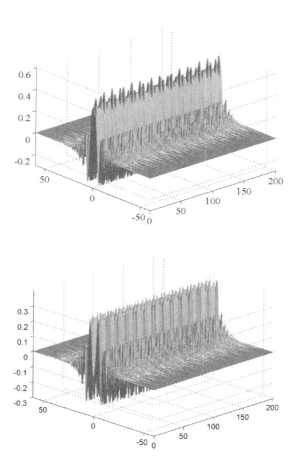

Fig. 12.3.   *The numerical solutions evolve with time, the real part (top) and the imaginary part (bottom).*

for Example 3.2. are shown in Fig. 12.3.

Example 3.3. We show an accuracy test for the 1D NLSEvc (2.1) with

$$\alpha(t) = \frac{1}{\alpha_0}\exp(\sigma t)\beta(t), \beta(t) = \beta_0 + \beta_1\cos(gt), \gamma(t) = \frac{\sigma}{2}. \qquad (3.12)$$

This problem has the chirp-less bright solitary wave solution[24]

$$\psi(x,t) = A_3\exp(\tfrac{\sigma}{2}t)\mathsf{sech}\{A_2[x - A_1\exp(\sigma t)(\tfrac{\beta_0}{\sigma} + \tfrac{\beta_1\sigma\cos(gt)+\beta_1 g\sin(gt)}{\sigma^2+g^2})] + A_4\}$$
$$\times \exp\{\mathrm{i}[A_1 x + \tfrac{1}{2}(A_2^2 - A_1^2)\exp(\sigma t)(\tfrac{\beta_0}{\sigma} + \tfrac{\beta_1\sigma\cos(gt)+\beta_1 g\sin(gt)}{\sigma^2+g^2})] + A_5\}, \qquad (3.13)$$

where α_0 is related to the initial peak power in system, β_0, β_1 and g describe Kerr nonlinearity. $\gamma(t)$ is the gain/loss function. $A_i(i = 1 - 5)$ are constants.

We set $\beta_0 = 0.1, \beta_1 = \alpha_0 = g = 1, A_1 = A_2 = A_3 = 1, A_4 = A_5 = 0$ and use the periodic boundary condition in $[-10, 15]$ as in Ref.,[24] while the initial condition is determined by the analytical solution (3.13) at $t = 0$. The intensity of numerical solutions in the temporal interval $[-10, 15]$ is plotted in Fig. 12.4 for different σ, which describes the nonlinear evolutionary behavior of the solitary wave. To test the LBM further the error evolution with time is plotted in Fig. 12.5 for $\Delta x = 10^{-2}$ and $\Delta t = 10^{-5}$. From Fig. 12.5 it can be found that the numerical results are in good agreement with the analytical ones.

4. Conclusion

In this paper the LB model for nD DE with source term has been applied directly to solve some important nonlinear complex equations by using complex-valued distribution function and relaxation time. Through the Chapman-Enskog expansion with small parameter ϵ in time and space, the nD DE can be exactly recovered to order $O(\epsilon^2)$. Unlike traditional numerical methods which solve the equations for macroscopic variables, the present model keeps the advantages of classical LB model, such as simplicity and symmetry of scheme, ease in coding, and intrinsical parallelism. As the application of the proposed LB model, simulations of three 1D nonlinear Schrödinger equation with variable coefficients are performed for accuracy test. Numerical results agree well with the analytical solutions. We found that to attain better accuracy the LB model for the test problems requires a relatively small time step Δt and $\Delta t = 10^{-4}$ is a proper choice. Since the Chapman-Enskog analysis shows that this kind of complex-valued LB model is only the direct *translation* of the classical LB model in complex-valued function, the LB model can be applied directly to other complex evolutionary equations or real ones with complex-valued solutions.

Although the preliminary work in this paper shows that the classical LB model has also potentials to simulate complex-valued nonlinear systems, some problems still need to be solved, such as how to improve the accuracy and efficiency of complex LB model.

References

1. Qian, Y. H., Succi, S., Orszag, S.: Recent advances in lattice Boltzmann computing. Annu. Rev. Comput. Phys. **3** (1995) 195–242
2. Dawson, S. P., Chen S. Y., Doolen,G. D.: Lattice Boltzmann computations for reaction-diffusion equations. J. Chem. Phys. **98** (1993) 1514–1523
3. Chopard, B., Droz, M.: Cellular automata modeling of physical systems. Cambridge University Press, Cambridge (1998)
4. Blaak, R., Sloot, P. M.: Lattice dependence of reaction-diffusion in lattice Boltzmann modeling. Comput. Phys. Comm. **129** (2000) 256–266
5. Van der Sman, R. G. M., Ernst, M. H.: Advection-diffusion lattice Boltzmann scheme for irregular lattices. J. Comput. Phys. **160**(2) (2000) 766–782

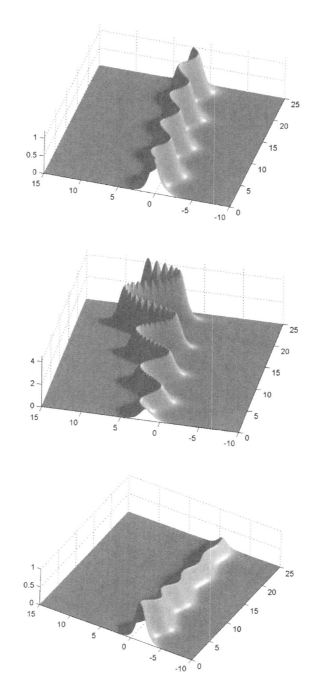

Fig. 12.4.   *The intensity evolve with time for $\sigma = 0$ (top), $\sigma = 0.06$ (middle) and $\sigma = -0.06$ (bottom).*

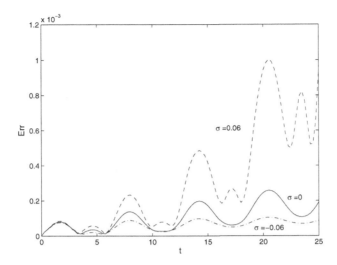

Fig. 12.5. *The global relative errors evolve with time, the problem three.*

6. Deng, B., Shi, B. C., Wang, G. C.: A new lattice Bhatnagar-Gross-Krook model for convection-diffusion equation with a source term. Chin. Phys. Lett. **22** (2005) 267–270

7. Yu, X. M., Shi, B. C.: A lattice Bhatnagar-Gross-Krook model for a class of the generalized Burgers equations. Chin. Phys. **25**(7) (2006) 1441–1449

8. Yan, G. W.: A lattice Boltzmann equation for waves. J. Comput. Phys. (2000) **161**(9) (2000) 61–69

9. Ginzburg, I.: Equilibrium-type and link-type lattice Boltzmann models for generic advection and anisotropic-dispersion equation. Advances in Water Resources. **28**(11) (2005) 1171–1195

10. Meyer, D. A.: From quantum cellular automata to quantum lattice gas. J. Stat. Phys. **85** (1996) 551–574

11. Succi, S., Benzi, R.: The lattice Boltzmann equation for quantum mechanics. Physica D **69** (1993) 327–332

12. Succi, S.: Numerical solution of the Schrödinger equation using discrete kinetic theory. Phys. Rev. E **53** (1996) 1969–1975

13. Boghosian, B. M., Taylor IV, W.: Quantum lattice gas models for the many-body Schrödinger equation. Int. J. Mod. Phys. C **8** (1997) 705–716

14. Yepez, J., Boghosian, B.: An efficient and accurate quantum lattice-gas model for the many-body Schrödinger wave equation. Comput. Phys. Commun. **146** (2002) 280–294

15. Yepez, J.: Quantum lattice-gas model for the Burgers equation. J. Stat. Phys. **107** (2002) 203–224

16. Vahala, G., Yepez, J., Vahala, L.: Quantum lattice gas representation of some classical solitons. Phys. Lett. A **310** (2003) 187–196

17. Vahala, G., Vahala, L., Yepez, J: Quantum lattice representations for vector solitons in external potentials. Physica A **362** (2006) 215–221

18. Zhong, L. H., Feng, S. D., Dong, P., et al.: Lattice Boltzmann schemes for the nonlinear Schrödinger equation. Phys. Rev. E. **74** (2006) 036704-1–9

19. Shi, B. C.: Lattice Boltzmann simulation of some nonlinear complex equations. Y. Shi et al. (Eds.): ICCS 2007, Part I, LNCS **4487** (2007) 818–825

20. Shi B. C., Guo Z. L.: Lattice Boltzmann model for nonlinear convection-diffusion equations. Phys. Rev. E **79** (2009) 016701-1–13

21. Guo, Z. L., Zheng, C. G., Shi, B. C.: Non-equilibrium extrapolation method for velocity and pressure boundary conditions in the lattice Boltzmann method. Chin. Phys. **11** (2002) 366–374

22. Hong, J. L., Liu, Y.: A Novel numerical approach to simulating nonlinear Schrödinger equations with varying coefficients. Appl. Math. Lett. **16** (2003) 759–765

23. Hong, J. L., Liu, X.-Y., Li, C.: Multi-symplectic Runge-Kutta-Nyström for nonlinear Schrödinger equations with variable coefficients. J. Comput. Phys. **226** (2007) 1968–1984

24. Zong, F.-D., Dai, C.-Q., Yang, Q., Zhang, J.-F.: Soliton solutions for variable co-efficient nonlinear Schrödinger equation for optical fiber and their application (In Chinese). Acta Physica Sinica. **55** (2006) 3805–3812

Appendix: Derivation of Macroscopic Equation

To derive the macroscopic equation (2.2), the Chapman-Enskog expansion in time and space is applied:

$$f_j = f_j^{eq} + \epsilon f_j^{(1)} + \epsilon^2 f_j^{(2)}, F = \epsilon F^{(1)}, \partial_t = \epsilon \partial_{t_1} + \epsilon^2 \partial_{t_2}, \nabla = \epsilon \nabla_1, \qquad (4.1)$$

where ϵ is a small parameter.

From Eqs. (4.1), (2.5) and (2.6), it follows that

$$\sum_j f_j^{(k)} = 0 (k \geq 1), \ \sum_j F_j^{(1)} = F^{(1)}, \ \sum_j \mathbf{c}_j F_j^{(1)} = \mathbf{0}, \qquad (4.2)$$

where $F_j^{(1)} = \omega_j F^{(1)}$.

Applying the Taylor expansion and Eq.(4.1) to Eq.(2.3), we have

$$O(\epsilon): D_{1j} f_j^{eq} = -\frac{1}{\tau \Delta t} f_j^{(1)} + F_j^{(1)}, \qquad (4.3)$$

$$O(\epsilon^2): \partial_{t_2} f_j^{eq} + D_{1j} f_j^{(1)} + \frac{\Delta t}{2} D_{1j}^2 f_j^{eq} = -\frac{1}{\tau \Delta t} f_j^{(2)} + \frac{\Delta t}{2} \partial_{t_1} F_j^{(1)}, \qquad (4.4)$$

where $D_{1j} = \partial_{t_1} + \mathbf{c}_j \cdot \nabla_1$.

Applying Eq.(4.3) to the left side of Eq.(4.4) and deleting the term $\frac{\Delta t}{2} \partial_{t_1} F_j^{(1)}$ on the both sides, we can rewrite Eq.(4.4) as

$$\partial_{t_2} f_j^{eq} + D_{1j}((1 - \frac{1}{2\tau}) f_j^{(1)}) + \frac{\Delta t}{2} \mathbf{c}_j \cdot \nabla_1 F_j^{(1)} = -\frac{1}{\tau \Delta t} f_j^{(2)}. \qquad (4.5)$$

Summing Eq.(4.3) and Eq.(4.5) over j and using Eq.(2.5) and Eq.(4.2), we have

$$\partial_{t_1} \varphi = F^{(1)}, \qquad (4.6)$$

$$\partial_{t_2} \varphi + \nabla_1 \cdot ((1 - \frac{1}{2\tau}) \sum_j \mathbf{c}_j f_j^{(1)}) = 0. \qquad (4.7)$$

Using Eqs. (4.3), (2.5) and (4.2), we have

$$\begin{aligned} \sum_j \mathbf{c}_j f_j^{(1)} &= -\tau \Delta t \sum_j \mathbf{c}_j (D_{1j} f_j^{eq} - F_j^{(1)}) \\ &= -\tau \Delta t (\nabla_1 \cdot (c_s^2 \varphi \mathbf{I})) = -\tau \Delta t c_s^2 \nabla_1 \varphi. \end{aligned} \qquad (4.8)$$

Then substituting Eq.(4.8) into Eq.(4.7), we obtain

$$\partial_{t_2}\varphi = \nabla_1 \cdot (\alpha\nabla_1\varphi), \tag{4.9}$$

where $\alpha = c_s^2(\tau - \frac{1}{2})\Delta t$. Therefore, combining Eq.(4.9) with Eq.(4.6), we have

$$\partial_t\varphi = \nabla \cdot (\alpha\nabla\varphi) + F. \tag{4.10}$$

The DE (2.2) is exactly recovered to order $O(\epsilon^2)$.

Perspectives in Mathematical Sciences
Interdisciplinary Mathematical Sciences, Volume 9, 2009
pp. 265–274

Chapter 13

Exponential Stability of Nonlocal Time-Delayed Burgers Equation

Yanbin Tang

School of Mathematics and Statistics,
Huazhong University of Science and Technology,
Wuhan, Hubei, 430074, P.R.China,
tangyb@mail.hust.edu.cn

Dedicated to Professor Youzhong Guo on the occasion of his 75th birthday

Burgers equation has been extensively studied and it has become one of the best known and most studied of all nonlinear partial differential equations. Even though Burgers equation, a one-dimensional version of Navier-Stokes, does not model any specific physical flow problem, it would be the first step to understand the turbulence exhibited in a flow.

The initial-boundary value problem of classical Burgers equation involving homogeneous Dirichlet data can be solved exactly using the Cole-Hopf transformation. Unfortunately, due to the introduction of time delay, it is very difficult to use the classical Cole-Hopf transformation to solve initial-boundary value problem of time-delay Burgers equation exactly. It means that the partial differential equation with time-delay will exhibit some quite different properties even if the small enough time-delay. In this paper, a nonlocal Burgers equation with a time delay is considered. Using the Liapunov function and uniform Gronwall inequality, the solution of the time-delayed nonlocal Burgers equation is shown to exhibit a delay-induced instability and be exponentially stable under small delays.

MSC: 35R10, 35B35, 35Q53.
Key words: Burgers equation with Time delay; Nonlocal Burgers equation; Gronwall inequality; Exponential stability.

1. Introduction

Because it describes such a wild array of physical phenomena, and lends itself to treatment by analytical methods, the Burgers equation

$$u_t = \nu u_{xx} - u u_x \qquad (1.1)$$

has been extensively studied, and it has become one of the best known and most studied of all nonlinear partial differential equations.[1,2,5,9,10] Even though Burgers equation, a one-dimensional version of Navier-Stokes, does not model any specific physical flow problem, it would be the first step to understand the turbulence exhibited in a flow.[10] The simplest derivation of Burgers equation occurs in the context of the kinematic-wave theory of traffic flow, especially, when the effects of vehicular diffusion are taken into account.[4,7,18] Assuming that there are no source/sink terms, it is not difficult to show the traffic density $\rho(x,t)$ satisfies the Burgers equation[8]

$$\rho_t = \nu\rho_{xx} - v_m(1 - \frac{2\rho}{\rho_s})\rho_x, \tag{1.2}$$

where the positive constants ν, v_m, ρ_s denote the diffusion coefficient, the maximum speed as $\rho \to 0$, the saturation value of the density, respectively. It has been proved that Burgers equation without delay is globally exponential stable at least in the norm of space H^1. The general theory of time delayed partial differential equations was described by C. Travis and G. Webb.[12–14] Some examples exhibit a delay-induced instability.[3,11,16,17,19] Jordan[8] considered the density transport equation of traffic flow

$$\rho_t(x, t + \lambda_0) = \nu\rho_{xx}(x,t) - v_m[1 - \frac{2\rho(x,t)}{\rho_s}]\rho_x(x,t), \tag{1.3}$$

a generalization of Eq.(1.1) that exhibits time delay, where the positive constant λ_0 denotes the reaction (or delay) time. Jordan[8] wanted to show that the initial-boundary value problem (IBVP) of delay Burgers equation (1.2) involving homogeneous Dirichlet data can be solved exactly using the Cole-Hopf transformation, like that of the classical case, and then to show that the delay induces the stability. Unfortunately, due to the introduction of delay, the classical Cole-Hopf transformation is not right to solve IBVP of Eq.(1.3) exactly, the formula (2.5) in Jordan[8] is wrong.

Does a small delay always destabilize Burgers equations? The answer is NO. Weijiu Liu[10] considered an initial-boundary value problem of the time-delayed Burgers equation

$$\begin{aligned} &u_t - \nu u_{xx} + u(t - \tau, x)u_x(t, x) = 0, \\ &u(t, 0) = u(t, 1) = 0, t > 0, \\ &u(s, x) = \varphi(s, x), x \in [0, 1], s \in [-\tau, 0], \end{aligned} \tag{1.4}$$

where $\nu > 0$ is the viscosity and $\tau > 0$ is the delay parameter, $\varphi(s, x)$ is an initial state in an appropriate function space. By the Liapunov function and uniform Gronwall inequality, Weijiu Liu[10] described the global existence and exponential stability of the solution of IBVP(1.4) if the delay parameter τ is sufficiently small, and gave an explicit estimate of the delay parameter in terms of the viscosity and initial data.

Theorem 1.1.[10]   *For any initial condition* $\varphi(t,x) \in C([-\tau,0], H_0^1(0,1))$, *IBVP(1.4) has a unique global mild solution* $u \in C([-\tau,\infty), H_0^1(0,1))$.

Theorem 1.2.[10]   *For any initial condition* $\varphi(t,x) \in C([-\tau,0], H_0^1(0,1))$, *there exists* $\tau_0 = \tau_0(\nu,\varphi)$, *for* $\tau < \tau_0$, *the solution of IBVP(1.4) satisfies*

$$\|u_x(t)\| \le \frac{k}{2}e^{-\omega t/2}, \forall t \ge 0,$$

where k is a constant depending only on ν, τ and φ.

Using a fixed point theorem and a comparison principle, Y. Tang and M. Wang[15] showed that the solution of IBVP(1.4) of time-delayed Burgers equation is exponentially stable under small delays. The result is more explicit, but also complements, the result given by Weijiu Liu.[10]

I.D. Chueshov[6] considered an initial-boundary value problem of the nonlocal Burgers equation

$$u_t - \nu u_{xx} + (\omega, u)u_x(t,x) = f(t,x), 0 < x < 1, t > 0$$
$$u(t,0) = u(t,1) = 0, t > 0, \tag{1.5}$$
$$u(0,x) = \varphi(x), x \in [0,1],$$

here $f(t,x)$ is continuous with the values in $L^2(0,1)$, that is, $f(t,x) \in C(R^+, L^2(0,1))$, and $\omega(x) \in L^2(0,1)$,

$$(\omega, u) = \int_0^1 \omega(x)u(t,x)dx,$$

$\varphi(0) = \varphi(1) = 0$, and described the global existence and unique of weak solution of IBVP(1.5) in the space $H_0^1(0,1)$.

Here we are interested in the global existence and exponential stability of the solution of the nonlocal time-delayed Burgers equation, we focus on the initial-boundary value problem of a nonlocal time-delayed Burgers equation

$$u_t - \nu u_{xx} + (\omega(x), u(t-\tau,x))u_x(t,x) = 0, 0 < x < 1, t > 0$$
$$u(t,0) = u(t,1) = 0, t > 0, \tag{1.6}$$
$$u(s,x) = \varphi(s,x), x \in [0,1], s \in [-\tau,0],$$

where $\varphi(0,0) = \varphi(0,1) = 0$. Using the Liapunov function and uniform Gronwall inequality, we show that the IBVP(1.6) is exponentially stable under small delays.

This paper is organized as follows. In Section 1.2 we prove that global existence of the solution. In Section 1.3 we prove the exponential stability of the solution.

2. Global existence for continuous initial data

In this section we study the IBVP(1.6) with initial data $\varphi \in C([-\tau,0], H_0^1(0,1))$. Define the linear operator A by

$$Aw = \nu w_{xx}, \quad D(A) = H^2(0,1)\bigcap H_0^1(0,1). \tag{2.1}$$

It's well known that A generates an analytic semigroup $T(t)$ on $L^2(0,1)$ which is compact for $t > 0$, and

$$||T(t)|| \le ke^{-\alpha t}, \quad \alpha > 0, k \ge 1.$$

We also define the nonlinear operator $F : C([-\tau,0], H_0^1(0,1)) \to L^2(0,1)$ by

$$F(\psi) = -(\omega, \psi(-\tau))\psi_x(0), \quad \forall \psi \in C([-\tau,0], H_0^1(0,1)),$$

then F is locally Lipschitz.

In fact, $\forall \varphi, \psi \in C([-\tau,0], H_0^1(0,1))$, we have

$$||F(\varphi) - F(\psi)||_{L^2} = ||(\omega, \psi(-\tau))\psi_x(0) - (\omega, \varphi(-\tau))\varphi_x(0)||_{L^2}$$
$$= ||(\omega, \psi(-\tau) - \varphi(-\tau))\psi_x(0) - (\omega, \varphi(-\tau))(\varphi_x(0) - \psi_x(0))||_{L^2}$$
$$\le |(\omega, \varphi(-\tau))|||\varphi_x(0) - \psi_x(0)||_{L^2} + |(\omega, \psi(-\tau) - \varphi(-\tau))|||\psi_x(0)||_{L^2}$$
$$\le ||\omega||_{L^2}||\varphi(-\tau)||_{L^2}||\varphi_x(0) - \psi_x(0)||_{L^2}$$
$$+ ||\omega||_{L^2}||\psi(-\tau) - \varphi(-\tau)||_{L^2}||\psi_x(0)||_{L^2},$$

$\forall u \in H_0^1(0,1)$, by the Poincaré inequality $||u_x||_{L^2} \ge \pi||u||_{L^2}$, we have

$$||F(\varphi) - F(\psi)||_{L^2}$$
$$\le \frac{1}{\pi}||\omega||_{L^2}[||\varphi(-\tau)||_{H_0^1}||\varphi_x(0) - \psi_x(0)||_{L^2} + ||\psi(-\tau) - \varphi(-\tau)||_{H_0^1}||\psi_x(0)||_{L^2}]$$
$$\le \frac{1}{\pi}||\omega||_{L^2}[||\varphi||_C||\varphi(0) - \psi(0)||_{H_0^1} + ||\psi||_C||\psi(-\tau) - \varphi(-\tau)||_{H_0^1}]$$
$$\le \frac{1}{\pi}||\omega||_{L^2}(||\varphi||_C + ||\psi||_C)||\varphi - \psi||_C,$$

therefore, $\forall \varphi, \psi \in C([-\tau,0], H_0^1(0,1)), ||\varphi||_C \le \rho, ||\psi||_C \le \rho$, we have

$$||F(\varphi) - F(\psi)||_{L^2} \le M(\rho)||\varphi - \psi||_C,$$

where $M(\rho) = 2\pi^{-1}||\omega||_{L^2}\rho$, and

$$||\varphi||_C = \sup_{-\tau \le s \le 0} ||\varphi(s,\cdot)||_{H_0^1}, \quad \forall \varphi(t,x) \in C([-\tau,0], H_0^1(0,1)).$$

Denote

$$u_t(s) = u(t+s), s \in [-\tau,0], \quad t \ge 0,$$

we then transform the problem (1.6) into the following integral equation

$$u(t) = \varphi(t), \quad t \in [-\tau,0], \tag{2.2}$$

$$u(t) = T(t)\varphi(0) + \int_0^t T(t-s)F(u_s)ds, \quad t > 0. \tag{2.3}$$

The continuous solutions of (2.2)(2.3) are called mild solutions of the IBVP(1.6). Then for every initial function $\varphi(s,x) \in C([-\tau,0], H_0^1(0,1))$, there exists a constant $T = T(\varphi) > 0$ such that the IBVP(1.6) has a unique mild solution $u(t,x)$ on $[-\tau, T]$ with $u(t,x) \in C([-\tau, T], H_0^1(0,1))$.

Furthermore, $\forall \tau > 0$, the solution of IBVP(1.6) does not blow up in finite time. In fact, for $0 \le t \le \tau$, we have

$$\frac{d}{dt}\int_0^1 u_x^2(t)dx$$

$$= 2\int_0^1 u_x(t)u_{xt}(t)dx$$

$$= -2\int_0^1 u_{xx}(t)u_t(t)dx$$

$$= -2\int_0^1 u_{xx}(t)[\nu u_{xx}(t) - (\omega, u(t-\tau))u_x(t,x)]dx$$

$$= -2\nu\int_0^1 u_{xx}^2(t)dx + 2(\omega, u(t-\tau))\int_0^1 u_{xx}(t)u_{xx}(t,x)dx$$

$$\le -2\nu\int_0^1 u_{xx}^2(t)dx + 2||\omega||_{L^2}||u(t-\tau)||_{L^2}\int_0^1 |u_{xx}(t)u_x(t,x)|dx$$

$$\le -2\nu\int_0^1 u_{xx}^2(t)dx + 2\pi^{-1}||\omega||_{L^2}||u(t-\tau)||_{H_0^1}\int_0^1 |u_{xx}(t)u_x(t,x)|dx$$

$$\le -2\nu\int_0^1 u_{xx}^2(t)dx + 2\pi^{-1}||\omega||_{L^2}||\varphi||_C\int_0^1 |u_{xx}(t)u_x(t,x)|dx$$

$$\le -2\nu\int_0^1 u_{xx}^2(t)dx + 2\nu\int_0^1 u_{xx}^2(t)dx + \frac{1}{2\nu\pi^2}||\omega||_{L^2}^2||\varphi||_C^2\int_0^1 u_x^2(t,x)dx$$

$$= \frac{1}{2\nu\pi^2}||\omega||_{L^2}^2||\varphi||_C^2\int_0^1 u_x^2(t,x)dx,$$

therefore we get

$$\int_0^1 u_x^2(t)dx \le \int_0^1 u_x^2(0)dx\, e^{\frac{1}{2\nu\pi^2}||\omega||_{L^2}^2||\varphi||_C^2 t}$$

$$\le ||\varphi||_C^2 e^{\frac{1}{2\nu\pi^2}||\omega||_{L^2}^2||\varphi||_C^2 \tau}$$

$$= M_0(||\varphi||_C, ||\omega||_{L^2}), \quad 0 \le t \le \tau,$$

where M_0 is a constant depending only on $||\varphi||_C$, $||\omega||_{L^2}$ and τ.

Similarly, we have

$$\int_0^1 u_x^2(t)dx \le M_n(||\varphi||_C, ||\omega||_{L^2}), \quad n\tau \le t \le (n+1)\tau.$$

where M_n is a constant depending only on $||\varphi||_C$, $||\omega||_{L^2}$, τ and n. Then we get the existence of the global solution of IBVP(1.6).

Theorem 2.1. *For any initial condition $\varphi(t,x) \in C([-\tau,0], H_0^1(0,1))$, IBVP(1.6) has a unique global mild solution $u \in C([-\tau,\infty), H_0^1(0,1))$.*

3. Decay estimate of solution

To describe the exponential stability, we need the following Gronwall inequality and notations.

Lemma 3.1.[10] *Let g, h and y be three positive and integrable function on (t_0, T) such that $y'(t)$ is integrable on (t_0, T). Assume that*

$$\frac{dy}{dt} \leq gy + h, \quad t_0 \leq t \leq T,$$

$$\int_{t_0}^{T} g(s)ds \leq c_1, \quad \int_{t_0}^{T} e^{\delta s} h(s)ds \leq c_2, \quad \int_{t_0}^{T} e^{\delta s} y(s)ds \leq c_3,$$

where δ, c_1, c_2, c_3 are positive constants. Then

$$y(t) \leq [c_2 + c\delta_3 + y(t_0)]e^{c_1}e^{-\delta(t-t_0)}, \quad t_0 \leq t \leq T.$$

For $\varphi(t, x) \in C([-\tau, 0], H_0^1(0, 1))$, denote

$$k = k(\varphi, \omega)$$

$$= \sup_{-\tau \leq s \leq 0} ||\varphi_x(s)||_{L^2} + 2||\varphi(0)||_{H_0^1} e^{[\frac{||\omega||_{L^2}^2}{2\nu^2}(\nu \int_{-\tau}^{0} ||\varphi_x(s)||_{L^2}^2 ds + ||\varphi(0)||_{L^2}^2)]}$$

$$\sigma = \sup\{\delta > 0 : ||\varphi(0)||_{H_0^1}^2 e^{[\frac{||\omega||_{L^2}^2}{\nu^2}(\nu \int_{-\tau}^{0} ||\varphi_x(s)||_{L^2}^2 ds + ||\varphi(0)||_{L^2}^2)]} \leq \frac{k^2}{4}, 0 \leq \tau \leq \delta\},$$

$$0 \leq \tau \leq \min\{\sigma, \frac{1}{\nu} \ln 2\}.$$

Theorem 3.1. *For any nontrivial initial condition $\varphi(x, t) \in C([-\tau, 0], H_0^1(0, 1))$, then $\exists \tau_0 = \tau_0(\nu, \varphi) = \min\{\sigma, \frac{1}{\nu} \ln 2\}$, for $\tau < \tau_0$, the solution $u(t, x)$ of IBVP(1.6) satisfies*

$$||u_x(t)||_{L^2}^2 \leq \frac{k^2}{4} e^{-\nu t}, \quad \forall t \geq 0, \tag{3.1}$$

therefore the trivial solution of IBVP(1.6) is asymptotically stable in $H_0^1(0, 1)$.

Proof. Denote

$$T_0 = \sup\{\delta : ||u_x(t)||_{L^2} \leq k, \quad 0 \leq t \leq \delta\}.$$

Since $||u_x(0)||_{L^2} = ||\varphi_x(0)||_{L^2} \leq k$ and $||u_x(t)||_{L^2}$ is continuous, there is a neighborhood $(0, t_0)$ such that $||u_x(t)||_{L^2} \leq k$ for $t \in (0, t_0)$, we have $T_0 > 0$. We shall prove that $T_0 = +\infty$. otherwise, if $T_0 < +\infty$, then we have

$$||u_x(t)||_{L^2} \leq k, \quad \forall - \tau \leq t < T_0; \qquad ||u_x(T_0)||_{L^2} = k. \tag{3.2}$$

Since $u(t,x)$ is the mild solution of IBVP(1.6), by direct computation, one has $\int_0^1 u(t)u_x(t)dx = 0$ and

$$\frac{d}{dt}\int_0^1 u^2(t)dx = 2\int_0^1 u(t)u_t(t)dx$$

$$= 2\nu \int_0^1 u(t)u_{xx}(t)dx - 2(\omega, u(t-\tau))\int_0^1 u(t)u_x(t,x)dx$$

$$= -2\nu \int_0^1 u_x^2(t)dx. \tag{3.3}$$

Due to Poincaré inequality $||u(t)||_{L^2}^2 \leq \pi^{-2}||u_x(t)||_{L^2}^2 \leq ||u_x(t)||_{L^2}^2$, it is clear that

$$\frac{d}{dt}\int_0^1 u^2(t)dx \leq -2\nu \int_0^1 u_x^2(t)dx,$$

then L^2−norm of the solution satisfies that

$$\int_0^1 u^2(t)dx \leq e^{-2\nu t}\int_0^1 u_x^2(0)dx$$

$$= ||\varphi(0)||_{L^2}^2 e^{-2\nu t}$$

$$\leq ||\varphi||_C^2 e^{-2\nu t}, \tag{3.4}$$

it means that the solution decays exponentially in $L^2(0,1)$.

On the other hand, from (3.3), we have

$$\frac{d}{dt}\int_0^1 u^2(t)dx + 2\nu \int_0^1 u_x^2(t)dx = 0,$$

hence

$$\frac{d}{dt}(e^{\nu t}\int_0^1 u^2(t)dx) = \nu e^{\nu t}\int_0^1 u^2(t)dx + e^{\nu t}\frac{d}{dt}\int_0^1 u^2(t)dx$$

$$= \nu e^{\nu t}\int_0^1 u^2(t)dx - 2\nu e^{\nu t}\int_0^1 u_x^2(t)dx.$$

Due to (3.4), we get

$$\frac{d}{dt}(e^{\nu t}\int_0^1 u^2(t)dx) + 2\nu e^{\nu t}\int_0^1 u_x^2(t)dx = \nu e^{\nu t}\int_0^1 u^2(t)dx$$

$$\leq \nu e^{-\nu t}\int_0^1 \varphi^2(0)dx,$$

integrating with respect to t from 0 to T_0,

$$e^{\nu T_0} \int_0^1 u^2(T_0)dx - \int_0^1 \varphi^2(0)dx + 2\nu \int_0^{T_0} [e^{\nu t} \int_0^1 u_x^2(t)dx]dt$$

$$\leq ||\varphi(0)||_{L^2}^2 \int_0^{T_0} e^{-\nu t}dt,$$

$$e^{\nu T_0} \int_0^1 u^2(T_0)dx + 2\nu \int_0^{T_0} [e^{\nu t} \int_0^1 u_x^2(t)dx]dt$$

$$\leq ||\varphi(0)||_{L^2}^2 (2 - e^{-\nu T_0}),$$

therefore we obtain

$$\nu \int_0^{T_0} [e^{\nu t} \int_0^1 u_x^2(t)dx]dt \leq ||\varphi(0)||_{L^2}^2, \tag{3.5}$$

and

$$\int_0^{T_0} [e^{\nu t} \int_0^1 u_x^2(t-\tau)dx]dt$$

$$= \int_{-\tau}^{T_0-\tau} [e^{\nu(s+\tau)} \int_0^1 u_x^2(s)dx]ds$$

$$= \int_{-\tau}^0 e^{\nu(s+\tau)} \int_0^1 u_x^2(s)dxds + \int_0^{T_0-\tau} e^{\nu(s+\tau)} \int_0^1 u_x^2(s)dxds$$

$$\leq e^{\nu\tau} \int_{-\tau}^0 ds \int_0^1 \varphi_x^2(s)dxds + e^{\nu\tau} \int_0^{T_0} e^{\nu s} \int_0^1 u_x^2(s)dxds$$

$$\leq e^{\nu\tau} \int_{-\tau}^0 ds \int_0^1 \varphi_x^2(s)dxds + \frac{1}{\nu} e^{\nu\tau} ||\varphi(0)||_{L^2}^2. \tag{3.6}$$

The last inequality comes from (3.5). Using the equation in (1.6) and Young's inequality, we have

$$\frac{d}{dt} \int_0^1 u_x^2(t)dx$$

$$= 2 \int_0^1 u_x(t)u_{xt}(t)dx$$

$$= -2 \int_0^1 u_{xx}(t)u_t(t)dx$$

$$= -2\nu \int_0^1 u_{xx}^2(t)dx + 2(\omega, u(t-\tau)) \int_0^1 u_{xx}(t)u_t(t)dx$$

$$\leq -2\nu \int_0^1 u_{xx}^2(t)dx + 2||\omega||_{L^2}||u(t-\tau)||_{L^2} \int_0^1 |u_{xx}(t)u_t(t|)dx$$

$$\leq -2\nu \int_0^1 u_{xx}^2(t)dx + 2\nu \int_0^1 u_{xx}^2(t)dx + \frac{1}{2\nu} ||\omega||_{L^2}^2 ||u(t-\tau)||_{L^2}^2 \int_0^1 u_x^2(t)dx$$

$$\leq \frac{1}{2\nu} ||\omega||_{L^2}^2 ||u_x(t-\tau)||_{L^2}^2 \int_0^1 u_x^2(t)dx.$$

Denote

$$y(t) = \int_0^1 u_x^2(t)dx, \quad g(t) = \frac{1}{2\nu}||\omega||_{L^2}^2 \int_0^1 u_x^2(t-\tau)dx, \quad h(t) = 0, \quad \delta = \nu,$$

$$c_1 = \frac{||\omega||_{L^2}^2}{2\nu^2}(\nu \int_{-\tau}^0 ||\varphi_x(s)||_{L^2}^2 ds + ||\varphi(0)||_{L^2}^2), \quad c_2 = 0, \quad c_3 = \frac{1}{\nu} \int_0^1 \varphi^2(0)dx,$$

where the constant c_1 comes from the following estimate

$$\int_0^{T_0} g(s)ds = \frac{||\omega||_{L^2}^2}{2\nu} \int_0^{T_0} ds \int_0^1 u_x^2(s-\tau)dx$$

$$\leq \frac{||\omega||_{L^2}^2}{2\nu} \int_0^{T_0} e^{\nu s}ds \int_0^1 u_x^2(s-\tau)dx$$

$$\leq \frac{||\omega||_{L^2}^2}{2\nu^2} e^{\nu\tau}(\nu \int_{-\tau}^0 ||\varphi_x(s)||_{L^2}^2 ds + ||\varphi(0)||_{L^2}^2).$$

The last inequality comes from (3.6). Then Lemma 3.1 implies that

$$\int_0^1 u_x^2(t)dx$$

$$\leq [\int_0^1 \varphi^2(0)dx + \int_0^1 \varphi_x^2(0)dx]e^{[\frac{||\omega||_{L^2}^2}{2\nu^2}e^{\nu\tau}(\nu \int_{-\tau}^0 ||\varphi_x(s)||_{L^2}^2 ds + ||\varphi(0)||_{L^2}^2)]}e^{-\nu t}$$

$$\leq ||\varphi(0)||_{H_0^1}^2 e^{[\frac{||\omega||_{L^2}^2}{\nu^2}(\nu \int_{-\tau}^0 ||\varphi_x(s)||_{L^2}^2 ds + ||\varphi(0)||_{L^2}^2)]}e^{-\nu t} \quad (0 \leq \tau \leq \tau_0)$$

$$\leq \frac{k^2}{4}e^{-\nu t} \quad (0 \leq \tau \leq \tau_0), \tag{3.7}$$

hence $||u_x(T_0)||_{L^2} \leq \frac{k}{2}e^{-\frac{1}{2}\nu T_0} < k$, this contradicts (3.2), so we have $T_0 = +\infty$, according to the definition of T_0, we get $||u_x(t)||_{L^2} \leq k, \forall t \geq 0$.

Therefore, (3.7) implies (3.1), the decay estimate of solution in H_0^1, then we prove Theorem 3.1. □

Remark For the initial-boundary value problem of the time-delayed nonlocal Burgers equation $u_t(t,x) - \nu u_{xx}(t,x) + (\omega(x), u(t-\tau, x))u_x(t,x) = 0$, Theorem 3.1 implies that the global mild solution of IBVP(1.6) decays exponentially in $H_0^1(0,1)$. Due to the result described in Theorem 2.1 and Theorem 3.1, we can see that although the nonlinear nonlocal term appears in the Burgers equation, the global asymptotic behavior of the solution does not change much. In other words, the asymptotic behavior of the time-delayed nonlocal Burgers equation is very similar to that of IBVP of the time-delayed Burgers equation $u_t(t,x) - \nu u_{xx}(t,x) + u(t-\tau, x)u_x(t,x) = 0$, see Weijiu Liu.[10]

Acknowledgments

The work was supported by the National Natural Science Foundation of China grant 10871078.

References

1. G. Avalos, I. Lasiecka and R. Rebarber, *Lack of time-delay robustness for stabilization of a structural acoustics model*, SIAM J. Control Optim., 37:5(1999), 1394-1418.

2. A. Balogh and M. Krstić, *Burgers equation with nonlinear boundary feedback: H^1 stability, well-posedness and simulation*, Mathematical Problems in Engineering, 6(2000), 189-200.

3. E. Beretta, F. Solimano and Y. Tang, *Analysis of a chemostat model for bacteria and virulent bacteriophage*, Discrete and Continuous Dynamical Systems, B2:4(2002), 495-520.

4. D.R. Bland, Wave Theory and Applications, Oxford University Press, Oxford,1988.

5. H. Choi, R. Temam, P. Moin and J. Kim, *Feedback control for unsteady flow and its application to the stochastic Burgers equation*, J. Fluid Mech., 253(1993), 509-543.

6. I.D. Chueshov, "Introduction to the theory of infinite-dimensional dissipative systems", Acta Scientific Publishing House, Kharkiv, Ukraine, 2002.

7. L. Debnath, "Nonlinear Partial Differential Equations for Scientists and Engineers", Birkhauser, Boston, MA,1997.

8. P.M. Jordan, *A note on Burgers equation with time delay: Instability via finite-time blow-up*, Physics Letters A, 372(2008), 6363-6367.

9. W. Liu and M. Krstić, *Backstepping boundary control of Burgers equation with actuator dynamics*, Systems and Control Letters, 41:4(2000), 291-303.

10. W. Liu, *Asymptotic behavior of solutions of time-delayed Burgers equation*, Discrete and Continuous Dynamical Systems, B2:1(2002), 47-56.

11. H. Smith, P. Waltman, "The theory of the chemostat", Cambridge University Press, Cambridge, UK, 1995.

12. C. Travis and G. Webb, *Existence and stability for partial functional differential equations*, Trans. Amer. Math Soc., 204(1974), 395-418.

13. C. Travis and G. Webb, *Partial differential equations with deviating arguments in the time variable*, J. Math. Anal. Appl., 56(1976), 397-409.

14. C. Travis and G. Webb, *Existence, stability and compactness in the $\alpha-$ norm for partial functional differential equations*, Trans. Amer. Math Soc., 240(1978), 129-143.

15. Y. Tang and M. Wang, *A Remark on Exponential Stability of Time-delayed Burgers Equation*. Discrete and Continuous Dynamical Systems, Series B, 12:1(2009), 219-225.

16. Y. Tang and L.Zhou, *Stability switch and Hopf bifurcation for a diffusive prey-predator system with delay*, J. Math. Anal. Appl., 334(2007), 1290-1307.

17. Y. Tang and L. Zhou, *Hopf Bifurcation and Stability of a Competition Diffusion System with Distributed Delay*, Publication of RIMS, 41:3(2005), 579-597.

18. G.B. Whitham, "Linear and Nonlinear Waves", Wiley, New York, 1974.

19. L. Zhou, Y. Tang and S. Hussein, *Stability and Hopf bifurcation for a delay competition diffusion system*, Chaos, Solitons and Fractals, 14:8(2002), 1201-1225.

Perspectives in Mathematical Sciences
Interdisciplinary Mathematical Sciences, Volume 9, 2009
pp. 275–291

Chapter 14

Bifurcation Analysis of the Swift-Hohenberg Equation with Quintic Nonlinearity and Neumann Boundary Condition

Qingkun Xiao and Hongjun Gao*

*Institute of Mathematics, School of Mathematical Science,
Nanjing Normal University, Nanjing 210097, P.R. China,
xiaoqingkun@hotmail.com*

*Institute of Mathematics, School of Mathematical Science,
Nanjing Normal University,
Nanjing 210097, P.R. China,
Center of Nonlinear Science, Nanjing University
Nanjing 210093, P.R. China,
gaohj@njnu.edu.cn*

This paper is concerned with the asymptotic behavior of the solutions $u(x, t)$ of the Swift-Hohenberg equation with quintic nonlinearity on a one-dimensional domain $(0, L)$ with Neumann boundary condition. With α and the length L of the domain regarded as bifurcation parameters, branches of nontrivial solutions bifurcating from the trivial solution at certain points are shown. Local behavior of these branches are also studied. Global bounds for the solutions $u(x, t)$ are established and the global attractor is investigated then. Finally, by a center manifold analysis, two types of structures in the bifurcation diagrams are presented when the bifurcation points are closer, and their stabilities are analyzed. Compared with our previous work, the differences between the Neumann boundary condition and Dirichlet boundary condition are shown. The impact of the initial function $u_0(x)$ on the linear stability of the trivial solution, and the impact of boundary conditions on the stability properties of the bifurcated solutions are shown.

1. Introduction

For many problems a comprehensive system of equations is either not known or too complicated to analyze. It is often appropriate to introduce phenomenological order-parameter models. An example of this sort is the fourth order model equation is Swift-Hohenberg equation first proposed in 1976 by Swift and Hohenberg[1] as a simple model for the Rayleigh-Bénard instability of roll waves. The Swift-Hohenberg equation plays an important role in bifurcation analysis. It's a useful model for the

---

*Corresponding author.

study of phenomena in various scientific areas from fluid dynamics and chemistry
to nonlinear optics, their model and its extensions have been the subject of many
analytical and numerical studies.[2–7]

Recently there are some other papers concerned with the Swift-Hohenberg equation. A rigorous numerical method for the study and verification of global dynamics
of the Swift-Hohenberg equation with cubic nonlinearity under periodic boundary
condition is presented.[8] Localized pattern formation with a large-scale mode is
investigated, and some numerical results are also given.[9] Stationary spatially localized hexagon patterns of the 2D Swift-Hohenberg equation with quadratic-cubic
nonlinearity are investigated, and the existence regions of fully localized hexagon
patches and of planar pulses consisting of a strip filled with hexagons which is embedded in the trivial state are traced out by numerical continuation techniques.[10]
The stationary Swift-Hohenberg equation has chaotic dynamics on a critical energy
level for a large (continuous) range of parameter values is proved by a computer
assisted, rigorous, continuation method.[11] Fronts are travelling waves in spatially
extended systems that connect two different spatially homogeneous rest states. If
the rest state behind the front undergoes a supercritical Turing instability, then
the front will also destabilize. On the linear level, however, the front will be only
convectively unstable since perturbations will be pushed away from the front as
it propagates. In,[12] the authors prove for a specific model system, in which the
Chafee-Infante equation is coupled to the Swift-Hohenberg equation, that the behavior described above carries over to the full nonlinear system. They also perform
numerical simulations to support their conclusions.

In,[5,6] the asymptotic behavior of the solutions of the Swift-Hohenberg equation
with cubic nonlinearity has been studied, two interesting types of structures in the
bifurcation diagram for (L, u) are exhibited. They demonstrate how these structures
arise naturally near certain bifurcation points, and that there are no others. They
also analyze their stability properties.

In our previous work,[13] we considered solutions of the Swift-Hohenberg equation
which reads

$$\frac{\partial u}{\partial t} = \alpha u - (1 + \frac{\partial^2}{\partial x^2})^2 u - u^5, \tag{1.1}$$

where the unknown function u is a real-valued function on the cylindrical domain
$Q = (0, L) \times R^+$, with the boundary conditions,

$$u = 0 \quad \text{and} \quad \frac{\partial^2 u}{\partial x^2} = 0, \quad \text{at} \quad x = 0, L. \tag{1.2}$$

and the initial condition

$$u(x, 0) = u_0(x), \quad \text{for} \quad 0 < x < L, \tag{1.3}$$

where the initial function u_0 is a smooth function satisfying

$$u_0(x) = -u_0(L - x), \quad \text{for} \quad 0 < x < L. \tag{1.4}$$

With the help of a center manifold analysis, two types of structures in the bifurcation diagrams are presented when the bifurcation points are closer, and their stabilities are analyzed.

In this paper, we consider the following Cauchy-Neumann problem for the Swift-Hohenberg equation in Q

$$\begin{cases} \frac{\partial u}{\partial t} = \alpha u - (1 + \frac{\partial^2}{\partial x^2})^2 u - u^5, & \text{for} \quad 0 < x < L, t > 0, \\ u' = 0, \quad u''' = 0, \quad \text{at} \quad x = 0, L, \\ u(x, 0) = u_0(x), \quad \text{for} \quad 0 < x < L, \end{cases} \quad (1.5)$$

However, we don't allow that the initial function $u_0(x)$ satisfies Eq. (1.4) in this paper.

The main motivation of this paper is to compare the results in this case with that in,[13] to show the impact of the initial function $u_0(x)$ on the linear stability of the trivial solution, and the impact of boundary conditions on the stability properties of the bifurcated solutions.

It's easy to see that Eq. (1.3) is a gradient system with corresponding Liapunov functional

$$J(u, L) = \frac{1}{L} \int_0^L \{\frac{1}{2}(u'')^2 - (u')^2 + \frac{1-\alpha}{2}u^2 + F(u)\}dx, \quad (1.6)$$

where

$$F(u) = \int_0^u s^5 ds = \frac{1}{6}u^6, \qquad F(0) = 0.$$

That is to say, if the stationary solutions of Eq. (1.3) are isolated, then $u(x, t)$ tends to one of these solutions as $t \to +\infty$.[14]

We view α and L as bifurcation parameters. It's easy to see that $u = 0$ is always a solution of Eq. (1.3). It's known that if $\alpha \le 0$, the trivial solution $u = 0$ is stable. When $\alpha > 0$, the case is much more complex.

In what follows, we consider Eq. (1.5) under the assumption $0 < \alpha < 1$.

We find that there exists critical lengths, nL_1 and nL_2, $n = 1, 2, \cdots$, where $L_1 = \frac{\pi}{\xi_+}$, $L_2 = \frac{\pi}{\xi_-}$, ξ_+ and ξ_- will be given in §2. At these points, branches of nontrivial solutions of Eq. (1.5) bifurcate from the trivial solution.

We will study structures of bifurcation curves in the (L, u)-plane when L_2 and $2L_1$ nearly coincide, i.e, when there exists $\alpha_1^* \in (0, 1)$ such that $2L_1(\alpha_1^*) = L_2(\alpha_1^*)$, $2L_1(\alpha) > L_2(\alpha)$ for $\alpha < \alpha_1^*$.

When $\alpha = \alpha_1^* + \varepsilon$ is fixed, where ε is positive and small, then $2L_1 < L_2$. It's easily verified that $\alpha_1^* = \frac{9}{25}$.

If $L \in (2L_1, L_2)$, we can set

$$L = 2L_1 + \delta L_{gap}, \quad 0 < \delta < 1. \quad (1.7)$$

where $L_{gap} = L_2 - 2L_1$.

By the center manifold theorem, we project Eq. (1.3) to the two-dimensional center manifold, we have the following theorem.

Theorem 1.1. *Suppose that $\alpha = \alpha_1^* + \varepsilon$ is fixed, ε is positive and small, $2L_1 < L_2$. Then if Eq. (1.6) holds,*

(1)the trivial solution is always unstable for $0 < \delta < 1$.

(2)when $0 < \delta < \frac{1}{13}$, the branches which bifurcate from $2L_1$ are unstable; when $\frac{1}{13} < \delta < 1$, they are stable.

(3)when $0 < \delta < \frac{7}{15}$, the branches which bifurcate from L_2 are stable; when $\frac{7}{15} < \delta < 1$, they are unstable.

(4)when $\frac{1}{13} < \delta < \frac{7}{15}$, the branch which connects these branches bifurcating from L_2 and $2L_1$ is unstable.

We also consider generalized cases, that is to say we focus on the situation when higher order bifurcation points coincide, i.e., when there exists an $\alpha^* \in (0,1)$ and positive integers m and n such that

$$nL_1(\alpha^*) = mL_2(\alpha^*),$$
$$nL_1(\alpha) > mL_2(\alpha), \quad \text{for} \quad \alpha < \alpha^*,$$

where $n \neq 2m$, $n \neq 3m$ and $n \neq 5m$. It's clearly that

$$\alpha^* = \alpha_{m,n}^* = (\frac{n^2 - m^2}{n^2 + m^2})^2.$$

When $\alpha = \alpha_{m,n}^* + \varepsilon$ is fixed, where $\varepsilon > 0$ is small, then $nL_1(\alpha) < mL_2(\alpha)$.

If $L \in (nL_1, mL_2)$, we can set

$$L = nL_1(\alpha) + \delta L_{gap}, \quad 0 < \delta < 1, \tag{1.8}$$

where

$$L_{gap} = mL_2(\alpha) - nL_1(\alpha).$$

In a similar manner, we will prove the following theorem.

Theorem 1.2. *Suppose that $\alpha = \alpha_{m,n}^* + \varepsilon$ is fixed, $\varepsilon > 0$ is small, $nL_1(\alpha) < mL_2(\alpha)$. Then if (1.7) holds,*

(1)the trivial solution is always unstable for $0 < \delta < 1$.

(2)when $0 < \delta < \frac{m^2}{m^2 + 3n^2}$, the branches which bifurcate from nL_1 are unstable; when $\frac{m^2}{m^2 + 3n^2} < \delta < 1$, the branches which bifurcate from nL_1 are stable.

(3)when $0 < \delta < \frac{3m^2}{3m^2 + n^2}$, the branches which bifurcate from mL_2 are stable; when $\frac{3m^2}{3m^2 + n^2} < \delta < 1$, the branches which bifurcate from mL_2 are unstable.

(4)when $\frac{m^2}{m^2 + 3n^2} < \delta < \frac{3m^2}{3m^2 + n^2}$, the branch which connects these branches bifurcating from mL_2 and nL_1 is unstable.

From the above theorems, we exhibit two types of structures in the bifurcation diagrams in the case when $n = 2m$ and the cases when $n \neq 2m$, $n \neq 3m$ and $n \neq 5m$.

Comparing with the results in,[13] we can observe that when $n = 2m$, the conclusion is different from the one under Dirichlet boundary condition, but when $n \neq 2m$,

$n \neq 3m$ and $n \neq 5m$, the results is the same under Dirichlet boundary condition and Neumann boundary condition.

The chapter is organized as follows. In §2, we study the linear stability of the trivial solution by Fourier expansion. Local behavior near L_1 and L_2 has been studied through implicit function theorem in §3. Then, some global bounds for the solutions are given in §4. In §5, asymptotic expressions for nontrivial stationary solutions for Eq. (1.5) as $\alpha \to 0^+$ have been considered. The proofs of Theorem 1.1 and Theorem 1.2 are given respectively in §6 and §7.

2. Linear Stability of the Trivial Solution

Firstly, we discuss the stability of the trivial solution, the linear problem about the trivial solution associated with Eq. (1.5) is

$$\begin{cases} \frac{\partial u}{\partial t} = \alpha u - (1 + \frac{\partial^2}{\partial x^2})^2 u, & \text{for} \quad 0 < x < L, \quad t > 0, \\ u' = 0, u''' = 0, & \text{at} \quad x = 0, L, \\ u(x, 0) = u_0(x), & \text{for} \quad 0 < x < L. \end{cases} \tag{2.1}$$

Since $u_0 \in L^2(0, L)$, the solution $u(x, t)$ of Eq. (2.1) can be written as a Fourier series,

$$u(x, t) = \sum_{n=1}^{\infty} a_n \varphi_n(x) e^{-\lambda_n t}, \tag{2.2}$$

where $\varphi_n(x)$ and λ_n are, respectively, the eigenfunctions and eigenvalues of the following eigenvalue problem

$$\begin{cases} \varphi^{(4)} + 2\varphi'' + (1 - \alpha)\varphi = \lambda\varphi, & \text{for} \quad 0 < x < L, \\ \varphi' = 0, \quad \varphi''' = 0, & \text{at} \quad x = 0, L, \end{cases} \tag{2.3}$$

and $a_n = (u_0, \varphi_n)$.

The eigenvalues λ_n are given by

$$\lambda_0 = 1 - \alpha, \quad \lambda_k = P(\frac{k\pi}{L}), k = 1, 2 \cdots, \quad \text{where} \quad P(\xi) = (\xi^2 - 1)^2 - \alpha.$$

and the corresponding eigenfunctions are

$$\varphi_0 = 1, \quad \varphi_k = \sqrt{2} \cos(\frac{k\pi x}{L}), \quad k = 1, 2 \cdots,$$

It's easy to see that $\lambda_0 > 0$ for arbitrary L. Since $\alpha \in (0, 1)$, $P(\xi)$ has two positive zeros ξ_- and ξ_+, and they are given by

$$\xi_- = (1 - \sqrt{\alpha})^{\frac{1}{2}}, \quad \xi_+ = (1 + \sqrt{\alpha})^{\frac{1}{2}}.$$

This implies that $\lambda_k < 0$, when $L \in (kL_1, kL_2)$, where $L_1 = \frac{\pi}{\xi_+}$, $L_2 = \frac{\pi}{\xi_-}$. If $L \in (0, L_1)$, then

$$\lambda_k = P(\frac{k\pi}{L}) > P(\xi_+) = 0, \quad \text{for all} \quad k \geq 1.$$

Thus, if $L < L_1$, the trivial solution is asymptotically stable.

We are interested in the cases when L is larger than L_1. Denote

$$I_k = \left\{ L > 0 : P(\frac{k\pi}{L}) \leq 0 \right\},$$

then $I_k = [kL_1, kL_2]$ for every $n \geq 1$.

For α small,

$$L_1(\alpha) \sim \pi - \frac{\pi}{2}\sqrt{\alpha}, \quad L_2(\alpha) \sim \pi + \frac{\pi}{2}\sqrt{\alpha}, \quad \text{as} \quad \alpha \to 0^+. \tag{2.4}$$

If α is small enough, the intervals I_k will not overlap, but be separated by intervals in which $\lambda_k > 0$, $k = 1, 2, \cdots$. Then there exists a gap between I_k and I_{k+1}, when

$$0 < \alpha < \alpha_k^* = 4(\frac{1}{2k+1} + (2k+1))^{-2}.$$

Therefore, when $0 < \alpha < \alpha_k^*$, there is a gap between I_k and I_{k+1}, we define

$$\Pi_k = (0, L_1) \bigcup (L_2, 2L_1) \bigcup \ldots \bigcup (kL_2, (k+1)L_1).$$

From Fourier series Eq. (2.2), we have the following theorem.

Theorem 2.1. *Let $u(t)$ be the solution of Eq. (1.3), in which $\alpha < \alpha_k^*$ for some $k \geq 1$. Then for all $L \in \Pi_k$, $u(t) \to 0$, as $t \to \infty$.*

We want to see the full-time dependent problem near those values of α and L where these bifurcation points approximately coincide with each other, and we have the following center manifold theorem.

Theorem 2.2. *Suppose that $\alpha = \alpha_1^*$, and so $L_2 = 2L_1$. Then $L = 2L_1$,*

(1) Eq. (1.3) has a two-dimensional center manifold X^c about the trivial solution and $X^c = span\{\varphi_1, \varphi_2\}$.

(2) $dimX^u = 0$, where X^u is the unstable manifold about the trivial solution.

3. Local Behavior near L_1 and L_2

For the sake of convenience, after a suitable rescaling of the spatial variable $L\bar{x} = x$, $\bar{u}(\bar{x}) = u(x)$, and then we omit the bar, the stationary equation of Eq. (1.3) then can be written as

$$\begin{cases} B(u, L) + L^4 u^5 = 0, & \text{for} \quad x \in (0, 1), \\ u' = 0, \quad u''' = 0, & \text{at} \quad x = 0, 1, \end{cases} \tag{3.1}$$

where

$$B(u, L) = u^{(4)} + 2L^2 u'' + L^4(1-\alpha)u. \tag{3.2}$$

The corresponding eigenvalue problem for the linear operator B is

$$\begin{cases} B(u, L) = \sigma u, & \text{for} \quad x \in (0, 1), \\ u' = 0, \quad u''' = 0, & \text{at} \quad x = 0, 1. \end{cases} \tag{3.3}$$

In addition, from Eq. (2.3) and Eq. (3.3), we obtain $\sigma_n = L^4 \lambda_n$, $n \geq 0$, where λ_n are the eigenvalues of Eq. (2.3), and the corresponding eigenfunctions ψ_n are given by

$$\psi_0(x) = 1, \quad \psi_k(x) = \sqrt{2}\cos(n\pi x), \quad k = 1, 2 \cdots$$

Besides Theorem 2.1 and Theorem 2.2, we want to get more information about the local behavior at the bifurcation point $(L, u) = (L_1, 0)$, and know the relation between u and L near this point, so we set

$$L = L_1(1 + \zeta), \quad u = \varepsilon(\psi_1 + w), \tag{3.4}$$

where ε is a small constant, ζ and w will turn out to be small, and substituting Eq. (3.4) into Eq. (3.2), we have

$$
\begin{aligned}
B(w, 2L_1) &= -2L_1^2[(1+\zeta)^2 - 1](\psi_2'' + w'') - L_1^4[(1+\zeta)^4 - 1](1 - \alpha)(\psi_2 + w) \\
&\quad - L_1^4(1+\zeta)^4 \varepsilon^4 (\psi_2 + w)^5 \\
&= h(\zeta, w, \gamma),
\end{aligned}
\tag{3.5}
$$

where $\gamma = \varepsilon^4$.

As in,[6] by Implicit Function Theorem, we have the following theorem.

Theorem 3.1. *Let $\alpha > 0$, there exists a unique branch in the (L, u)-plane of nontrivial solutions of Eq. (3.1) which bifurcates from the trivial solution at L_1. Its local behavior is given by*

$$\|u\|_{L^2}^4 \sim \frac{8}{5\pi}\sqrt{\alpha}(1 + \sqrt{\alpha})^{\frac{3}{2}}(L - 2L_1), \quad as \quad L \to L_1^+.$$

Similarly, we can prove the following theorem about the local behavior near L_2.

Theorem 3.2. *Let $0 < \alpha < 1$, there exists a unique branch in the (L, u)-plane of nontrivial solutions of Eq. (3.1) which bifurcates from the trivial solution at L_2. Its local behavior is given by*

$$\|u\|_{L^2}^4 \sim \frac{8}{5\pi}\sqrt{\alpha}(1 - \sqrt{\alpha})^{\frac{3}{2}}(2L_2 - L), \quad as \quad L \to L_2^-.$$

Local behavior of $J(u, L)$ near L_1 and L_2 can also be established.

Theorem 3.3. *On the branches defined in Theorem 3.1 and Theorem 3.2, the local behavior of $J(u, L)$ is respectively given by*

$$J(u, L) \sim -\frac{8\sqrt{10}}{15\pi^{\frac{3}{2}}}\alpha^{\frac{3}{4}}(1 + \sqrt{\alpha})^{\frac{9}{4}}(L - L_1)^{\frac{3}{2}} \quad as \quad L \to L_1^+,$$

$$J(u, L) \sim -\frac{8\sqrt{10}}{15\pi^{\frac{3}{2}}}\alpha^{\frac{3}{4}}(1 - \sqrt{\alpha})^{\frac{9}{4}}(L_2 - L)^{\frac{3}{2}} \quad as \quad L \to L_2^-.$$

4. Global Bounds for Solutions of Eq. (1.5)

In this section, we give some a priori estimates for the solutions of Eq. (1.5). Define

$$\mu(L) = \min\{\lambda_n(L), n = 0, 1, 2, \cdots\}. \tag{4.1}$$

Let $u(t) = u(x, t)$ be the solution of Eq. (1.5). As in,[13] we have the same estimates for $u(t)$.

Theorem 4.1. *Let $u(t)$ be the solution of Eq. (1.5), and μ is defined by Eq. (4.1).*
(a) Suppose that $\mu > 0$, then

$$\|u\|_{L^2} \leq \|u_0\|_{L^2} \exp(-\mu t) \quad for \quad 0 \leq t < \infty, \tag{4.2}$$

and

$$u(t) \to 0 \quad as \quad t \to \infty \quad in \quad L^2(0, L).$$

(b) Suppose that $\mu = 0$, then

$$\|u(t)\|_{L^2} \leq \frac{\|u_0\|_{L^2}}{\sqrt[4]{1 + 4\|u_0\|_{L^2}^4 t}} \quad for \quad 0 \leq t < \infty, \tag{4.3}$$

and

$$u(t) \to 0 \quad as \quad t \to \infty \quad in \quad L^2(0, L).$$

(c) Suppose that $\mu < 0$, then

$$\|u(t)\|_{L^2} \leq \max\{\|u_0\|_{L^2}, \sqrt[4]{|\mu|}\} \quad for \quad 0 \leq t < \infty, \tag{4.4}$$

and

$$\limsup_{t \to \infty} \|u(t)\|_{L^2} \leq \sqrt[4]{|\mu|}.$$

Remark From Theorem 4.1, we obtain that, if $\mu \geq 0$, the origin then is a global attractor in $L^2(0, L)$. If $\mu < 0$, the global attractor is the ball B_R in $L^2(0, L)$ with radius $R = \sqrt[4]{|\mu|}$ and centered at the origin.

Bounds for $\|u_x\|_{L^2}$ can also be established.

Theorem 4.2. *Let $u(t)$ be the solution of Eq. (1.5), and μ is defined by Eq. (4.1).*
(a) If $\mu > 0$, then

$$\|u_x(t)\|_{L^2} = O(\exp(-\mu t)) \quad as \quad t \to \infty.$$

(b) If $\mu = 0$, then

$$\|u_x(t)\|_{L^2} = O(t^{-\frac{1}{4}}) \quad as \quad t \to \infty.$$

(c) If $\mu < 0$, then

$$\limsup_{t \to \infty} \|u_x(t)\|_{L^2} \leq \sqrt[4]{2|\mu|(1 + \alpha)}.$$

5. Stationary Solutions in the Limit as $\alpha \to 0^+$

In this section, we construct asymptotic expressions for nontrivial stationary solutions for Eq. (1.5) as $\alpha \to 0^+$. The stationary solution of Eq. (1.5) is the solution of the following problem

$$\begin{cases} v^{(4)} + 2v'' + (1 - \alpha)v + v^5 = 0, & \text{for} \quad 0 < x < L, \\ v' = 0, \quad v''' = 0, & \text{at} \quad x = 0, L. \end{cases} \tag{5.1}$$

We choose $L \in (kL_1, kL_2)$, where k is an arbitrary positive integer, and α is small enough such that $[kL_1, kL_2]$ and $[(k+1)L_1, (k+1)L_2]$ are disjoint. That is to say

$$0 < \alpha < \alpha_k^* = 4(\frac{1}{2k + 1} + (2k + 1))^{-2}.$$

We set

$$x = Ly \quad \text{and} \quad w(y) = v(x),$$

then Eq. (5.1) becomes

$$\begin{cases} w^{(4)} + 2L^2 w'' + L^4((1 - \alpha)w + w^5) = 0, & \text{for} \quad x \in (0, 1), \\ w' = 0, \quad w''' = 0, & \text{at} \quad x = 0, 1. \end{cases} \tag{5.2}$$

It follows from Theorem 3.1 and Theorem 3.2 that, as $L \to L_1^+$ and $L \to L_2^-$, the solutions scales like $\sqrt[4]{\alpha}$. Therefore, we suppose $w(y)$ has the form

$$w(y) \sim \alpha^{\frac{1}{4}} w_1(y) + \alpha^{\frac{3}{4}} w_2(y) + \alpha^{\frac{5}{4}} w_3(y) + \alpha^{\frac{7}{4}} w_4(y) + \cdots \tag{5.3}$$

We write

$$L = kL_1 + k\delta(L_2 - L_1) = k[L_1 + \delta L_{gap}], \quad 0 < \delta < 1,$$

from Eq. (2.4), we write $L_1(\alpha)$, $L_2(\alpha)$ and $L_{gap}(\alpha)$ into Fourier series of $\alpha^{\frac{1}{4}}$.

Therefore, we have

$$L(\alpha) = k\pi\{1 + \frac{1}{2}(2\delta - 1)\alpha^{\frac{1}{2}} + \frac{3}{8}\alpha + \frac{5}{16}(2\delta - 1)\alpha^{\frac{3}{2}} + \cdots\}, \tag{5.4}$$

and equate the coefficients of $\alpha^{\frac{1}{4}}$, $\alpha^{\frac{3}{4}}$, $\alpha^{\frac{5}{4}}$, $\alpha^{\frac{7}{4}}$ equal to zero, we obtain the following equations

$O(\alpha^{\frac{1}{4}}) : L_1 w_1 = 0,$

$O(\alpha^{\frac{3}{4}}) : L_1 w_2 + L_2 w_1 = 0,$

$O(\alpha^{\frac{5}{4}}) : L_1 w_3 + L_2 w_2 + L_3 w_1 + k^4\pi^4 w_1^5 = 0,$

$O(\alpha^{\frac{7}{4}}) : L_1 w_4 + L_2 w_3 + L_3 w_2 + L_4 w_1 + k^4\pi^4(2(2\delta - 1)w_1^5 + 5w_1^4 w_2) = 0,$

where

$$L_1 z = z^{(4)} + 2k^2\pi^2 z'' + k^4\pi^4 z,$$

$$L_2 z = 2k^2\pi^2(2\delta - 1)(z'' + k^2\pi^2 z),$$

$$L_3 z = 2k^2\pi^2(\frac{1}{4}(2\delta - 1)^2 z'' + \frac{3}{4}z'') + k^4\pi^4(\frac{3}{2}(2\delta - 1)^2 z + \frac{1}{2}z),$$

$$L_4 z = 2k^2\pi^2(2\delta - 1)z'' + k^4\pi^4(\frac{3}{2}(2\delta - 1)z + \frac{1}{2}(2\delta - 1)^3 z),$$

and $w_j(y)$, $j = 1, 2, 3, 4$, all need to satisfy the boundary condition

$$w'_j(y) = 0, \quad w'''_j(y) = 0 \quad \text{at} \quad y = 0, 1.$$

This implies that

$$w_1(y) = \gamma_1 \cos(k\pi y),$$

where γ_1 is to be determined. Since $L_2 w_1 = 0$, the equation for w_2 reduces to $L_1 w_2 = 0$, it follows that

$$w_2(y) = \gamma_2 \cos(k\pi y),$$

and γ_2 is also to be determined. From the solvability condition, we have

$$(w_1, L_1 w_3) = -(w_1, L_2 w_2 + L_3 w_1 + k^4 \pi^4 w_1^5) = 0. \tag{5.5}$$
$$(w_1, L_1 w_4) = -(w_1, L_2 w_3 + L_3 w_3 + L_4 w_1 + k^4 \pi^4 (2(2\delta - 1) w_1^5 + 5 w_1^4 w_2))$$
$$= 0. \tag{5.6}$$

This yields

$$\gamma_1(\delta) = \sqrt[4]{\frac{32}{5} \delta(1 - \delta)}, \quad \delta \in (0, 1). \tag{5.7}$$

$$\gamma_2(\delta) = \frac{3}{8}(1 - 2\delta) \sqrt[4]{\frac{32}{5} \delta(1 - \delta)}, \quad \delta \in (0, 1). \tag{5.8}$$

We then obtain the expansion of $w(y)$

$$w(y, \delta) = \sqrt[4]{\frac{32}{5} \delta(1 - \delta)} \{ \sqrt[4]{\alpha} + \frac{3}{8}(1 - 2\delta) \alpha^{\frac{3}{4}} \} \cos(k\pi y) + \cdots \tag{5.9}$$

Then we consider the stability of the above asymptotic solution by setting

$$u(x, t) = c(t) \sqrt[4]{\alpha} v_1(x) + \cdots, \quad v_1(x) = \cos(\frac{k\pi x}{L}).$$

Project this solution onto the local center manifold $X^c = span\{\cos(\frac{k\pi x}{L})\}$, and from Eq. (5.4), we expand to leading order in α, then we obtain

$$\dot{c} = -\frac{1}{8} c(5c^4 - 32\delta(1 - \delta)).$$

It's clearly that $c(t) = \gamma_1$ is one solution of full time dependent PDE. Linearizing this equation about γ_1 yields

$$\dot{d} = -16\delta(1 - \delta)d, \tag{5.10}$$

from Eq. (5.9), we conclude that this solution is stable for small δ.

6. Stationary Solutions in the Limit as $\alpha \to \alpha_1^{*+}$

In this section, we analyze the case when the bifurcation points L_2 and $2L_1$ are close together, i.e, when $\alpha = \alpha_1^* + \varepsilon$ where ε is positive and small. We will compare the results in this section with that in.[13]

When $\alpha = \alpha_1^*$, the Eq. (1.5) has a two-dimensional center manifold X^c about the trivial solution

$$X^c = span\{\varphi_1, \varphi_2\} = span\{\cos(\frac{\pi x}{L}), \cos(\frac{2\pi x}{L})\}.$$

In our analysis, we fix α close to α_1^*, and use L as a bifurcation parameter close to $L^* = L_2(\alpha^*)$.

We write

$$u(x, t) = a(t)\varphi_1(x) + b(t)\varphi_2(x),$$

and project the differential equation onto X^c. It follows that

$$\frac{1}{L} \int_0^L \{u_t + u_{xxxx} + 2u_{xx} + (1 - \alpha)u + u^5\}\varphi_1 dx = 0,$$

$$\cdot \ \frac{1}{L} \int_0^L \{u_t + u_{xxxx} + 2u_{xx} + (1 - \alpha)u + u^5\}\varphi_2 dx = 0.$$

This yields a pair of differential equations for $a(t)$ and $b(t)$, then we obtain the following system

$$\begin{cases} \dot{a}(t) = -P(\frac{\pi}{L})a - \frac{5}{8}a^5 - \frac{35}{2}a^3b^2 - \frac{15}{2}ab^4, \\ \dot{b}(t) = -P(\frac{2\pi}{L})b - \frac{35}{4}a^4b - 15a^2b^3 - \frac{5}{2}b^5. \end{cases} \tag{6.1}$$

Since the original equation is a gradient system, and the reduced system as well, the corresponding Liyapunov functional is given by

$$V(a, b) = \frac{1}{2}P(\frac{\pi}{L})a^2 + \frac{1}{2}P(\frac{2\pi}{L})b^2 + \frac{5}{12}a^6 + \frac{35}{8}a^4b^2 + \frac{15}{4}a^2b^4 + \frac{5}{12}a^6, \tag{6.2}$$

$$\dot{a} = -V_a(a, b) \quad \text{and} \quad \dot{b} = -V_b(a, b),$$

$$\frac{dV}{dt} = V_a\dot{a} + V_b\dot{b} = -(V_a^2 + V_b^2) \le 0.$$

We will consider the structures of the set of stationary solutions of Eq. (6.1) in the (δ, u) plane, where $u = (a, b)$, and discuss their stabilities. They are defined by the pair of equations

$$V_a(a, b) = 0 \quad \text{and} \quad V_b(a, b) = 0.$$

We fix

$$\alpha = \alpha_1^* + \varepsilon, \quad \varepsilon > 0, \quad \text{and} \quad \alpha_1^* = \frac{9}{25},$$

then $2L_1 < L_2$ and we write

$$L = 2L_1 + \delta L_{gap}, \quad \text{where} \quad L_{gap} = L_2 - 2L_1.$$

Note that $L_{gap} > 0$ and $L_{gap}(\varepsilon) \to 0$ as $\varepsilon \to 0$, and we expand $2L_1$ and L_2 in powers of ε,

$$2L_1 = \frac{2\pi}{\sqrt{1 + \sqrt{\alpha_1^* + \varepsilon}}} = \frac{\pi\sqrt{10}}{2}\{1 - \frac{25}{96}\varepsilon + O(\varepsilon^2)\} \quad \text{as} \quad \varepsilon \to 0,$$

$$L_2 = \frac{\pi}{\sqrt{1 - \sqrt{\alpha_1^* + \varepsilon}}} = \frac{\pi\sqrt{10}}{2}\{1 + \frac{25}{24}\varepsilon + O(\varepsilon^2)\} \quad \text{as} \quad \varepsilon \to 0,$$

$$L_{gap}(\varepsilon) = \frac{125}{192}\pi\sqrt{10}\varepsilon + O(\varepsilon^2) \quad \text{as} \quad \varepsilon \to 0.$$

We rescale the variables according to

$$(a, b) \to \sqrt[4]{\varepsilon}(a, b) \quad \text{and} \quad t \to \frac{1}{\varepsilon}t. \tag{6.3}$$

Then we obtain the leading order equation

$$\begin{cases} \dot{a}(t) = g_1(a, b) = a\{\frac{5}{4}(1 - \delta) - \frac{5}{2}a^4 - \frac{35}{2}a^2b^2 - \frac{15}{2}b^4\}, \\ \dot{b}(t) = g_2(a, b) = b\{5\delta - \frac{35}{4}a^4 - 15a^2b^2 - \frac{5}{2}b^4\}. \end{cases} \tag{6.4}$$

6.1. *Stationary solutions*

The stationary solutions of Eq. (6.4) are the points where the null clines by Γ_1 and Γ_2 are defined by

$$\Gamma_i = \{(a, b) : g_i(a, b) = 0\} \quad \text{for} \quad i = 1, 2$$

intersect.

Obviously, Γ_1 consists of two components

$$\Gamma_1^1 = \{(a, b) : a = 0\}, \quad \Gamma_1^2 = \{(a, b) : \frac{5}{2}a^4 + \frac{35}{2}a^2b^2 + \frac{15}{2}b^4 = \frac{5}{4}(1 - \delta)\}.$$

And Γ_2 consists of two components

$$\Gamma_2^1 = \{(a, b) : b = 0\}, \quad \Gamma_2^2 = \{(a, b) : \frac{35}{4}a^4 + 15a^2b^2 + \frac{5}{2}b^4 = 5\delta\}.$$

Note that both Γ_1 and Γ_2 are invariant under the transformation $(a, b) \to (-a, -b)$, under $(a, b) \to (a, -b)$ and $(a, b) \to (-a, b)$. In what follows we shall discuss only the null clines in the first quadrant. We can immediately see that,

$$\Gamma_1^1 \bigcap \Gamma_2^1 = \{(0, 0)\}, \quad \Gamma_1^1 \bigcap \Gamma_2^2 = \{(a, b) : a = 0, b = \sqrt[4]{2\delta}\},$$

$$\Gamma_2^1 \bigcap \Gamma_1^2 = \{(a, b) : a = \sqrt[4]{\frac{1}{2}(1 - \delta)}, b = 0\}.$$

Then we next focus on the points of intersection of Γ_1^2 and Γ_2^2.

In Γ_1^2, we write

$$a^2 = r_1(\theta)\cos\theta, \quad b^2 = r_1(\theta)\sin\theta, \quad \theta \in (0, \frac{\pi}{2}).$$

In Γ_2^2, we write

$$a^2 = r_2(\theta)\cos\theta, \quad b^2 = r_2(\theta)\sin\theta, \quad \theta \in (0, \frac{\pi}{2}).$$

then we obtain two equations. At the points where the null clines intersect, we have $r_1 = r_2$, we divide it by $\sin^2\theta$ and write $x = \cot\theta$, then we obtain the following equation

$$(1 - \frac{7\gamma}{4})x^2 + (7 - 3\gamma)x + 3 - \frac{\gamma}{2} = 0, \tag{6.5}$$

where $\gamma = \frac{1-\delta}{2\delta}$.

Notice that x must be positive. When $\gamma = \frac{4}{7}$, $x = -\frac{19}{37}$, we omit this case. When $\gamma \neq \frac{4}{5}$, the discrimination of Eq. (6.5) is

$$\triangle = (7 - 3\gamma)^2 - 4(1 - \frac{7\gamma}{4})(3 - \frac{\gamma}{2})$$

$$= \frac{11\gamma^2}{2} - 19\gamma + 37 > 0, \tag{6.6}$$

then we have

$$x = \frac{3\gamma - 7 \pm \sqrt{\triangle}}{2(1 - \frac{7\gamma}{4})}. \tag{6.7}$$

If $\frac{4}{7} < \gamma < 6$, Eq. (6.5) admits one positive solution; for other γ, Eq. (6.5) admits no positive solutions.

From $\gamma = \frac{4}{7}$, we have $\delta = \frac{7}{15}$; and from $\gamma = 6$, $\delta = \frac{1}{13}$.

Therefore, for $\frac{1}{13} < \delta < \frac{7}{15}$, $\Gamma_1^2 \bigcap \Gamma_2^2$ has one point P in the first quadrant; for other δ, the intersection is empty in the first quadrant.

6.2. *Stability*

We discuss the stabilities of the stationary solutions of Eq. (1.5) by means of an analysis of Eq. (6.1).

It's easy to see that the origin O is always unstable.

When $\delta < \frac{1}{13}$, $A = (0, \sqrt[4]{2\delta})$ is a saddle; when $\delta > \frac{1}{13}$, A is a stable node.

When $\delta < \frac{7}{15}$, $B = (\sqrt[4]{\frac{1}{2}(1-\delta)}, 0)$ is a stable node; when $\delta > \frac{7}{15}$, B is a saddle. We denote

$$\rho(a,b) = \frac{5}{4}(1 - \delta) - \frac{25}{2}a^4 - \frac{105}{2}a^2 b^2 - \frac{15}{2}b^4,$$

$$\sigma(a,b) = 5\delta - \frac{35}{4}a^4 - 45a^2 b^2 - \frac{25}{2}b^4.$$

The characteristic equation associated with P is given by

$$\lambda^2 - (\rho + \sigma)\lambda + \rho\sigma - a^2 b^2(35a^2 + 30b^2)^2 = 0.$$

By the definition of Γ_1^2 and Γ_1^2, and after simple calculation, it can be verified that

$$\rho\sigma - a^2b^2(35a^2 + 30b^2)^2$$
$$= a^2b^2(10a^2 + 35b^2)(30a^2 + 10b^2) - a^2b^2(30a^2 + 35b^2)^2 < 0.$$

That's to say, the two corresponding eigenvalues have different signs, so the points in $\Gamma_1^2 \bigcap \Gamma_2^2$ are saddles.

Then next we translate these results to the stabilities of the stationary solutions of Eq. (1.3).

We draw a conclusion that the trivial solution is always unstable.

When $0 < \delta < \frac{1}{13}$, the branches which bifurcate from $2L_1$ are unstable; when $\frac{1}{13} < \delta < 1$, they are stable.

When $0 < \delta < \frac{7}{15}$, the branches which bifurcate from L_2 are stable; when $\frac{7}{15} < \delta < 1$, they are unstable.

When $\frac{1}{13} < \delta < \frac{7}{15}$, the branch which connects these branches bifurcating from L_2 and $2L_1$ is unstable.

Remark If we fix α close to α_1^*, and use L be close to $L^* = 2L_2(\alpha_1^*)$, we can also obtain above results. Comparing with the results in,[13] we find that even for the same L near $2L_2(\alpha_1^*)$, the stability of the bifurcated solution from $2L_2$ may be different, as well as the stability of the branch which connects these branches bifurcating from $2L_2$ and $4L_1$. However, stability of the branches which bifurcate from $4L_1$ is the same under the two boundary conditions.

7. Another Structure in the Bifurcation Diagrams

In this section, we focus on the cases when higher order bifurcation points coincide, i.e., when there exists an $\alpha^* \in (0, 1)$ and positive integers m and n such that

$$nL_1(\alpha^*) = mL_2(\alpha^*). \tag{7.1}$$

Since $L_1 < L_2$, it follows that $n > m$.

It's easy to see that Eq. (7.1) occurs when

$$\alpha^* = \alpha_{m,n}^* = \left(\frac{n^2 - m^2}{n^2 + m^2}\right)^2. \tag{7.2}$$

When $n = 2m$, Eq. (6.1) becomes $2L_1 = L_2$, therefore, $\alpha^* = \alpha_1^*$. It's the case discussed in §6. So here, we consider the cases when $n \neq 2m$. First we generalize the center manifold theorem.

Theorem 7.1. *Suppose that Eq. (7.1) holds for some α^*, then if $L=nL_1(\alpha^*)$, we have the following:*

(a) There exists a two-dimensional center manifold X^c spanned by the eigenfunctions φ_m and φ_n;

(b) $dim X^u = n - m - 1$, where X^u is the unstable manifold about the trivial solution.

Projecting the Swift-Hohenberg equation onto $X^c = span\{\varphi_m, \varphi_n\}$ and writing

$$u(x,t) = a(t)\varphi_m(x) + b(t)\varphi_n(x), \tag{7.3}$$

we then obtain the following:

When $n = 3m$, the new system is given by

$$\begin{cases} \dot{a} = -P(\frac{m\pi}{L})a - \frac{5}{2}a^5 - \frac{25}{4}a^4b - 15a^3b^2 - \frac{15}{2}a^2b^3 - \frac{15}{2}ab^4, \\ \dot{b} = -P(\frac{n\pi}{L})b - \frac{5}{2}b^5 - 15a^2b^3 - \frac{15}{2}a^3b^2 - \frac{15}{2}a^4b - \frac{5}{4}a^5. \end{cases} \tag{7.4}$$

When $n = 5m$, the new system is given by

$$\begin{cases} \dot{a} = -P(\frac{m\pi}{L})a - \frac{5}{2}a^5 - \frac{5}{4}a^4b - 15a^3b^2 - \frac{15}{2}ab^4, \\ \dot{b} = -P(\frac{n\pi}{L})b - \frac{5}{2}b^5 - 15a^2b^3 - \frac{15}{2}a^4b - \frac{1}{4}a^5. \end{cases} \tag{7.5}$$

When $n \neq 3m$, $n \neq 5m$, the new system is given by

$$\begin{cases} \dot{a} = -P(\frac{m\pi}{L})a - \frac{5}{2}a^5 - 15a^3b^2 - \frac{15}{2}ab^4, \\ \dot{b} = -P(\frac{n\pi}{L})b - \frac{5}{2}b^5 - 15a^2b^3 - \frac{15}{2}a^4b. \end{cases} \tag{7.6}$$

We set $\alpha = \alpha^*_{m,n} + \varepsilon(\varepsilon > 0)$, then $nL_1(\alpha) < mL_2(\alpha)$, and write

$$L = nL_1(\alpha) + \delta L_{gap},$$

where

$$nL_1(\alpha) = nL_1(\alpha^*) - \frac{\sqrt{2}\pi}{16} \frac{(m^2 + n^2)^{\frac{5}{2}}}{(n^2 - m^2)n^2}\varepsilon,$$

$$mL_2(\alpha) = mL_2(\alpha^*) + \frac{\sqrt{2}\pi}{16} \frac{(m^2 + n^2)^{\frac{5}{2}}}{(n^2 - m^2)m^2}\varepsilon,$$

$$L^*_{m,n} = nL_1(\alpha^*) = mL_2(\alpha^*) = \pi\sqrt{\frac{m^2 + n^2}{2}},$$

$$L_{gap} = \frac{\sqrt{2}\pi}{16} \frac{(m^2 + n^2)^{\frac{7}{2}}}{(n^2 - m^2)m^2n^2}\varepsilon + O(\varepsilon^2),$$

and we have

$$L = nL_1(\alpha) + \delta L_{gap} = L^*_{m,n}\{1 + \frac{(m^2 + n^2)^2\varepsilon}{8n^2(n^2 - m^2)}(\frac{m^2 + n^2}{m^2}\delta - 1) + O(\varepsilon^2)\},$$

$$P(\frac{m\pi}{L}) = -\frac{m^2 + n^2}{n^2}(1 - \delta)\varepsilon + O(\varepsilon^2),$$

$$P(\frac{n\pi}{L}) = -\frac{m^2 + n^2}{m^2}\delta\varepsilon + O(\varepsilon^2).$$

We rescale the variables according to

$$(a, b) \rightarrow \sqrt[4]{\varepsilon}(a, b) \quad \text{and} \quad t \rightarrow \frac{1}{\varepsilon}t. \tag{7.7}$$

When $n \neq 3m$, $n \neq 5m$, we obtain the leading order equations

$$\begin{cases} \dot{a}(t) = a\{\frac{m^2 + n^2}{n^2}(1 - \delta) - \frac{5}{2}a^4 - 15a^2b^2 - \frac{15}{2}b^4\}, \\ \dot{b}(t) = b\{\frac{m^2 + n^2}{m^2}\delta - \frac{15}{2}a^4 - 15a^2b^2 - \frac{5}{2}b^4\}. \end{cases} \tag{7.8}$$

Here we consider the case when $n \neq 3m$, $n \neq 5m$.

Note that the null clines both Γ_1 and Γ_2 are invariant under the transformation $(a, b) \rightarrow (-a, -b)$, under $(a, b) \rightarrow (a, -b)$ and $(a, b) \rightarrow (-a, b)$. In what follows we shall discuss only the null clines in the first quadrant.

Because the reduced system is the same as the one given in,[13] therefore, the conclusion is the same.

Then we have the following results about the stabilities of the stationary solutions of Eq. (1.5).

The trivial solution is always unstable.

When $0 < \delta < \frac{m^2}{m^2 + 3n^2}$, the branches which bifurcate from nL_1 are unstable; when $\frac{m^2}{m^2 + 3n^2} < \delta < 1$, the branches which bifurcate from nL_1 are stable.

When $0 < \delta < \frac{3m^2}{3m^2 + n^2}$, the branches which bifurcate from mL_2 are stable; when $\frac{3m^2}{3m^2 + n^2} < \delta < 1$, the branches which bifurcate from mL_2 are unstable.

And the branch which connects these branches bifurcating from mL_2 and nL_1 is unstable.

Remark We can observe that when $n \neq 2m$, $n \neq 3m$ and $n \neq 5m$, we have the same conclusion as in,[13] however, when $n = 3m$, the reduced system is different from the one in,[13] we may expect that there are different results under this case between two boundary conditions. Because of complexity of Eq. (7.4) and Eq. (7.5), now we have not yet obtained the situation when $n = 3m$ or $n = 5m$, we will investigate this two cases elsewhere for further discussion.

Acknowledgments

This work was supported by the NSFC Grant No. 10871097, and National Basic Research Program of China (973 Program) No. 2007CB814800.

References

1. J. B. Swift and P. C. Hohenberg, Hydrodynamic fluctuations at the convective instability. *Phys. Rev. A.* **15**(1977), 319–328.
2. P. Collet and J. P. Eckmann, *Instabilities and fronts in extended systems, Princeton Series in Physics.* Princeton University Press, Princeton, NJ, 1990.
3. M. F. Hilali, S. Metens, P. Borckmans and G. Dewel, Pattern selection in the generalised Swift-Hohenberg model. *Phys. Rev. E.* **51**(1995), 2046-2052.
4. H. Sakaguchi and B. Malomed, Grain boundaries in two-dimensional traveling-wave patterns. *Phys. D.* **118**(1998), 250-260.
5. L. A. Peletier and V. Rottschäfer, Pattern selection of solutions of the Swift-Hohenberg equation. *Phys. D.* **194**(2004), 95-126.
6. L. A. Peletier and J. F. Williams, Some Canonical Bifurcations in the Swift-Hohenberg Equation. *SIAM J. Appl. Dynamical Systems,* **6**(2007), 208-235.
7. C. J. Budd and R. Kuske, Localized periodic patterns for the non-symmetric generalized Swift-Hohenberg equation, *Phys. D.* **208**(2005), 73-95.
8. S. Day, Y. Hiraoka, K. Mischaikow and T. Ogawa, Rigorous Numerics for Global

Dynamics: A Study of the Swift-Hohenberg Equation. *SIAM J. Appl. Dynamical Systems.* **4**(2005), 1-31.

9. J. H. P. Dawes, Localized Pattern Formation with a Large-Scale Mode: Slanted Snaking. *SIAM J. Appl. Dynamical Systems.* **7**(2008), 186-206.

10. D. J. B. Lloyd, B. Sandstede, D. Avitabile and Alan R. Champneys, Localized Hexagon Patterns of the Planar Swift-Hohenberg Equation. *SIAM J. Appl. Dynamical Systems*, **7**(2008), 1049-1100.

11. J. B. Van Den Berg and J. P. Lessard, Chaotic Braided Solutions via Rigorous Numerics: Chaos in the Swift-Hohenberg Equation. *SIAM J. Appl. Dynamical Systems*, **7**(2008), 988-1031.

12. A. Ghazaryan and B. Sandstede, Nonlinear convective instability of Turing-unstable fronts near onset: a case study. *SIAM J. Appl. Dynamical Systems*, **6**(2007), 319-347.

13. Qingkun Xiao and Hongjun Gao, Bifurcation analysis of the Swift-Hohenberg equation with quintic nonlinearity, to appear in *International Journal of Bifurcation and Chaos*, 2009.

14. J. K. Hale, *Asymptotic Behavior of Dissipative Systems.* American Mathematical Society, Providence, 1988.

Perspectives in Mathematical Sciences
Interdisciplinary Mathematical Sciences, Volume 9, 2009
pp. 293–306

Chapter 15

A New GL Method for Mathematical and Physical Problems

Ganquan Xie and Jianhua Li

GL Geophysical Laboratory
GLganquan@gmail.com

In this chapter, a new GL method for solving ordinary and partial differential equations is proposed. These equations may govern the electromagnetic fields, macro- and micro-physical, chemical, and financial problems in science and engineering. The domain can be finite, infinite, or part of an infinite domain with a curved surface. The differential equation may be held in an infinite domain which includes a finite inhomogeneous domain. The inhomogeneous domain is divided further into finite sub-domains. The solution of the differential equation can be represented as an explicit recursive sum of the integrals in the inhomogeneous sub-domains. An explicit relationship between forward modeling and inversion is found. The analytical solution of the equation in the infinite homogeneous domain is called an initial global field. The global field is updated by local scattering field successively, sub-domain by sub-domain. Once all sub-domains are scattered and the finite updating process is finished in all the sub-domains, the solution of the equation is obtained. Such a method is called the Global and Local field method, in short the GL method. The GL method is totally different from Finite Element Method (FEM) and Finite Difference Method (FD). The GL method directly assembles the inverse matrix and gets the solution successively sub-domain by sub-domain. There is no big matrix equation needed to be solved in the GL method which overcomes the difficulties in FEM and FD in solving big matrix equations. When the FEM and FD are used to solve a differential equation in an infinite domain, the artificial boundary and absorption boundary condition are necessary and difficult. The error reflections from the artificial absorption boundary condition downgrade the accuracy of the forward solution and damage the inversion resolution. The GL method resolves the artificial boundary difficulty encountered in the FEM and FD methods. There is no artificial boundary and no absorption boundary condition for infinite domains in the GL method. A triangle integral equation of Green's functions is proposed and several theorems for the theoretical analysis of the GL method are established. A numerical discretization of the GL method is presented. The numerical solution of the GL method is shown to converge to the exact solution when the maximum diameter of the sub-domain is going to zero. An error estimation of the GL method for solving the wave equation is presented. The simulations show that the GL method is accurate, fast, and stable for solving elliptic, parabolic, and hyperbolic equations. The GL method has wide applications in the 3D electromagnetic (EM) fields, 3D elastic and plastic etc seismic fields, acoustic fields, flow fields, and quantum fields. The GL method software for the above 3D EM etc fields is developed by the authors in the GL Geophysical Laboratory.

1. Introduction

Analytical method and numerical method are the two ways for solving differential equations. Since the beginning of history, they have been developed independently from each other. Analytical methods are only suitable for uniform medium with isotopic entire infinite space, layered media with finite space, and the region with regular geometrical shape, such as spheroidal region and so on. There are many journals and books which publish analytical methods. Analytic methods are not suitable for solving differential equations in the region with an irregular geometrical shape; but the numerical methods may be made available to do so. However, the existing numerical methods, such as finite element method (FEM), finite difference method (FD) and the Born approximation (BA) and so on, have several difficulties and limitations:

1) In FEM and FD methods, solving certain large scale matrix equation is necessary and difficult, which is hard to by-pass.

2) When using FEM and FD methods to solve differential equations in the infinite domain, the artificial boundary and its absorption condition are necessary and difficult.

3) The origin, north pole and south pole points are singular points, and can not be removed, in cylindrical and spherical coordinates.

4) The process for solving big matrix equation occurring in finite element method and finite difference method is hard to run parallelly.

5) Born approximation can only be used for low-frequency and low-contrast materials.

6) The formats of FEM and FD for high-order differential equations are rather complicated.

The new GL method proposed by us, combining the advantages of both analytical and numerical methods, provides a way to overcome the above-mentioned difficulties in FD method and FEM method:

1) Fundamentally different from the way in combining the integral big matrix in finite element method, the GL method directly assembles inverse matrix and gets solution successively subdomain by subdomain. There is no big matrix equation needed to solve for the GL method in the GL method which overcomes the difficulties in FEM and FD in solving big matrix equations. No matter how large the region is, the GL method needs only to solve some small, such as certain 3×3 or 6×6 matrix equations. For some scalar differential equation problems, the use of the GL method does not require to solve any matrix equation.

2) Since the introduction of the Sommerfeld radiation condition, more than 80 years have been spent in researching the artificial boundary and its absorption

and radiation conditions. Regarding to both FEM method and FD method, the structures of necessary artificial boundary and its absorption condition are difficult. Many mathematicians, physicists and engineers have made their contributions. Professor Feng Kang proposed the higher-order radiation and absorption conditions. Professors Yin Long-an and Shao Xiumin have done an outstanding study on artificial boundary conditions. In the 45-minute invited speeches at previous conferences of International Congress of Mathematicians, there is no shortage of the subject related to the artificial boundary and its absorption condition. But there are reflection errors included in many absorb conditions, computation proved that the reflection error indeed exist in angular point due to the PML artificial absorption boundary condition. The error reflections from the artificial absorption boundary condition downgrade the accuracy of the forward solution and damage the inversion resolution. Different from all numerous past efforts, the new GL method here no longer needs any artificial boundary, absorbs the boundary condition, and overcomes the difficulties for constructing artificial boundary and absorption boundary condition.

4) Without needing to solve any big matrix equation, the GL method is a parallel algorithm for its own right. On the other hand, the use of the FEM and FD methods involves the process of solving big matrix equations which is hard to run parallelly.

5) The Born approximation can only be used for low-frequency and low-contrast mediums. Our GL method does not have such restrictions.

6) The FD method in solving higher order differential equation does not increase the degree of complexity, when non-uniform medium coefficients only appear in low-order terms.

Our GL method may be used to solve the elliptic type, parabolic type and hyperbolic type partial differential equations, defined in certain infinite regions which may contain finite regions occupied by inhomogeneous media. Such infinite regions include entire spaces, infinite laminated regions, or partially infinite regions bounded by some surfaces. These differential equations may describe macrophysics and microphysics phenomena with electromagnetic fields and other fields in mechanics, chemistry, finance study and engineering. The solution regions of differential equations may be infinite which contain some bounded and finite homogeneous-medium subregions. The nonhomogeneous-medium region is divided into the sum of certain subdomains. In the above infinite region including non-uniform medium, we can explicitly express the solution of the differential equation in a recursive sum of certain integrals for inhomogeneous subdomains. We can transfer the non-linear recursive relationship consisting of forward and backward calculations into an explicit procedure.

In the homogeneous infinite region, the analytical solution of differential equation is known as the initial general field. The global field is updated by local scattering field, subdomain by subdomain successively. Once all subdomains are scattered and the finite updating process is finished in all the subdomains, then the solution of the equation is obtained. We call it the global and local fields method, referred to as the GL method. The GL method does not require an artificial boundary nor an absorption boundary condition in order to find the exact solution of a differential equation successively subdomain by subdomain. The FEM and FD methods are simply numerical methods. However, the GL method unifies analytic method and numerical method in a compatible way. We have proposed and developed the formulations for solution fields and for integral equations of the Green functions. We have proposed a numerically discrete GL method. When the diameters of each subdomains tend to zero, we have proved that the numerical solutions of wave equation solved by GL method converge to exact solutions. An error estimate of the GL method solution of a wave equation has also been obtained. If the trapezoidal integral is used, the rate of convergence is $O(h^2)$. If the Gauss integral is used, the upper rate of convergence is $O(h^4)$. And in using the FEM and FD methods to solve a differential equation in an infinite region, the artificial boundary need to depart very far from the non-homogeneous domain, which consumes a lot of units and grids outside the non-homogeneous domain. But the GL method only requires to calculate the non-homogeneous domain, thereby greatly saves the memory and speeds up the computation.

The GL method has relaxed the restrictions on geometry of the grids and the subdomains. Arbitrary convex or concave polyhedrons, partial balls, partial cylinders and so on can all be used as subdomains in the same computational region. The GL method can be an algorithm with network, semi-net or non-network algorithm. Many experiments indicate that, the solutions of wave equations obtained by the GL method converge rapidly to the exact solutions, but the solutions by FEM are not the case. Solutions of elliptic, parabolic and hyperbolic equations of the GL simulation results show that, our GL method has the advantages of being accurate, rapidly convergent and stable. The GL method can be widely used in three-dimensional problems arising in electromagnetic fields, elasto-plastic mechanics fields, seismic wave fields, acoustic fields, fluid fields, quantum fields, nano-materials, electromagnetic stirring, forest exploitation and exploration, and geophysics. A GL software aimed at computing three-dimensional electromagnetic fields has been studied and developed.

2. Integral Equation

2.1. *Wave Equation*

We consider the following three-dimensional wave equation in frequency domain

$$\Delta u(r) + \frac{\omega^2}{c^2(r)} u(r) = s(r), r \in R^3 \tag{2.1}$$

Where r is a space variable, $r = (x, y, z)$, $r \in R^3$, $c(r)$ is the coefficient, the so-called wave velocity, namely a nonhomogeneous media function on finite domain

$\Omega \subset R^3$, is a constant in $r \in R^3 \backslash \Omega$, $c(r) = c_b \, c_b$, ω is the frequency, $u(r) = u(r, \omega)$ is the unknown wave field function, $\Delta = \Delta_r$ is the Laplacian or Laplace operator, $S(r)$ is a source function known in the finite domain, we consider the Delta function point source item specially, $s(r) = \delta(r - r_s)$, r_s is a finite source. If in entire space R^3 ,$c(r) = c_b$, The solution $u_b(r)$ of equation (2.1) is called the incident ray in background medium, $u_b(r) = e^{-ik_b|r-r_s|}/4\pi|r - r_s|$, $k_b = \omega/c_b$,and $k(r) = \omega/c(r)$, moreover the solution of equation (2.1) meets the following distant field radiation condition,

$$\lim_{r \to \infty} r \left(\frac{\partial u}{\partial n} - iku \right) = 0, \tag{2.2}$$

And $ru(r)$ is finite at infinity.

2.2. Green Function

The Green function of wave equation (2.1) satisfies the following relation,

$$\Delta G(r, r') + \frac{\omega^2}{c^2(r)} G(r, r') = \delta(r - r'), \quad r \in R^3, r' \in R^3, \tag{2.3}$$

And the following distant field radiation condition

$$\lim_{r' \to \infty} r' \left(\frac{\partial G(r', r)}{\partial n_{r'}} - ikG(r', r) \right) = 0, \tag{2.4}$$

With $r'G(r, r')$ be finite at infinity. Where r is the virtual source, r' is the virtual acceptance point, $\delta(r - r')$ the pulse Delta function. If $c(r) = c_b$ in entire space R^3, The Green function is called the Green function in background medium

$$G_b(r, r') = e^{-ik_b|r-r'|}/4\pi|r - r'|.$$

2.3. Integral Equation

Theorem 1 Let Ω be a finite and bounded domain with non-uniform medium, embedded in uniform infinite medium. Point source r_s is also finite, then the wave equation (2.1) is equivalent to the following integral equation

$$u(r) = u_b(r) - \omega^2 \int_{\Omega} \left(\frac{1}{c^2(r')} - \frac{1}{c_b^2} \right) G(r', r) u_b(r') dr' \tag{2.5}$$

where $u_b(r)$ is the incident ray in background medium, $G(r', r)$ is Green function of equation (2.1).

This proof of the theorem is rather long, may refer to author's papers [11], [6], [19], [20].

Theorem 2 Let Ω be a finite and bounded domain with non-uniform medium, embedded in uniform infinite medium. If the right term of equation (2.1) is the Delta function, then equation (2.5) is the integral equation of Green function.

This proof is similar to that of Theorem 1, see references [11-12].

3. New GL method Combining Global and Local fields

Here, we proposed our GL method Combining Global and Local fields as follows:
  1) Let domain Ω be divided into a set of N-subdomains $\{\Omega_k\}$, such that

$$\Omega = \cup_{k=1}^{n}\Omega_k.$$

  2) Let $u_0(r) = u_b(r), G_0(r',r) = G_b(r',r)$, $u_b(r)$ be the analytic incident ray in uniform background medium, $G_b(r',r)$ an analytic Green function for uniform background medium. By inductive description, suppose that, u_{k-1} and $G_{k-1}(r',r)$ already known in the $(k-1)th$ step computing Ω_{k-1}, We solve Green function integral equation(5) in subdomain. Ω_k to obtain $G_k(r',r)$.
  3) In subdomain Ω_k use the following integral formula,

$$u_k(r) = u_{k-1}(r) - \omega^2 \int\limits_{\Omega_k} \left(\frac{1}{c^2(r')} - \frac{1}{c_b^2}\right) G_k(r',r)u_{k-1}(r')dr', \qquad (3.1)$$

we obtain $u_k(r)$.
  4) For $k = 1, 2, \ldots, N$, steps 2) and 3) forming a finite iteration, $u_N(r)$ is our solution of GL method combined Global and Local fields.

4. Basic Theory of GL

4.1. *Fundamental Theorem*

Theorem 3 If the Green function is obtained via computing by GL process 1)-4) in Section 3, then the function $G_n(r',r)$ satisfies

$$\Delta_{r'}G_n(r',r) + \frac{\omega^2}{c^2(r')}G_n(r',r) = \delta(r'-r), \qquad (4.1)$$

And radiation condition 4), $G_n(r',r)$ is Green function of the equation (2.1). Proof of this theorem is rather longer, we avoid to state it.
Theorem 4 If the solution function $u_n(r)$ and Green function $G_n(r'',r)$ are obtained via computing by GL process 1)-4) in Section 3, then the function $u_n(r)$ satisfies the following integral equation:

$$u_n(r) = u_b(r) - \omega^2 \int\limits_{\Omega} \left(\frac{1}{c^2(r')} - \frac{1}{c_b^2}\right) G_n(r',r)u_b(r')dr', \qquad (4.2)$$

And distant field radiation condition 2), $u_n(r)$ is the exact solution of wave equation (2.1).
  We use mathematical induction process to prove this theorem. Obviously, when $N = 0$, this theorem is tenable. Assume for any $m-1$, $\tilde{\Omega}_{m-1} = \cup_{k=1}^{m-1}\Omega_k$, integral equation (5.2) is valid, then we have

$$u_{m-1}(r) = u_b(r) - \omega^2 \int\limits_{\tilde{\Omega}_{m-1}} \left(\frac{1}{c^2(r')} - \frac{1}{c_b^2}\right) G_{m-1}(r',r)u_b(r')dr'. \qquad (4.3)$$

$m - th$ step iteration of GL Method, gives

$$u_m(r) = u_{m-1}(r) - \omega^2 \int_{\Omega_m} \left(\frac{1}{c^2(r')} - \frac{1}{c_b^2} \right) G_m(r', r) u_{m-1}(r') dr'. \qquad (4.4)$$

From equation (4.3) iterating $u_{m-1}(r)$ to equation(4.4), and after some derivation and finishing, we have

$$
\begin{aligned}
u_n(r) &= u_b(r) - \omega^2 \int_{\Omega_n} \left(\frac{1}{c^2(r')} - \frac{1}{c_b^2} \right) G_n(r', r) u_b(r') dr' \\
&\quad - \omega^2 \int_{\cup_{k=1}^{n-1} \Omega_k} \left(\frac{1}{c^2(r'')} - \frac{1}{c_b^2} \right) G_n(r'', r) u_b(r'') dr'' \\
&= u_b(r) - \omega^2 \int_{\cup_{k=1}^{n} \Omega_k} \left(\frac{1}{c^2(r'')} - \frac{1}{c_b^2} \right) G_n(r'', r) u_b(r'') dr'' \\
&= u_b(r) - \omega^2 \int_{\Omega} \left(\frac{1}{c^2(r')} - \frac{1}{c_b^2} \right) G_n(r', r) u_b(r') dr'.
\end{aligned}
\qquad (4.5)
$$

The equation (4.5) is really the equation (4.2).

We already proved that $u_n(r)$ satisfies the integral equation (4.2). Its function $G_n(r'', r)$ satisfies (4.1) and the distant field radiation condition 4), $G_n(r'', r)$ is Green function of the equation (2.1). From Theorem 1, also satisfies the radiation condition 2), therefore, it is the exact solution of wave equation (2.1). Theorem 4 is proven completely.

Notice: From the fundamental Theorem 3 and Theorem 4, our GL finite iterative solution has nothing to do with the rank order of subdomain Ω_k. In third section, what we proposed is the exact solution by GL method. Theorem 3 and Theorem 4 have proven our GL method's solution is the exact solution of wave equation (2.1).

4.2. GL Numerical Method

In third, what we proposed is the exact GL method. The step 2) and step 3) in the GL method, we use the numerical method to calculate integral. We obtain the GL numerical method. Essentially, in GL method, step 2) is using the local scattering inverse matrix on Ω_k to modify the global inverse matrix; The step 3) is using the local scattering field on Ω_k to modify the global field. When we use the trapezoidal numerical integration to calculate the integral, the GL numerical method is simply the algebra iterative process, there is no need to solve the matrix equation. Yet we still use the Gauss numerical integration to calculate integral, there are only a small matrix equation no more than 10 orders need to solve.

5. GL Method for Solving Parabolic Type Partial Differential Equation

The GL method is suitable for solving wave equation, is also suitable for elliptic and parabolic type equation.

Theorem 5 Let Ω be a finite and bounded domain filled with non-uniform media, point source r_s is also finite, then the following parabolic equation,

$$\Delta u(r) - \frac{i\omega}{c^2(r)} u(r) = \delta(r - r_s), r \in R^3, \tag{5.1}$$

is equivalent to the following integral equation,

$$u(r) = u_b(r) + i\omega \int_\Omega \left(\frac{1}{c^2(r')} - \frac{1}{c_b^2} \right) G(r', r) u_b(r') dr', \tag{5.2}$$

where $u_b(r)$ is the background medium incident ray, $G(r', r)$ is the Green function of equation (5.1), they meet the following radiation conditions 2) and 4) separately, where $k_b = \sqrt{-i\omega}/c_b$.

The proof is very similar to it of Theorem 1.

6. GL Method for Solving One-dimensional Wave Equation

In Section 2, we described the GL method for solving three-dimensional wave Equation (2.1), now, we use it to solve one-dimensional wave equation

$$\frac{\partial^2 u}{\partial x^2} + \frac{\omega^2}{c^2(x)} u = \delta(x - x_s), x \in R. \tag{6.1}$$

The Green function of equation (6.1) satisfies

$$\frac{\partial^2 G(x', x)}{\partial x'^2} + \frac{\omega^2}{c^2(x')} G(x', x) = \delta(x' - x). \tag{6.2}$$

Let $c = c_b$, Equations (6.1) and (6.2) will turn into their background medium equations, respectively.

Theorem 6 For one-dimensional wave equation (6.1) and its Green function equation (6.2), Theorem 1 and Theorem 2 are valid.

The proof process of Theorem 6 is similar to those of Theorem 1 and Theorem 2. It is worth to note that; in case of one-dimensional, when x tends to infinite, the Green function of wave field does not decay to zero. What fortunately is, due to distant field radiation condition, $\lim_{x \to \infty} (\partial u/\partial x - iku) = 0$, caused the item of boundary condition still tend to zero. The Theorem 6 is proved.

7. The Simulation of GL Method for Solving One-dimensional Wave Equation

7.1. *Discretization of GL Method*

The discretization of the GL method for solving one-dimensional wave equation is as follows:

1) Let the interval $[a, b]$ be divided into n sub-intervals, shown as in (7.1),

$$a = x_1 < x_2 < \cdots < x_{k-1} < x_k < x_{k+1} < \cdots < x_n < x_{n+1} = b. \tag{7.1}$$

2) Let $u_0(x) = u_b(x)$, and $G_0(x', x) = G_b(x', x)$, shown as follows,

$$u_0(x) = u_b(x) = -\frac{e^{-ik_b|x-x_s|}}{2ik_b}, \quad \omega > 0, \tag{7.2}$$

$$G_0(x', x) = G_b(x', x) = -\frac{e^{-ik_b|x'-x|}}{2ik_b}, \quad \omega > 0. \tag{7.3}$$

According to induction method, assume that u_{k-1} and $G_{k-1}(x', x)$ has been calculated in the $(k-1)th$ step. We solve the Green function integral equation (7.2) in the interval $[x_{k-1}, x_k]$ to get $G_k(x', x)$. Using the GL approach to solve the one-dimensional wave equation, we have achieved a very high accuracy solution. We have proved that the GL numerical solution converges to the exact solution.

3) The GL iterative formula,

$$u_k(x) = u_{k-1}(x) - \omega^2 \int_{x_k}^{x_{k+1}} ((\frac{1}{c^2(x)} - \frac{1}{c_b^2})G_k(x', x)u_{k-1}(x'))dx', k = 1, 2, \ldots, n.$$

$$\tag{7.4}$$

4) $u_n(x)$ is a GL solution of wave equation (6.1).

7.2. *Convergence*

Theorem 7 Suppose in Section 7, two steps GL iteration 2) and 3) for integral equation (), the trapezoidal integral is applied to calculate numerical integrations in $G_{k-1}(x'', x')$ and $u_{k-1}(x')$, we may prove: when $\max|x_{k+1} - x_k| = l \to 0$, then GL numerical solution $u_n^h(x)$ is convergent to its exact solution. Moreover, we have $|u_n^h(x) - u(x)| \le ch^2$, where c is a fixed constant, independent of h. The proof of Theorem 7 is omitted For simplicity.

Theorem 8 Under conditions of Theorem 7, if Gaussian numerical integration is used in the numerical integration, then we have the super convergence estimate,

$$|u_n^h(x) - u(x)| \le ch^4.$$

The proof of Theorem 8 is omitted.

7.3. *Computational Simulation of GL Method*

In the bounded interval $[0, 6]$ with inhomogeneous medium, we use the new GL method 1)-4), to solve one-dimensional wave equation() in a discrete way; in the interval $[0, 6] = [0, 2] \cup [2, 4] \cup [4, 6]$, $c_b = 3 \times 10^7 m/s$; in subinterval $[0, 2]$, $c_1 = 3 \times 10^7 m/s$; in subinterval $[2, 4]$, $c_2 = 21,276,593m/s$; in subinterval $[4, 6]$, $c_3 = 17,336,030m/s$. A dipole source is located at $x_s = 2$, 128 frequencies are used to calculate the wave field; the minimum frequency is 1 Hz, the highest frequency is 3.14×108 Hz. We using GL method obtained the precise wave field result. Our GL wave field and the exact wave field are shown in Figure 15.1 to Figure 15.6. 33 nodes were used in calculating; on PC, the CPU time is 11 seconds. For high-frequency waves, we the wave field calculated by GL method, is matched precisely to the one-dimensional exact wave field; however, the FEM solutions of one-dimensional wave field is deviated very far from the one-dimensional exact wave field.

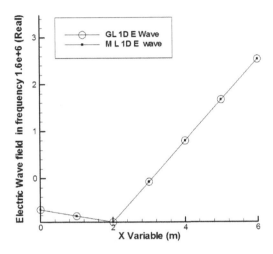

Fig. 15.1.   GL and Exact Electric wave with freq. $1.6e^6$ Hz

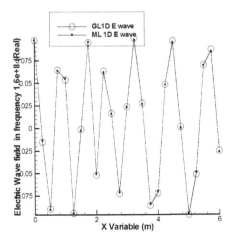

Fig. 15.2.   GL and Exact Electric wave with freq. $1.6e^8$ Hz

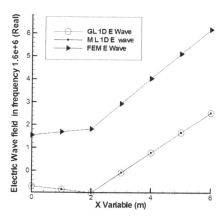

Fig. 15.3. GL, Exact and FEM Electric wave with freq. $1.6e^6$Hz

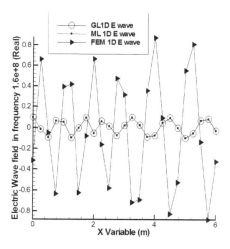

Fig. 15.4. GL, Exact and FEM Electric wave with freq. $1.6e^8$Hz

8. Memories and Discussion

Thirty-two years ago, my (G. Q. Xie's) article "3-D finite element method for elastic questions" was published in the journal *Mathematics in Practice and Theory*, Vol.1 (1975) [5]. Professor Feng Kang's article proposed difference schemes based on variational principle (1965) [1]. The article of Professors Huang Hongzhi, Wang Jinyen, Cui Junzi and Lin Zongkai, researched and developed variational difference schemes for two-dimensional elastic problem (1966)[2]. Our article studied and developed three dimensional finite-element method [5]. These were the early finite-element studies, independent of the works of western scholars. We cherish the memory of

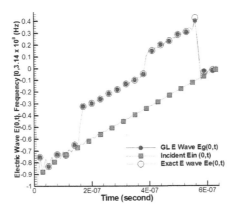

Fig. 15.5.   GL and Exact Electric wave E(0,t) in time domain

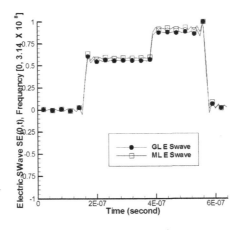

Fig. 15.6.   GL and Exact Scat. Electric S-wave SE(0,t) on time

Professor Feng Kang and fellow professors and colleagues. The name "Finite element" was first seen in the article written by Professor Cui Junzhi published in a Chinese scholarly journal (1972)[3]. Professor Lin Qun comprehensively and systematically developed and led research work for the theory of finite element methods (1999) [4]. Professor Guo Youzhong made a good summary and refinement of the mathematical methods for physical science (1993)[6]. In 1972, Professor Li Jainhua and I completed the research of "3-D finite element method for elastic questions" in Hunan Institute of Computer Science. In Spring 1973, we two, together with Professors Wang Xianru and Gu Yinghua, with our own three-dimensional finite-element program, obtained some high accuracy results. In 1999, we published a super-relaxation SOR method, a new secret recipe [28]. Since 1973, we have intended to "completely change the customary way in finite-element method, only to solve a big matrix gathered by matrices," and started a long dream. In 1982, for the problem of finite domain, we found a new method which "directly assembles the inverse matrix and gets the solution successively subdomain by subdomain." In 1994, for the problem over an infinite domain, we proposed a "global integral and local differential, forward and backward algorithm GILD", which realizes not only "directly assembling the inverse matrix and getting the solution successively subdomain by subdomain.", but also, using a method of coupling the integral equation with the Green function kernel equations and Galerkin equation", overcomes the difficulties brought about by the artificial boundary and absorption boundary condition [7-8]. In 2002, we also proposed and developed the AGILD method [9-10] and, our dream came true: We established the global and the local field GL method [11-12].

9. Application and Conclusion

The application of the GL method is very wide [13-26], especially suitable for parallel algorithm hardening. We have put forward the differential and integral equations for elastic displacement field, acoustic field, and electromagnetic field [7-8], [26-28]. We have established and developed the GILD and GL models for many applications including three-dimensional and two-dimensional elastic wave field [20-22], electromagnetic field, earthquake wave fields, sound wave fields [9-12], petroleum and mineral resource geophysics [10], forest mining, quantum nano electro-magnetism QEM materials [15-16], fluids, temperature fields, physical-chemical and financial mathematics [18]. From 2002 to the present, we have devoted ourselves to developing and applying the GL algorithm software [11-28].

The GL method is an original piece of research work. The references quoted here are mostly the articles written by the authors.

References

1. Feng Kang, On the array of difference based on variation principle, Applied Mathematics and Computational Mathematics, 1965, vol.4.
2. Huang Hongzhi, Wang Jinyen, Cui Junzi, Lin Zongkai, On the array of difference based on variation principle of elasticity theory, Applied Mathematics and Computational Mathematics, 1966, vol.3.
3. Cui Junzi, Variation method and finite element method, Mathematics in Practice and Theory, 1972, vol.1.
4. Lin Qun, Yan Xinxin, High accuracy finite element method and estimate, Science Press, 1999.
5. Xie Ganquan, 3-D finite element method for elastic questions, Mathematics in Practice and Theory, 1975, vol.1, pp.8-41.
6. Guo Youzhong, Li Qingxi, Mathematical methods for physical science, Wuhan University Publishing House, 1993.
7. Xie Ganquan, Li J., Majer E., Zuo D., Oristaglio M., 3-D electromagnetic modeling and nonlinear inversion: Geophysics, 2000, 65, no.3, pp.804-8228.

8. Xie Ganquan, Li J., New parallel SGILD modeling and inversion, Physics D, 1999, v.133, pp.477-487.

9. Li Jianhua, Xie G., Xie L., Xie L., New Stochastic AGLID EM Modeling and Inversion, PIERS Online, 2006, vol.2, no.5, pp.490-494.

10. Li Jianhua, Xie G., Oristaglio, M., Xie L., Xie F., A 3D-2D AGILD EM Modeling and Inversion Imaging, PIERS Online, 2007, vol.3, no.4, pp.423-429.

11. Xie Ganquan, Xie F., Xie L., and Li J. New GL method and its advantages for resolving historical difficulties, Progress In Electromagnetics Research, PIER, 2006, no.63, pp.141-152, 2006.

12. Xie Ganquan, Li J., Xie L., Xie F., A GL Metro Carlo EM Inversion, Journal of Electromagnetic Waves and Applications, 2006, vol.20, no.14, pp.1991-2000, 2006.

13. Xie Ganquan, Li J., Xie L., Xie F., GL EM Modeling and Inversion Based On the New EM Integral Equation. Report of GLGEO patent, 2005, No.GLP05001, pp.38-96.

14. Li Jianhua, Xie G., Xie L., Xie F., A 3D GL EM Modeling and Inversion for Forest Exploration and Felling, PIERS Online, 2007, vol.3, no.4, pp.402-410.

15. Xie Ganquan, Li Jianhua, Li J., Xie F., 3D and 2.5D AGLID EMS Stirring Modeling in the Cylindrical Coordinate System, PIERS Online, 2006, vol.2, no.5, pp.505-509.

16. Xie Ganquan, Li Jianhua, Xie Feng,, Xie Lee, 3D GL EM and Quantum Mechanical Coupled Modeling for the Nanometer Materials, PIERS Online, 2007, vol.3, no.4, pp.418-422.

17. Xie Ganquan, Li J., Xie L., Xie F., Li J., The 3D GL EM-Flow-Heat-Stress Coupled Modeling, PIERS Online, 2007, vol.3, no.4, pp.411-417.

18. Li Jianhua, Xie G., Xie L., A new GL Method for Solving Differential Equation in Electromagnetic and Phys-Chemical and Financial Mathematics, 2007, to be publish in PIERS online.

19. Xie Ganquan, Lin C.C., Li J., GILD EM modeling in nanometer material using magnetic field integral equation. Published in J. MATHMATICA APPLICATA, 2003, no.16, pp.149-156.

20. Xie Ganquan, Li J., Lin C.C., New SGILD EM modeling and inversion in Geophysics and Nano-Physics: Three Dimensional Electromagnetics, 2002, vol.2, pp.193-213.

21. Lin Chien-Chang, Xie Ganquan, Li Jianhua, Deformation analysis for materials using GILD modeling method, The Chinese Journal of Mechanics-Series A, 2003, vol.19, no.1, pp.73-81.

22. Xie Ganquan, Xie F., and Li J., New GL and GILD Superconductor Electromagnetic Modeling, PIERS Online, 2005, vol.1, no.2, pp.173-177.

23. Xie Ganquan, Li J., Li Jianhua, New AGILD EMS Electromagnetic Field Modeling, PIERS Online, 2005, vol.1, no.2, pp.168-172..

24. Xie Ganquan, Li J., Xie F., 2.5D AGILD Electromagnetic Modeling and Inversion, PIERS Online, 2006, vol.2, no.4, pp.390-394.

25. Xie Ganquan, A new iterative method for solving the coefficient inverse problem of wave equation, Communication on pure and applied math., 1986, vol.39, pp.307-322.

26. Xie Ganquan, Chen Y. M., A modified pulse spectrum technique for solving inverse problem of two-dimensional elastic wave equation. Appl. Numer. Math. 1985, vol.1, no.3, pp.217-237.

27. Xie Ganquan, Li J., Nonlinear integral equation of coefficient inversion of acoustic wave equation and TCCR iteration: Science In China, 1987, vol.32, no.5, pp.513-523.

28. Xie Ganquan, Li Jianhua, New parallel GILD-SOR modeling and inversion for E-O-A strategic simulation: IMACS series book in computational and applied math., 1999, V.5, pp.123-138.

Perspectives in Mathematical Sciences
Interdisciplinary Mathematical Sciences, Volume 9, 2009
pp. 307–352

Chapter 16

Harmonically Representing Topological Classes

Yisong Yang

*Department of Mathematics,
Polytechnic Institute of New York University,
Brooklyn, New York 11021, USA*
yyang@math.poly.edu

Dedicated to Professor Youzhong Guo on the occasion of his 75th birthday

The purpose of this article is to present a survey on some profound constructs which connect a broad range of subjects including topological classes and their harmonic representatives, calculus of variations and partial differential equations, and soliton solutions, especially the Yang–Mills instantons, in quantum field theory. Throughout, the analytic technicalities involved will be kept to the minimum but the beautiful interaction of different areas of mathematics and physics will be emphasized.

1. Introduction

This article is based on some lectures notes written for the 2006 summer graduate school at Fudan University for which the main theme then was to report some latest work on representing the Chern–Pontryagin classes over the $4m$-sphere by the Yang–Mills instantons, under the general title "Topics in Mathematical Physics". Such a subject gave me an opportunity to present to the participants a lively interplay of various important areas in contemporary mathematics and physics such as topology, geometry, analysis, quantum mechanics, and gauge field theory. The lectures were divided into four related but self-contained sections, which will be observed too in this article: Harmonic maps, Hodge theory, and instantons – In this section, we review some examples of harmonic maps between Riemannian manifolds, the Hodge theory for the harmonic representation of de Rham cohomology, the Yang–Mills instantons representing the second Chern–Pontryagin class in four dimensions, and their main applications. Quantum tunneling, imaginary time, instantons, and Liouville-type equations – In this section, we start from a simplified discussion of a one-dimensional quantum tunneling phenomenon and the motivation of using imaginary time so that we can work on an Euclidean spacetime. We review various

Yang–Mills solutions in four dimensions. In particular, we present Witten's solution via the integrable Liouville equation. Atiyah–Singer index theorem and calculation of dimension of moduli space – In this section, we give an elementary introduction to the Atiyah–Singer index theorem. We begin by reviewing some classical examples. We then show how to use it to compute the dimension of the moduli space of the Yang–Mills instantons and recover the result of Atiyah, Hitchin, and Singer in four dimensions. Topological classes and instantons in all $4m$ dimensions and nonlinear elliptic equations – In this section, we first present the general Yang–Mills theory of Tchrakian in arbitrary $4m$ dimensions. We will see that as in the classical four dimensions, the energy-minimizing Yang–Mills fields realizing a prescribed value of the Chern–Pontryagin class are self-dual or anti-self-dual. We then present the resolution of the existence problem by a variational study of a quasilinear elliptic partial differential equation. We conclude that the top Chern–Pontryagin classes over S^{4m} can all be represented harmonically by self-dual or anti-self-dual Yang–Mills fields.

We will also discuss many open problems which are closely or loosely related to our main theme.

2. Harmonic Maps, Hodge Theory, and Instantons

In all of our discussion here, the manifolds are assumed to be smooth, Riemannian, orientable, connected, without boundary, and compact (unless otherwise stated).

Example 1. Harmonic Maps Between Compact Riemannian Manifolds

Consider elastic deformation represented by a single deflection variable u. Then the normalized energy stored in the solid due to this deformation is given by

$$\int e(u)\,\mathrm{d}x = \int |\nabla u|^2\,\mathrm{d}x$$

In general, let $\phi : (M, g) \to (N, h)$ be a differentiable map and consider the energy

$$E(\phi) = \int_M e(\phi)\,\mathrm{d}V_g$$
$$e(\phi) = |\mathrm{d}\phi|^2 = g^{ij}\partial_i\phi^a\partial_j\phi^b h_{ab}.$$

Harmonic maps are the critical points of $E(\phi)$ which satisfy an elliptic equation of divergence form.

Problem. *Let $\phi_0 : M \to N$ be given. Can ϕ_0 be deformed in the sense of homotopy equivalence into a harmonic map?*

Major Theorems

1.1. $\dim(M) = 1$. So $M = S^1$ and harmonic maps are closed geodesics of N.

Theorem of Closed Geodesics: Yes. More precisely, $\pi_1(N)$ can be represented by geodesics of minimum energy.

This theorem (also referred to as Hilbert's theorem[37]) is so classical that it is often stated without referring to its original contributors. According to Bott,[19] it may be traced back to several people including Hadamard, Cartan, etc. In,[19] Bott gives an elementary proof based on using geodesic polygons.

An Important Application - Synge's Theorem:[87] A compact, orientable, and even dimensional manifold with a positive sectional curvature must be simply connected.

The proof uses the relation between the second variation of the energy functional and the sectional curvature and shows that, if a closed geodesic is nontrivial, one can always deform it to achieve a lower energy, hence arriving at a contradiction.

1.2. The sectional curvature of (N, h) is nonpositive.

Theorem of Eells and Sampson: Yes, any map is homotopic to a harmonic map which has minimum energy in its homotopy class.

Method 1: Heat flow (Eells and Sampson[36])

$$\frac{\partial \phi}{\partial t} = -\text{grad}E(\phi) = \tau(\phi),$$

$$\phi|_{t=0} = \phi_0$$

Method 2: Perturbation and calculus of variation (Uhlenbeck[96])

$$E_\varepsilon(\phi) = \int_M \left(e(\phi) + \varepsilon |d\phi|^p \right) dV_g \quad (p > m)$$

E_ε satisfies the Palais–Smale (PS) condition over $W^{1,p}(M, N)$. Find a critical point of E_ε in each connected component of $W^{1,p}(M, N)$ and pass to the $\varepsilon \to 0$ limit when the sectional curvature is nonpositive.

If the sectional curvature is negative, then the harmonic map in each homotopy class is unique.[48]

A weak existence theorem without any condition on curvature.

Theorem of Eells and Ferreira:[35] Suppose $\dim(M) \geq 3$. Then for any $\phi_0 : (M, g_0) \to (N, h)$, one can find a conformal metric g and a harmonic map $\phi : (M, g) \to (N, h)$ such that ϕ is homotopy equivalent to ϕ_0.

Method: Minimizing the functional

$$\int_M (1 + e(\phi))^p \, dV_{g_0}$$

for $p > m$ and taking $g = (1 + e(\phi))^{2(p-1)/(m-2)} g_0$.

1.3. $M = N = S^n$.

Theorem of Smith:[82] Every element of the homotopy group $\pi_n(S^n) = \mathbb{Z}$ has a harmonic representative for $n \leq 7$.

For $n = 2$, the solutions are known explicitly and carry minimum energy.[10] For $3 \leq n \leq 7$, the energy has infimum 0 which can easily be seen by a rescaling argument, and hence does not achieve its absolute minimum in any class of degree $k \neq 0$. For $n \geq 8$, the situation is not very well understood.

Method: Symmetry reduction and solution to ODEs.

Existence of n-harmonic maps between n-spheres:[109] Any map from S^n into itself is homotopy equivalent to a smooth critical point of the (conformal) n-energy

$$\int_{S^n} |\mathrm{d}\phi|^n$$

However, it is not clear whether the solution is an energy minimizer (except for $n = 2$) because the method of[109] is similar to that of Smith.[82]

1.4. $M = T^2, N = S^2$, any metrics.

Theorem of Eells and Wood:[38] All classes with degree $k \neq \pm 1$ have harmonic representatives. The classes with $k = \pm 1$ have no harmonic representative.

The proof for nonexistence follows from an index theorem type argument (see later part of this lecture series).

Slightly a bit later, Wood[106] obtained a stronger result concerning the reversed direction of the above nonexistence theorem: Let M and N be two Riemann surfaces so that the genus of N is q. Then, for $q > 0$, the only harmonic maps from M into N are constant maps. In particular, there are no nontrivial harmonic maps from M into T^2.

1.5. $M = S^m, N = S^n, m \neq n$. It is well known that $\pi_m(S^n) = 0$ when $m < n$. So in this case the problem is trivial. For $m > n$, the problem is complicated.

The simplest situation is when $M = S^3, N = S^2$, and the homotopy classes are represented by the Hopf invariants in $\pi_3(S^2) = \mathbb{Z}$. It is not hard to see[108] that any Hopf invariant which is the square, i.e., $Q = k^2$, can be harmonically represented. To see this, one uses the Hopf map $H : S^3 \to S^2$ which has unit Hopf number, $Q(H) = 1$. Let $f : S^2 \to S^2$ be a harmonic map of degree k. Then $\phi = f \circ H : S^3 \to S^2$ is harmonic and

$$Q(\phi) = Q(f \circ H) = \deg(f)^2 Q(H) = k^2$$

as expected.

There is no general theory yet in any of those nontrivial settings. A very interesting situation would be that for the general Hopf fibration

$$S^{4n-1} \to S^{2n}, \quad n \geq 1$$

Note that except some isolated cases such as $S^3 \to S^2$ and $S^{11} \to S^6$, we are no longer facing an infinite cyclic group ($\mathbb{Z}$).

Here are some more known examples that all elements in the homotopy groups have harmonic representatives:[37]

$$\pi_7(S^3) = \mathbb{Z}_2, \quad \pi_7(S^5) = \mathbb{Z}_2, \quad \pi_9(S^6) = \mathbb{Z}_{24},$$
$$\pi_{15}(S^9) = \mathbb{Z}_2, \quad \pi_{n+1}(S^n) = \mathbb{Z}_2 \quad (3 \leq n \leq 8)$$

Stability Theorem of Xin and Leung:[60,107,108] For $n \geq 3$, a stable harmonic map from S^n into any Riemannian manifold N or from any compact manifold M into S^n must be constant.

Method: A calculation of the second variation of the energy.

Excursion/Invitation into Physics

1. Cosmic strings generated from harmonic maps

Consider the Einstein equations over a $(3+1)$-dimensional spacetime of metric signature $(+---)$:

$$R_{\mu\nu} - \frac{1}{2}g_{\mu\nu}R = -8\pi G T_{\mu\nu}$$

where $T_{\mu\nu}$ is the energy-momentum tensor. In the context of static cosmic strings,[30,98,99,112,113] these equations reduce to a scalar equation

$$K_g = 8\pi H$$

where K_g is the Gauss curvature of an unknown compact Riemann surface M and $H \geq 0$ is the energy density (or Hamiltonian) of any physical model. Integrating the above equation and using the Gauss–Bonnet theorem, we have

$$2\pi\chi(M) = \int_M K_g \, dV_g \geq 0$$

where $\chi(M)$ is the Euler characteristic of M which has the expression $\chi(M) = 2-2q$ (q = genus = number of handles attached to S^2). Hence $q = 0$ or 1 and the latter case is trivial. So $q = 0$ and $M = S^2$.

If physics is induced from the nonlinear σ-model (Heisenberg's ferromagnetism), we arrive at the harmonic maps from S^2 to S^2 which can be used to generate an important class of explicit cosmic string solutions.[30,114]

2. The Skyrme model for elementary particles (baryon-meson scattering)

Let A be an $n \times n$ matrix and define $\sigma_i(A)$ to be the coefficients of the characteristic polynomial of A determined by the expansion

$$\det(A + \lambda I) = \sum_{i=0}^{n} \sigma_i(A) \lambda^{n-i}$$

Suppose $\dim(M) = \dim(N) = n$. For a map $\phi : M \to N$, the geometrized Skyrme energy is of the form

$$E(\phi) = \int_M (\sigma_1(g^{-1}\phi^*h) + \sigma_{n-1}(g^{-1}\phi^*h)) \, dV_g$$

One would like to prove the existence of a minimizer among the topological class $\deg(\phi) = k$. Note that $e(\phi) = \sigma_1(g^{-1}\phi^*h)$ and the additional term $\sigma_{n-1}(g^{-1}\phi^*h)$ is called the Skyrme term.

In the original setting of Skyrme, $M = \mathbb{R}^3$, $N = SU(2) = S^3$, and the energy functional is written as

$$E(\phi) = \int_{\mathbb{R}^3} \Big(\sum_{1\leq i\leq 3} |\partial_i\phi|^2 + \sum_{1\leq i<j\leq 3} |\partial_i\phi \wedge \partial_j\phi|^2 \Big) \, dx$$

and the topological degree (baryon number) has the integral representation

$$\deg(\phi) = \frac{1}{2\pi^2} \int_{\mathbb{R}^3} \det(\phi, \partial_1\phi, \partial_2\phi, \partial_3\phi)(x) \, \mathrm{d}x$$

The basic question is again the existence of solutions of the constrained minimization problem

$$E_k = \inf\{E(\phi) \,|\, E(\phi) < \infty, \ \deg(\phi) = k\}$$

The only result we know is that the problem has solutions for $k = \pm 1$.[63]

With radial symmetry, we know that E has a critical point in any degree class.[40,111] It is an important open question whether the minimizers at $k = \pm 1$ are all radially symmetric.

3. Faddeev knots

In this situation, we are interested in the existence of a minimizer for the Faddeev energy functional which governs maps from S^3 into S^2 and contains a Skyrme-like term in addition to the quadratic term giving rise to harmonic maps. We know that minimizers exist at the unit Hopf charge $Q = \pm 1$ among other things. See.[63,64]

Like harmonic maps, the regularity issue for both the Skyrme and Faddeev problems are difficult and unsettled.

Example 2. Hodge Theory

This is even a more classical theory than the work on harmonic maps. It can be established by using either elliptic theory of PDEs[14,31,49,59,100] or heatflow approach,[70] and the latter is perhaps one of the earliest heatflow successes in differential topology.

Let (M, g) be a compact oriented manifold of dimension n and $\Omega^k(M)$ be the space of all degree k differential forms on M. Then, the de Rham complex

$$\mathrm{d} : 0 \to \mathbb{R} \to \Omega^0(M) \to \Omega^1(M) \to \cdots \to \Omega^n(M) \to 0$$

gives us the de Rham cohomology group

$$H^k(M) = \ker(\mathrm{d} : \Omega^k(M) \to \Omega^{k+1}(M))/\mathrm{d}\Omega^{k-1}(M)$$

On the other hand, using the Hodge star $* : \Omega^k(M) \to \Omega^{n-k}(M)$, which is an isometry and satisfies $* * \alpha = (-1)^{k(n-k)}\alpha$ on any k-form α, we can express the volume element of (M, g) as $*1 = \mathrm{d}V_g$ which allows us to define an inner product on $\Omega^k(M)$ by

$$\langle \alpha, \beta \rangle = \int_M \alpha \wedge *\beta, \quad \alpha, \beta \in \Omega^k(M)$$

Let δ be the adjoint of d such that $\langle \mathrm{d}\alpha, \beta \rangle = \langle \alpha, \delta\beta \rangle$. Then $\delta = (-1)^{nk+n+1} * \mathrm{d}* : \Omega^{k-1}(M) \to \Omega^k(M)$, induces the Laplace–Beltrami operator

$$\Delta = \mathrm{d}\delta + \delta\mathrm{d} = (-1)^{nk+n+1}(\mathrm{d} * \mathrm{d} * + (-1)^n * \mathrm{d} * \mathrm{d}) : \Omega^k(M) \to \Omega^k(M)$$

In particular, when $n = $ even, we have

$$\Delta = -(*\mathrm{d} * \mathrm{d} + \mathrm{d} * \mathrm{d}*)$$

Basic properties: Δ commutes with d, δ, and $*$; $\langle \Delta \omega, \omega \rangle = \langle d\omega, d\omega \rangle + \langle \delta \omega, \delta \omega \rangle$, etc.

The harmonic forms are the members in the kernel of $\Delta : \Omega^k(M) \to \Omega^k(M)$,

$$\mathcal{H}^k(M) = \{\omega \in \Omega^k(M) \,|\, \Delta \omega = 0\}$$

Since $\Delta \omega = 0$ if and only if $d\omega = 0$ and $\delta \omega = 0$, we have the natural inclusion

$$i : \mathcal{H}^k(M) \to H^k(M)$$

The Hodge Theorem: The above inclusion is in fact an isomorphism. In other words, each cohomological class in the de Rham group $H^k(M)$ has a unique harmonic representative.

Some of the immediate consequences of the Hodge Theorem includes:

Finite Dimensionality of Cohomology: $H^k(M)$ is finitely dimensional.
Proof: It follows from elliptic theory that $\dim \mathcal{H}^k(M) < \infty$.

Poincaré Duality: It is well known that the Poincaré bilinear form

$$P : H^k(M) \times H^{n-k}(M) \to \mathbb{R}, \quad P([\alpha], [\beta]) = \int_M \alpha \wedge \beta$$

is nonsingular and thus defines an isomorphism between $H^{n-k}(M)$ and the dual space of $H^k(M)$. That is,

$$(H^k(M))^* \cong H^{n-k}(M)$$

Using the Hodge Theorem, the above result is straightforward: The commutativity of the Hodge dual $*$ with Δ implies that $*$ defines an isomorphism between $\mathcal{H}^k(M)$ and $\mathcal{H}^{n-k}(M)$.

Calculation of Top Cohomology: $H^n(M) \cong \mathbb{R}$.
Proof: This follows from $H^0(M) = \mathbb{R}$ and the Poincaré duality.

Vanishing Euler Characteristic of Manifolds of Odd Dimensions: The obvious pairing in odd dimensions leads to the immediate conclusion

$$\chi(M) = \sum_{k=0}^{n} (-1)^k \dim(H^k(M)) = 0$$

Example 3. Instantons and Chern–Pontryagin Classes

Consider the $(3+1)$-dimensional Minkowski spacetime defined by the line element

$$ds^2 = (dx^0)^2 - \sum_{j=1}^{3} (dx^j)^2$$

The main ingredient in the context of instantons is to make the time coordinate imaginary so that the Minkowski spacetime becomes the Euclidean space $\mathbb{R}^4$,

$$x^0 = ix^4, \quad ds^2 = \sum_{\mu=1}^{4} (dx^\mu)^2$$

This change of variable is also called the Wick transformation in quantum field theory. In the next section, we shall briefly explain the physical meaning of such a transformation, but for now, we only discuss its mathematical contents.

The Vacuum Maxwell Equations

Let A_μ be a real-valued vector field over $\mathbb{R}^4$. The Euclidean space version of the electromagnetic field is the curvature tensor $F_{\mu\nu}$ induced from A_μ:

$$F_{\mu\nu} = \partial_\mu A_\nu - \partial_\nu A_\mu$$

and the associated total action (energy) is given by

$$E(A) = \int_{\mathbb{R}^4} F_{\mu\nu}^2$$

The Maxwell equations are the Euler–Lagrange equations of the above functional:

$$\partial_\mu F_{\mu\nu} = 0$$

It is well known that all finite action solutions are trivial, $F_{\mu\nu} = 0$, and such a property is analogous in spirit with the theorems of Liouville and Bernstein for nonlinear PDEs.

Using differential forms and the Hodge theory, we can formulate a simple 'non-analytic' proof of the above fact. In fact, we replace the gauge potential A_μ and electromagnetic field $F_{\mu\nu}$ by a connection 1-form A and the induced curvature F respectively, i.e., $A = A_\mu \mathrm{d}x^\mu$ and $F = \mathrm{d}A = F_{\mu\nu} \mathrm{d}x^\mu \wedge \mathrm{d}x^\nu$. Then

$$E(A) = \int_{\mathbb{R}^4} |\mathrm{d}A|^2 \mathrm{d}x = \langle \mathrm{d}A, \mathrm{d}A \rangle = \int_{S^4} \mathrm{d}A \wedge *\mathrm{d}A$$

Note that the conformal structure of the energy allows us to work either on S^4 or S^4. Now the Maxwell equations become

$$\mathrm{d} * \mathrm{d}A = 0$$

which implies that $F = \mathrm{d}A$ is harmonic. Since $H^2(S^4) = 0$, we have $F = 0$ immediately as expected.

In 3 Euclidean dimensions, one needs to consider the addition of the matter sector which leads to the Abelian Higgs model or the Ginzburg–Landau theory and is the simplest gauge field theory. There is a similar Bernstein type theorem which says that all finite energy static solutions are gauge-equivalent to the trivial ones. It is very easy to formulate a topological proof of this statement but its analytic proof has not been seen yet. In 2 Euclidean dimensions, we arrive at the classical Ginzburg–Landau vortex model for superconductivity and there has been a lot of work in this area.

The Yang–Mills Equations

In the classical model of Yang and Mills,[110] one considers the simplest non-Abelian symmetry group $SU(2)$ whose associated Lie algebra $su(2)$ is generated by the 2×2 matrices t_1, t_2, t_3 satisfying the commutation relation

$$[t_a, t_b] \equiv t_a t_b - t_b t_a = \varepsilon_{abc} t_c, \quad a, b, c = 1, 2, 3,$$

where the symbol ε_{abc} is skewsymmetric with respect to permutation of subscripts and $\varepsilon_{123} = 1$. In fact, in terms of the Pauli matrices σ_a $(a = 1, 2, 3)$,

$$\sigma_1 = \begin{pmatrix} 0 & 1 \\ 1 & 0 \end{pmatrix}, \quad \sigma_2 = \begin{pmatrix} 0 & -i \\ i & 0 \end{pmatrix}, \quad \sigma_3 = \begin{pmatrix} 1 & 0 \\ 0 & -1 \end{pmatrix},$$

these generators are realized by the relation $t_a = \sigma_a/2i$ $(a = 1, 2, 3)$.

Let $A = (A_\mu)$ $(\mu = 1, 2, 3, 4)$ be an $su(2)$-valued gauge field over the Euclidean space $\mathbb{R}^4$. Then A_μ may be represented by

$$A_\mu = A_\mu^a t_a.$$

In analogy to the Maxwell electromagnetic field, the field strength tensor or curvature $F_{\mu\nu}$ induced from A_μ is defined by

$$F_{\mu\nu} = \partial_\mu A_\nu - \partial_\nu A_\mu + [A_\mu, A_\nu].$$

In view of $2\mathrm{Tr}\,(t_a t_b) = -\delta_{ab}$, we can define the total energy as

$$E(A) = -\frac{1}{2} \int_{\mathbb{R}^4} \mathrm{Tr}\,(F_{\mu\nu}^2)\,\mathrm{d}x$$

so that the associated Euler–Lagrange equations are

$$\partial_\mu F_{\mu\nu} + [A_\mu, F_{\mu\nu}] = 0$$

which are the vacuum Yang–Mills equations, generalizing the electromagnetic Maxwell equations. The solutions of these equations are called the Yang–Mills fields. It is interesting to note that these equations can be rewritten as the classical Maxwell equations in the presence of an external source current,

$$\partial_\mu F_{\mu\nu} = j_\nu$$

if we identify j_ν with $-[A_\mu, F_{\mu\nu}]$. In other words, the Yang–Mills equations contain a self-induced current as source to sustain "electromagnetism".

Gauge Invariance and Topological Characterization

It is important to notice that there is a gauge symmetry

$$A_\mu \mapsto U A_\mu U^{-1} - (\partial_\mu U) U^{-1}, \quad F_{\mu\nu} \to U F_{\mu\nu} U^{-1}, \quad U \in SU(2)$$

so that the finite-energy condition implies that the gauge field near infinity is a pure gauge, or

$$A_\mu \sim -(\partial_\mu U) U^{-1}$$

on spheres near infinity of $\mathbb{R}^4$. Since $SU(2)$ is topologically a 3-sphere as well, U gives rise to an element in

$$\pi_3(S^3)$$

In other words, the boundary condition of the gauge field near infinity is topological and is given by the Brouwer degree of the map U restricted to any sphere, say S^3_∞, near infinity.

Analytically, the degree of $U : S^3_\infty \to S^3$ can be represented as an integral,

$$
\begin{aligned}
Q &= \deg(U) \\
&= -\frac{1}{24\pi^2} \int_{S^3_\infty} \epsilon_{\mu\nu\alpha\beta} \, \mathrm{Tr} \, \{(U^{-1}\partial_\nu U)(U^{-1}\partial_\alpha U)(U^{-1}\partial_\beta U)\} \, \mathrm{d}S_\mu \\
&= -\frac{1}{4\pi^2} \int_{S^3_\infty} K_\mu \, \mathrm{d}S_\mu \\
&= -\frac{1}{4\pi^2} \int_{\mathbb{R}^4} \partial_\mu K_\mu \, \mathrm{d}x
\end{aligned}
$$

On the other hand, in terms of the gauge field A_μ, a lengthy calculation shows that

$$
K_\mu = \epsilon_{\mu\nu\alpha\beta} \, \mathrm{Tr} \, \left(\frac{1}{2} A_\nu \partial_\alpha A_\beta - \frac{1}{3} A_\nu A_\alpha A_\beta \right)
$$

which is a Chern–Simons term.[29] Therefore $\partial_\mu K_\mu$ is the classical second Chern form given by

$$
\partial_\mu K_\mu = \frac{1}{4} \, \mathrm{Tr} \, \{F_{\mu\nu} \tilde{F}_{\mu\nu}\}, \quad \tilde{F}_{\mu\nu} = \frac{1}{2} \epsilon_{\mu\nu\alpha\beta} F_{\alpha\beta}
$$

In summary, we see that the gauge field carries a topological index given by the second Chern class,

$$
Q = c_2 = -\frac{1}{16\pi^2} \int_{\mathbb{R}^4} \mathrm{Tr} \, \{F_{\mu\nu} \tilde{F}_{\mu\nu}\} \, \mathrm{d}x
$$

Words on compactness: Although the setting is on $\mathbb{R}^4$, the conformal structure of both the energy and topology implies that the setting can be viewed as being placed over S^4, fixing the standard metric on S^4 and using stereographic projection as coordinates.

Words on smooth extension to S^4: This is guaranteed by the removable singularity theorem of Uhlenbeck[97] under the natural condition $E(A) < \infty$. In other words, a finite energy solution over $\mathbb{R}^4$ can be smoothly extended to a solution over the full S^4.

Therefore, from now on, we can work interchangeably over $\mathbb{R}^4$ and S^4 according to convenience. In particular, we often omit writing these spaces out when there is no risk of confusion.

So far, we have not touched the issue of harmonicity yet for the Yang–Mills fields. Here we present a quick discussion on this. This structure is better seen when we use differential forms to reformulate the problem. When we do this, we also prepare ourselves for the higher dimensional extensions of the Yang–Mills theory.

The Yang–Mills Fields, Differential Forms, and Harmonicity

Let A be an $su(2)$-valued connection 1-form and F be the induced curvature,

$$
F = \mathrm{d}A + A \wedge A
$$

The Yang–Mills energy is

$$
E(A) = -\int \mathrm{Tr} \, (F \wedge *F) = \langle F, F \rangle = \|F\|^2
$$

and the Chern class is

$$c_2(F) = -\frac{1}{8\pi^2} \int \text{Tr}\,(F \wedge F)$$

In terms of the connection 1-form A, the connection D operates on an $su(2)$-valued p-form ω according to the relation

$$D\omega = \mathrm{d}\omega + A \wedge \omega + (-1)^{p+1}\omega \wedge A$$

The Yang–Mills equation (or the Euler–Lagrange equation of the energy) is

$$D(*F) = 0$$

Recall the Bianchi identity

$$DF = 0$$

Then we see that the Yang–Mills field F is necessarily "harmonic" with respect to the connection D:

$$\Delta_D F = -(*D * D + D * D*)F = 0$$

It is not hard to see that the converse is also true because, like before,

$$\langle \Delta_D F, F \rangle = \|D * F\|^2$$

so that F is harmonic if and only if F satisfies the Yang–Mills equation.

Hence, the problem that, for a given second Chern class $c_2 = k$, find a solution of the Yang–Mills equation is equivalent to the problem of finding a harmonic representative among this given second Chern class.

The Self-Dual Equation and Minimization of Energy

The Yang–Mills equation has an obvious first integral reduction due to the Bianchi identity:

$$F = \pm * F$$

which is called the self-dual or anti-self-dual reduction of the Yang–Mills field. In fact, all known explicit solutions are the solutions of these equations.

In order to explore the meaning the self-dual or anti-self-dual solutions, we rewrite the Yang–Mills curvature F as

$$F = F^+ + F^-, \quad F^\pm = \frac{1}{2}(F \pm *F)$$

Then F^+ is self-dual, $*F^+ = F^+$, and F^- is anti-self-dual, $*F^- = -F^-$. It can also be checked that F^+ and F^- are orthogonal, $\langle F^+, F^- \rangle = 0$.

With the above preparation, we have

$$E(A) = \|F^+\|^2 + \|F^-\|^2$$
$$8\pi^2 c_2(F) = \|F^+\|^2 - \|F^-\|^2$$

Therefore, we obtain the lower bound

$$E(A) = 2\|F^{\mp}\|^2 \pm (\|F^+\|^2 - \|F^-\|^2)$$
$$= 2\|F^{\mp}\|^2 + 8\pi^2|c_2(F)|$$
$$\geq 8\pi^2|k|$$

The lower bound is saturated if and only if $F^- = 0$ or $F^+ = 0$ for $k = |k|$ or $k = -|k|$. That is, the topological energy minimum

$$E_k = \min\{E(A) \,|\, c_2(F) = k\} = 8\pi^2|k|$$

is attained if and only if F is self-dual or anti-self-dual. The set of these possible energy minima, or the set of classical soliton masses,

$$\mathcal{M} = \{E_k \,|\, k = 0, \pm 1, \pm 2, \cdots\} = \{0, 8\pi^2, 16\pi^2, \cdots\}$$

is called the energy (mass) spectrum of the classical Yang–Mills theory. Note that the energy (mass) of a nontrivial solution is at least $8\pi^2$. In other words, there can be no classical soliton with an energy (mass) below the "energy (mass) gap" $\Delta \equiv 8\pi^2$. One of the seven Clay Institute Millennium Prize Problems concerns the energy (mass) gap of the energy (mass) spectrum of the *quantum* Yang–Mills theory.

Clay Institute Millennium Prize Problem on Quantum Yang–Mills Theory and Its Mass Gap: Prove that for any compact gauge group, quantum Yang–Mills theory on $\mathbb{R}^4$ exists and has a mass gap $\Delta > 0$ (quoted from[54]).

Thus, the problem has two components: (i) Quantize a classical Yang–Mills theory. (ii) Establish a positive mass gap for the mass spectrum of the theory.

Recall that, even when quantizing a 1D single-particle Newtonian motion, a lot of machinery is needed and the quantized theory cannot be made completely accurate except for a few extremely simple cases.

Saddle Point Solutions ("Sphalerons")

It had long been an outstanding open question whether the Yang–Mills equation has a finite-energy solution which is not self-dual or anti-self-dual, thus nonminimal. Note that, when the gauge group is enlarged to $SU(3)$, or the spacetime is altered to $S^3 \times S^1$, there are solutions which are indeed not self-dual or anti-self-dual.[22,47]

For $k = 0$, L. Sibner, R. Sibner, and Uhlenbeck[80] proved the existence of a nonminimal solution using the min-max approach developed earlier by Taubes[89] for the Yang–Mills–Higgs equations in three dimensions.

For all $k \neq \pm 1$, Sadun and Segert[77] proved the same conclusion.

For $k = \pm 1$, whether or not there exists a nonminimal Yang–Mills solution is still an open question.

The main strategy in the work of Taubes[89] is an application of the Lyusternik–Shnirelman theory, which is an infinitely dimensional Morse theory, and a construction of noncontractible loops in the configuration space of the $SU(2)$ Yang–Mills–Higgs fields. Some difficult issues include noncompactness, gauge ambiguity, and infinite dimensions. An important open problem in this direction is the existence of nonminimal solutions in the Weinberg–Salam electroweak theory for which the gauge group is slightly larger, $G = SU(2) \times U(1)$. Klinkhamer and Manton[58,68] indeed constructed noncontractible loops but detailed analytic issues remain to be settled.

3. Quantum Tunneling, Imaginary Time, Instantons, and Liouville-Type Equations

Modern physics contains a great amount of concepts that go against our routine intuition. For example, in quantum mechanics, the Heisenberg uncertainty principle implies that there is no such notion of a rest particle whatsoever and quantum fluctuations are everywhere. An important phenomenon is quantum tunneling. For example, imagine that you run into a rigid wall and you know for sure that you will be bounced back. Quantum tunneling predicts that, if your mass is small enough so that it is comparable to the Planck constant, $\hbar$, there is a considerable probability that you will end up on the side of the wall, without breaking the wall or losing energy. So, virtually, you have passed the wall through a tunnel. Likewise, when you run into a deep cliff, there is also a probability that you get bounced back. This tunneling phenomenon is fundamental for semiconductor devices and nuclear fission processes. For example, we can mention the celebrated alpha decay theory developed in 1928 by Gamow which explains the physical mechanism of radioactive elements. On the other hand, the concept of imaginary time was introduced by Feynman as a technical convenience for the calculation of transition amplitude through path integrals (see below). In 1983, Hawking and Hartle introduced it in quantum cosmology in order to eliminate spacetime singularities associated with the beginning of time and curvature blowup, thereby replacing the Minkowski spacetime with the Euclidean spacetime.

Consider the 1D motion of a particle of mass m given by the action

$$S(x) = \int \mathcal{L}(x, \dot{x}) \, dt = \int \left\{ \frac{m}{2} \dot{x}^2 - V(x) \right\} dt$$

where $V(x)$ is the potential energy. The classical motion is described by the equation

$$m\ddot{x} = -V'(x)$$

Suppose that $a < b$ are two isolated absolute minimum points of $V(x)$ with $V(a) = V(b) = 0$. Then these are the two stable equilibria of the classical equation which are the ground states and stay isolated even under small perturbations. Quantum mechanically, however, this is not the case. In other words, there is a considerable probability that the state $x = a$ goes through a phase transition to become state $x = b$ which is measured by something called the transition amplitude which is proportional to the path integral

$$\int \mathcal{D}(x) \mathrm{e}^{\mathrm{i}S(x)/\hbar}$$

where the integration is taken over all possible paths ending at $x = a$ and $x = b$ whose precise mathematical formulation still bothers mathematicians today. However, we are not concerned about this and only note that this would give us a positive value for the transition amplitude. In other words, the phase transition from the state $x = a$ to the state $x = b$ is possible quantum mechanically.

Since $\hbar$ is small, the factor $\mathrm{e}^{\mathrm{i}S(x)/\hbar}$ is highly oscillatory. In order to overcome this difficulty, Feynman replaced the real time variable t by the imaginary time variable τ through $t = -\mathrm{i}\tau$ so that the path integral becomes a real one,

$$\int \mathcal{D}(x) \mathrm{e}^{-S_E(x)/\hbar}$$

where S_E is the Euclidean version of the action given by

$$S_E(x) = \int \left\{ \frac{m}{2} \left(\frac{\mathrm{d}x}{\mathrm{d}\tau} \right)^2 + V(x) \right\} \mathrm{d}\tau$$

It is interesting to note that in terms of the imaginary time τ the classical motion is now governed by an up-side-down potential $-V(x)$,

$$m \frac{\mathrm{d}^2 x}{\mathrm{d}\tau^2} = -(-V'(x))$$

It will be instructive to look at two explicit cases.

1. Double potential well case: $V(x) = \frac{\lambda}{8}(x^2 - 1)^2$ and there are only two ground states, $x = -1$ and $x = 1$. It is easily seen that there is a solution that connects these ground states,

$$x(-\infty) = -1, \quad x(\infty) = 1$$

and minimizes the action S_E. In fact, we have after integrating by parts,

$$S_E(x) = \frac{m}{2} \int \left(\frac{\mathrm{d}x}{\mathrm{d}\tau} + \frac{1}{2}\sqrt{\frac{\lambda}{m}}(x^2 - 1) \right)^2 + \frac{2}{3}\sqrt{\lambda m}$$

so that the minimal action is attained with

$$\min S_E = \frac{2}{3}\sqrt{\lambda m}$$

at the solution of the first-order self-dual equation

$$\frac{\mathrm{d}x}{\mathrm{d}\tau} = \frac{1}{2}\sqrt{\frac{\lambda}{m}}(1 - x^2)$$

which is in fact equivalent to the original second-order equation.

Here are some obvious observations: (i) the leading-order contribution to the transition amplitude is given by

$$\mathrm{e}^{-2\sqrt{\lambda m}/3\hbar}$$

which becomes significant only when the particle mass m is comparable to the Planck constant $\hbar$ otherwise it is insignificant and classical physics dominates; (ii) the transition in terms of imaginary time can be realized classically, although such a transition cannot be made classically in terms of real time by analyzing the equation of motion and conservation of energy; (iii) in the leading-order calculation of the transition amplitude, it is more important to know the existence of an action minimizer and the associated action minimum than the explicit form of an action minimizer.

2. Infinitely-many potential well case: For example, we consider the sine-Gordon model, $V(x) = \lambda(1 - \cos x)$. There are infinitely many ground states given by $x_N = 2N\pi, N = 0, \pm 1, \pm 2, \cdots$. We are interested in quantum tunneling between any two of these states,

$$x(-\infty) = 2N\pi, \quad x(\infty) = 2(N + k)\pi$$

Therefore we are led to asking the question whether there exists a classical solution in terms of imaginary time to realize the above boundary condition. Such a solution is necessarily an action minimizer. It is not hard to prove that[65] such a solution exists if and only if

$$k = \pm 1$$

An interesting implication of this result is that, for the sine-Gordon model, the most likely quantum tunneling happens between "neighboring" ground states.

The Self-Dual Yang–Mills Instantons a la BPST and 't Hooft

Likewise, when we consider the $SU(2)$ Yang–Mills theory describing nuclear interactions, the ground states can be characterized topologically by homotopy classes representing mappings from S^3, which is the compactified physical space $\mathbb{R}^3$, into $SU(2)$, which is a 3-sphere as a manifold. The Yang–Mills instantons of the second Chern number k then describes the quantum tunneling between the ground state of homotopy class n_1 and that of homotopy class n_2 so that $k = n_1 - n_2$. We shall not discuss this physical process any further but only remark that, again, the explicit forms of the solutions realizing these topological numbers are not important for the calculation of the leading-order tunneling amplitude because these self-dual solutions enable us to evaluate their minimal energy values exactly, and, these results are nonperturbative.

Therefore, in the subsequent presentation, we shall focus on mathematics.

The boundary condition gives us a hint to choose the gauge field A_μ to be

$$A_\mu(x) = \frac{x^2}{x^2 + \lambda^2} U^{-1}(x) \partial_\mu U(x)$$

where $\lambda > 0$ is a parameter and the group element $U \in SU(2)$ may be specified to be

$$U(x) = \frac{x_\mu \omega_\mu}{|x|}$$

with the 2×2 ω-matrices defined by

$$\omega_a = \mathrm{i}\sigma_a, \quad a = 1, 2, 3; \quad \omega_4 = \begin{pmatrix} 1 & 0 \\ 0 & 1 \end{pmatrix}$$

Introduce the 't Hooft tensors

$$\eta_{\mu\nu} = -\frac{1}{4}(\omega_\mu^\dagger \omega_\nu - \omega_\nu^\dagger \omega_\mu), \quad \overline{\eta}_{\mu\nu} = -\frac{1}{4}(\omega_\mu \omega_\nu^\dagger - \omega_\nu \omega_\mu^\dagger)$$

It is straightforward to examine that these tensors are either self-dual or anti-self-dual,

$$\eta_{\mu\nu} = *\eta_{\mu\nu}, \quad \overline{\eta}_{\mu\nu} = -*\overline{\eta}_{\mu\nu}$$

We need to represent the gauge field in terms of the 't Hooft tensors so that self-duality becomes apparent to achieve,

$$A_\mu(x) = \frac{2x_\nu}{x^2 + \lambda^2}\eta_{\mu\nu}$$

Consequently, we obtain the self-dual curvature 2-tensor,

$$F_{\mu\nu}(x) = \frac{4\lambda^2}{(x^2 + \lambda^2)^2}\eta_{\mu\nu}$$

One of the interesting features of this solution is that its energy density peaks at the origin $x = 0$ with a level determined by λ. In other words, this solution looks like a particle, or an instanton, located at $x = 0$ with a size specified by a parameter.

Of course, we may represent $\eta_{\mu\nu}$ in terms of the standard basis, $\{t_a\}_{a=1,2,3}$, of the Lie algebra $su(2)$, in the form

$$\eta_{\mu\nu} = \eta_{\mu\nu}^a t_a$$

Various properties of the real-valued tensors $\eta_{\mu\nu}^a$ are stated in,[93] of which, the most useful one for our purpose here is

$$\eta_{\mu\nu}^a \eta_{\mu\nu}^b = 4\delta_{ab}$$

Inserting the 2-tensor $F_{\mu\nu}$ into the Chern integral and using $\mathrm{Tr}\,(t_a t_b) = -\delta_{ab}/2$, we have

$$c_2 = \frac{6}{\pi^2} \int_{\mathbb{R}^4} \frac{\lambda^4}{(x^2 + \lambda^2)^4}\,\mathrm{d}x = 1$$

Hence we have constructed an instanton of unit topological charge, $c_2 = 1$. This one-instanton solution was discovered by Belavin, Polyakov, Schwartz, and Tyupkin[11] and is known as the BPST solution.[1]

We then show that the above method may be generalized to obtain instantons of an arbitrary topological charge, $c_2 = k$. To this end, we rewrite

$$U^{-1}\partial_\mu U = \frac{2x_\nu}{x^2}\eta_{\mu\nu}$$

On the other hand, define

$$\tilde{A}_\mu(x) = \left(\partial_\nu \ln\left[1 + \frac{\lambda^2}{x^2}\right]\right)\eta_{\mu\nu}.$$

We thus obtain the relation

$$\tilde{A}_\mu = \left(\frac{2x_\nu}{x^2 + \lambda^2} - \frac{2x_\nu}{x^2}\right)\eta_{\mu\nu} = A_\mu - U^{-1}\partial_\mu U$$

$$= A_\mu + (\partial_\mu U^{-1})U = UA_\mu U^{-1} + U\partial_\mu U^{-1}$$

In other words, the gauge fields A_μ and $\tilde{A}_\mu$ are equivalent. Consequently, the field strength tensor induced from $\tilde{A}_\mu$ is also self-dual and we get a gauge-equivalent self-dual instanton. Hence we may write the obtained solution in the form

$$A_\mu = (\partial_\nu \ln f)\eta_{\mu\nu}$$

where $f = 1 + \lambda^2/x^2$. At the first glance, this procedure does not lead to any new solutions. However, it suggests that we may obtain more solutions if we simply use the above as an ansatz for which f is a positive-valued function to be determined by our self-duality requirement. It turns out that a general choice of f is

$$f(x) = 1 + \sum_{j=1}^{k} \frac{\lambda_j^2}{(x - p_j)^2}, \quad \lambda_j > 0, \quad p_j \in \mathbb{R}^4, \quad j = 1, 2, \cdots, k$$

which contains $5k$ continuous parameters and is called the 't Hooft solution.[1,93] In fact this solution describes k instantons located at the points $p_1, p_2, \cdots, p_k$ with sizes determined by the parameters $\lambda_1, \lambda_2, \cdots, \lambda_k$. It can be examined that $c_2 = k$ (we omit the details). The 't Hooft instantons have been extended by Jackiw and Rebbi[51] and Ansourian and Ore[4] into a form containing $5k + 4$ parameters which is the most general explicit self-dual solution known, although, according to a result[6,79] based on the Atiyah–Singer index theorem,[9] the number of free parameters of a general self-dual instanton in the class $c_2 = k$ is $8k - 3$. This conclusion was first arrived at by physicists[21,52] using plausible arguments: $4k$ parameters determine the positions and k parameters the sizes of the instantons as in the 't Hooft solution case, $3k$ parameters determine the asymptotic orientations of the instantons in the internal space $SU(2) = S^3$ from which the 3 parameters originated from the global $SU(2)$ gauge equivalence must be subtracted. For a general construction of 4-dimensional Yang–Mills instantons, see.[5,7]

Witten's Instantons

Witten's instanton[102] is symmetric with respect to rotation of the spatial coordinates x_i $(i = 1, 2, 3)$ and is of the form

$$A_i^a = \epsilon_{iaj}\frac{x_j}{r^2}(1 - \phi_2(r, t)) + \frac{1}{r^3}(\delta_{ia}r^2 - x_i x_a)\phi_1(r, t) + \frac{x_i x_a}{r^2}a_1(r, t),$$

$$A_4^a = \frac{x_a}{r}a_2(r, t), \quad a, i, j = 1, 2, 3$$

where $r^2 = x_1^2 + x_2^2 + x_3^2$, $t = x_4$ is the temporal coordinate, and a_1, a_2, ϕ_1, ϕ_2 are real-valued functions. Thus, the field strength tensor becomes

$$
F_{4i}^a = -\left(\frac{\partial \phi_2}{\partial t} + a_2 \phi_1\right)\frac{\epsilon_{iaj} x_j}{r^2} + \left(\frac{\partial \phi_1}{\partial t} - a_2 \phi_2\right)\frac{(\delta_{ai} r^2 - x_a x_i)}{r^3}
$$
$$
+ \left(\frac{\partial a_1}{\partial t} - \frac{\partial a_2}{\partial r}\right)\frac{x_a x_i}{r^2},
$$
$$
\frac{1}{2}\epsilon_{ijj'} F_{jj'}^a = -\frac{\epsilon_{iaj'} x_{j'}}{r^2}\left(\frac{\partial \phi_1}{\partial r} - a_1 \phi_2\right) - \left(\frac{\partial \phi_2}{\partial r} + a_1 \phi_1\right)\frac{(\delta_{ai} r^2 - x_a x_i)}{r^3}
$$
$$
+ (1 - \phi_1^2 - \phi_2^2)\frac{x_a x_i}{r^4}
$$

Inserting these, we obtain the reduced expressions for the total energy

$$
E = \frac{1}{4}\int_{\mathbb{R}^3} \mathrm{d}x \int_{\mathbb{R}} \mathrm{d}t\{F_{\mu\nu}^a F_{\mu\nu}^a\}
$$
$$
= 4\pi \int_{-\infty}^{\infty} \mathrm{d}t \int_0^{\infty} \mathrm{d}r\left\{|D_i \phi|^2 + \frac{1}{4}r^2 f_{ij}^2 + \frac{1}{2r^2}(1 - |\phi|^2)^2\right\}
$$

and the Chern class

$$
c_2 = -\frac{1}{16\pi^2}\int_{\mathbb{R}^4} \mathrm{Tr}\,(F_{\mu\nu} * F_{\mu\nu})
$$
$$
= -\frac{1}{2\pi}\int_{-\infty}^{\infty} \mathrm{d}t \int_0^{\infty} \mathrm{d}r\left\{(1 - |\phi|^2)f_{12} - \mathrm{i}(D_1 \phi \overline{D_2 \phi} - \overline{D_1 \phi} D_2 \phi)\right\},
$$

where now ϕ is a complex field defined by $\phi = \phi_1 + \mathrm{i}\phi_2$, ∂_1 and ∂_2 denote $\partial/\partial r$ and $\partial/\partial t$, respectively, $f_{ij} = \partial_i a_j - \partial_j a_i$ $(i, j = 1, 2)$, $D_i \phi = \partial_i \phi + \mathrm{i}a_i \phi$ $(i = 1, 2)$.

The above in fact can be viewed as a Ginzburg–Landau theory over the Poincaré hyperbolic half space $\mathbb{R}_+^2 = \{(t, r) \mid -\infty < t < \infty, 0 < r < \infty\}$ equipped with the line element

$$
\mathrm{d}s^2 = r^{-2}(\mathrm{d}r^2 + \mathrm{d}t^2)
$$

In terms of these, the self-dual equation becomes a vortex equation,

$$
D_1 \phi + \mathrm{i}D_2 \phi = 0,
$$
$$
r^2 f_{12} = |\phi|^2 - 1
$$

and knowledge on superconducting vortices tells us that the energy density peaks at the spots where ϕ vanishes. Suppose that

$$
p_1, \cdots, p_k \in \mathbb{R}_+^2
$$

are zeros of ϕ. Then the substitution $u = \ln|\phi|$ gives us the following scalar nonlinear elliptic equation with sources,

$$
\Delta u = \frac{1}{r^2}(\mathrm{e}^{2u} - 1) + 2\pi \sum_{s=1}^k \delta_{p_s}, \quad x \in \mathbb{R}_+^2
$$

We now use the method of Witten[102] to construct the solutions explicitly. We momentarily neglect the singular source term and consider

$$r^2 \Delta u = e^{2u} - 1.$$

It is seen that we arrive at the Liouville equation

$$\Delta v = e^{2v}$$

after the transformation

$$u = \ln r + v.$$

All the solutions of the Liouville equation can be expressed explicitly (integrable). In terms of the complex variable $z = r + it$, we have the representation

$$v(z) = \ln \left(\frac{2|F'(z)|}{1 - |F(z)|^2} \right)$$

The solution is free of singularities if $F'(z) \neq 0$ and $|F(z)| < 1$.

Returning to the original function u, we have

$$u(z) = \ln \left(\frac{2r|F'(z)|}{1 - |F(z)|^2} \right), \quad z = r + it.$$

We only remark that we can choose F suitable to get the solutions of the equation realizing k vortex points $p_1, \cdots, p_k$ and such solutions belong to the topological class $c_2 = k$. These vortex points give rise to $2k$ free parameters in the Poincaré half space. Since there are 4 choices of the imaginary time variable, we obtain a total of $8k$ parameters. Discounting again the 3 parameters describing global gauge symmetry, we arrive again at the miracle number, $8k - 3$. Although we do not justify here whether this number count is accurate, we see that Witten's construction does lead to a quite general description of solutions.

Excursion to the Liouville-Type Equations in Physics

Although the Liouville equation can be solved exactly using any of those well developed methods including Liouville's method,[66] the Bäcklund transformation,[69] the inverse scattering transformation,[3] the method of separation of variables,[61] etc., a small variation of it often spoils its integrability. Below are some important examples of such variations.

1. The well-known Ginzburg–Landau self-dual vortex equation[53] for superconductivity is of the form

$$\Delta u = e^u - 1 + 4\pi \sum_{s=1}^{k} \delta_{p_s}(x)$$

Jaffe and Taubes[53] ask the question whether the solutions of this equation can be obtained in closed forms. Using the Painlevé tests, Schiff[78] argues that this equation is nonintegrable and the answer to the question of Jaffe–Taubes should be negative.

However, the Painlevé tests can only be regarded as giving a compelling evidence which is not sufficient to draw conclusion.

A milder question is to replace the Laplacian Δ by a Laplace–Beltrami operator and ask the question that for what metric the equation becomes integrable. We have seen that a supportive example is the equation of Witten for which the equation is integrable when the metric is Poincar'e's hyperbolic metric. However, there is no general picture towards this direction at all at this time.

2. The Abelian relativistic Chern–Simons vortex equation is lightly more complicated and takes the form

$$\Delta u = e^u(e^u - 1) + 4\pi \sum_{s=1}^{k} \delta_{p_s}(x)$$

The work of Schiff[78] also shows that this equation is nonintegrable. There are two types of boundary conditions at infinity. The first type is given by

$$\lim_{|x|\to\infty} u(x) = 0$$

and is called topological. Topological solutions[85] resemble the Ginzburg–Landau vortices and the solutions may not be unique as evidenced already in the compact setting.[88] The associated magnetic and electric charge is quantized and given by

$$Q = k$$

The second type of boundary condition is given by

$$\lim_{|x|\to\infty} u(x) = -\infty$$

and the charge is continuous,

$$Q = k + \alpha$$

It can be shown[28,86] that for any $\alpha \in (k + 2, \infty)$, the equation has a radially symmetric nontopological solution when $p_1 = \cdots = p_k$. For nonradially symmetric solutions where the vortex points do not coincide, we encounter a difficult problem and there is only some partial progress available. Notably, Chae and Imanuvilov[25] obtained solutions for α small and Chan, Fu, and Lin[27] obtained solutions for α large. The mathematical importance of this problem is that the technical issues associated to it are not well developed and its complete solution invites new ideas.

3. Systems of nonlinear PDEs are much harder problems but occur frequently in theoretical physics. For example, the study of a nonrelativistic condensed-matter problem leads to the system of the coupled "Ginzburg–Landau" vortex equations[34]

$$\Delta u = \lambda(e^v - 1) + 4\pi \sum_{i=1}^{M} \delta_{p_i}(x),$$

$$\Delta v = \lambda(e^u - 1) + 4\pi \sum_{i=1}^{N} \delta_{q_i}(x)$$

Any work on this system (over $\mathbb{R}^2$ or a compact domain which can either be a closed 2-surface or a bounded planar domain subject to the Dirichlet condition) will be interesting. Besides, it will also be interesting to study the integrability of the following "Liouville" system,

$$\Delta u = \pm e^v, \quad \Delta v = \pm e^u$$

Note that this system may be viewed as a "radical root" of the Liouville equation in the sense that the Liouville equation is recovered when $u = v$.

To see some detailed structure of the equations, we use the new variables $f = u + v$ and $g = u - v$ and we see that we have a variational functional of the form

$$\int (|\nabla f|^2 - |\nabla g|^2 + \text{nonlinear terms})$$

Note that a similar, but easier, problem that occurs in the relativistic Chern–Simons theory[33,57] containing two species of superconducting bosons and involves a nonlinear system of the form

$$\Delta u = \lambda e^v (e^u - 1) + 4\pi \sum_{i=1}^{M} \delta_{p_i}(x),$$

$$\Delta v = \lambda e^u (e^v - 1) + 4\pi \sum_{i=1}^{N} \delta_{q_i}(x)$$

This system has a similar variational structure and has recently been studied in[62] using some new techniques (indefinite minimization and domain expansion).

4. Atiyah–Singer Index Theorem and Calculation of Dimension of Moduli Space

An important question concerning the solutions of equations is to determine how large the solution space is. Interestingly, sometimes a question like this has its clue in the geometric and topological setting of the problem. For example, a nonlinear example concerning the counting of critical points of a differentiable function over a manifold is the Morse theory. Here we may mention the Morse inequality: Let f be a differentiable function over a compact manifold M so that the critical points of f are all nondegenerate. A critical point of f is said to have index k if the Hessian of f has exactly k negative eigenvalues. Using C_k to denote the number of the critical points of f of index k. Then we have

$$\sum_{i=0}^{k}(-1)^i C_i \geq \sum_{i=0}^{k}(-1)^i \dim H^i(M), \quad k = 0, 1, \cdots, n = \dim(M)$$

and the equality holds at $k = n$,

$$\sum_{i=0}^{n}(-1)^i C_i = \sum_{i=0}^{n}(-1)^i \dim H^i(M) = \chi(M)$$

In particular, we have

$$C_k \geq \dim H^k(M), \quad k = 0, 1, \cdots, n$$

which gives the following lower bound for the total number of critical points,

$$\sum_{i=0}^{n} C_i \geq \sum_{i=0}^{n} \dim H^i(M) \geq 2$$

This is of course true because f attains its maximum and minimum points in M. When more information on the topological structure of M is known, we may derive further information on the critical points of f. As an example, take $M = T^2$ (2-torus). Since $\dim H^1(M) = 2$, we have $C_1 \geq 2$, which implies that f has at least two saddle points as well.

A very general theory concerning linear equations is the Atiyah–Singer index theorem which can be expressed symbolically as:

Theorem (Atiyah–Singer). Let $L(f) = 0$ be a linear differential equation. Then

$$\text{Analytic Index of } L \; = \; \text{Topological Index of } L$$

The left-hand side is often a measurement of the dimension of the solution space of $L(f) = 0$ and the right-hand side is often a quantity that accounts for the global geometric and topological properties of the domain and range spaces of the operator L. In this section, we are concerned with the calculation of the dimension number of the moduli space of the Yang–Mills instantons which can be carried out by using the Atiyah–Singer index theorem.

The following excerpts are quoted from Wikipedia, a free online encyclopedia:

"In the mathematics of manifolds and differential operators, the Atiyah–Singer index theorem is an important unifying result that connects topology and analysis. It deals with elliptic differential operators (such as the Laplacian) on compact manifolds. It finds numerous applications, including many in theoretical physics.

"When Michael Atiyah and Isadore Singer were awarded the Abel Prize by the Norwegian Academy of Science and Letters in 2004, the prize announcement explained the Atiyah–Singer index theorem in these words:

"Scientists describe the world by measuring quantities and forces that vary over time and space. The rules of nature are often expressed by formulas, called differential equations, involving their rates of change. Such formulas may have an 'index,' the number of solutions of the formulas minus the number of restrictions that they impose on the values of the quantities being computed. The Atiyah–Singer index theorem calculated this number in terms of the geometry of the surrounding space."

In the book of Yu,[115] we read the following words of S. S. Chern:

"Even if there will be no research results, it is worthwhile to study the Atiyah–Singer index theorem."

Examples of Index Calculations and Historical Preludes

Let $T : V \to W$ be a linear operator between two finitely dimensional vector spaces with $\dim(V) = m$ and $\dim(W) = n$. Let $K \subset V$ and $R \subset W$ be the kernel and range of T, respectively, and K' and R' be their complements in the corresponding spaces,

$$V = K \oplus K', \quad W = R \oplus R'$$

Recall that R' measures the set in W that the operator T misses and is called the cokernel of T. The dimension of R' is the thing that measures "the number of restrictions that they impose on the values of the quantities being computed" stated above in the Abel Prize citations. The index of T is then defined by

$$\text{index}(T) = \dim(\ker(T)) - \dim(\text{coker}(T))$$

Since $T : K' \to R$ is an isomorphism, we quickly get

$$\text{index}(T) = (\dim(K) + \dim(K')) - (\dim(R) + \dim(R'))$$
$$= m - n$$

which is independent of the operator itself but only the domain and range space dimensions.

Similarly, for an operator between two infinitely dimensional vector spaces, we can define its index so long as its kernel and cokernel are of finite dimensions. Such an operator is called Fredholm. If T is a self-adjoint Fredholm operator from a Hilbert space into itself so that

$$\langle T\alpha, \beta \rangle = \langle \alpha, T\beta \rangle$$

then $\ker(T) = \text{coker}(T)$ and $\text{index}(T) = 0$. In particular, the Laplace operator Δ acting on differential forms over an oriented manifold has zero index.

An interesting situation is a calculation of the 'radical root' of Δ, which is

$$D = \text{d} + \delta : \Omega(M) \to \Omega(M)$$

Of course, D is self-adjoint and there is nothing special so far. However, when we restrict D to the space of all even-order forms, we get an operator $\mathcal{D}$ called the Gauss–Bonnet operator,

$$\mathcal{D} = D : \Omega^{\text{even}}(M) \to \Omega^{\text{odd}}(M)$$

and we see immediately that

$$\ker(\mathcal{D}) = \oplus_k \mathcal{H}^{2k}(M), \quad \text{coker}(\mathcal{D}) = \oplus_k \mathcal{H}^{2k+1}(M)$$

Consequently, we see that we can express the index of $\mathcal{D}$ by the Euler characteristic of the underlying manifold M,

$$\text{index}(\mathcal{D}) = \sum_k (-1)^k \dim \mathcal{H}^k(M) = \chi(M)$$

If $\dim(M) = \text{odd}$, then $\chi(M) = 0$ and there is nothing interesting; if $\dim(M) = \text{even} = 2n$, recall that the Chern–Gauss–Bonnet theorem says that the integral of the Pfaffian $Pf(\Omega)$, a $2n$-form constructed from the $so(2n)$-valued curvature 2-form of the Levi-Civita connection of a compact Riemannian manifold M of dimension $2n$, gives rise to the Euler characteristic of M,

$$\int_M Pf(\Omega) = (2\pi)^n \chi(M)$$

Now consider a complex manifold of complex dimension m and real dimension $n = 2m$. We use $\Omega^{p,q}(M)$ to denote the complex exterior differential forms having bases spanned by p factors of dz_k and q factors of $d\overline{z}_k$. Then the natural decomposition

$$df = \frac{\partial f}{\partial z_k} dz_k + \frac{\partial f}{\partial \overline{z}_k} d\overline{z}_k = \partial f + \overline{\partial} f$$

on a complex-valued function f gives us the Dolbeault operator

$$\overline{\partial} : \Omega^{p,q}(M) \to \Omega^{p,q+1}(M)$$

Since $\overline{\partial}^2 = 0$, we have a complex called the Dolbeault complex which gives rise to the Dolbeault cohomology over the complex field,

$$H^{p,q}(M) = \ker\left(\overline{\partial} : \Omega^{p,q}(M) \to \Omega^{p,q+1}(M)\right)/\overline{\partial}\Omega^{p,q-1}(M)$$

Like before, we have

$$\text{index}(\overline{\partial}) = \sum_{q=0}^{m} (-1)^q \dim H^{0,q}(M)$$

This quantity is also called the holomorphic Euler characteristic of M.

Using $T_c(M)$ to denote the complex tangent bundle spanned locally by $\partial/\partial z_k$ ($k = 1, 2, \cdots, n$) and $\text{td}(T_c(M))$ the Todd class whose specific form does not concern us here, we recall that the Hirzebruch–Riemann–Roch theorem (1954) states that

$$\text{index}(\overline{\partial}) = \int_M \text{td}(T_c(M))$$

which generalizes the original Riemann–Roch theorem for curves and surfaces. A much more extended form of this theorem is the Grothendieck–Hirzebruch–Riemann–Roch theorem[18] dated 1958.

Motivated by the above and other similar relations connecting analysis, geometry, and topology, Gelfand proposed the following problem.

Gelfand Problem: Let L be an operator between the smooth sections of the vector bundles E and F over a Riemannian manifold M so that $\ker(L)$ and $\text{coker}(L)$ are both of finite dimensions and the index of L is well defined. Can the index of L be expressed in terms of certain topologically invariant quantities of M, E, F, L?

We next study the index theorem of Atiyah and Singer[8,9,39,74] which solves the Gelfand problem affirmatively in the elliptic situation.

A (Very) Soft Introduction to the Atiyah–Singer Index Theorem

Let $\{E_k\}$ be a finite sequence of vector bundles over M and

$$D_k : C^\infty(E_k) \to C^\infty(E_{k+1})$$

be a differential operator between the corresponding spaces of smooth sections. When $D_{k+1}D_k = 0$ or image$(D_k) \subset \ker(D_{k+1})$ $(\forall k)$, we say that the sequence is a complex.

Let $\delta_k : C^\infty(E_{k+1}) \to C^\infty(E_k)$ be the dual of D_k and set

$$\Delta_k = \delta_k D_k + D_{k-1}\delta_{k-1} : C^\infty(E_k) \to C^\infty(E_k)$$

be the induced Laplacian. The complex is elliptic if Δ_k is an elliptic operator for any k. For an elliptic complex (E, D), we can define the cohomology space

$$H^k(E, D) = \ker(D_k)/\text{image}(D_{k-1})$$

and we can show that there holds the generalized Hodge theorem,

$$H^k(E, D) \cong \ker(\Delta_k) \equiv \mathcal{H}^k(E, D)$$

In particular, finite dimensionality property holds as before for any k.

The index of the elliptic complex (E, D) is given by

$$\text{index}(E, D) = \sum_k (-1)^k \dim H^k(E, D) = \sum_k (-1)^k \dim \mathcal{H}^k(E, D)$$

which looks like a direct generalization of the Euler characteristics for the de Rham complex and the Dolbeault complex.

In order to see that the above is indeed an operator index, we recall a standard device in topology called the "rolling up". Define

$$F_0 = \oplus_k E_{2k}, \quad F_1 = \oplus_k E_{2k-1}$$

and

$$A = \oplus_k(D_{2k} + \delta_{2k-1}), \quad A^* = \oplus_k(\delta_{2k} + D_{2k-1})$$

then $A : C^\infty(F_0) \to C^\infty(F_1)$ and $A^* : C^\infty(F_1) \to C^\infty(F_0)$ are dual to each other. With the associated Laplacians

$$\Box_0 = A^*A = \oplus_k \Delta_{2k}, \quad \Box_1 = AA^* = \oplus_k \Delta_{2k-1}$$

we arrive at

$$\text{index}(A) = \dim \ker \Box_0 - \dim \ker \Box_1$$
$$= \sum_k (-1)^k \dim \ker \Delta_k = \text{index}(E, D)$$

Finally, the Atiyah–Singer index theorem may be stated in a single-line formula,

$$\text{index}(A) = \text{index}(E, D) = \int_{\psi(M)} \text{ch}(\Sigma(A)) \wedge \rho^*\text{td}(TM)$$

where $\Sigma(A)$ is the symbol bundle constructed from A, F_0, F_1, M, $\text{ch}(\Sigma(A))$ is the Chern character of $\Sigma(A)$, $\text{td}(TM)$ is the Todd class of the tangent bundle TM, and $\rho : \psi(M) \to M$ is the compactified tangent bundle of M obtained from using the unit disk bundle and unit sphere bundle of T^*M. The detailed structure of these constructions do not concern us here. Instead, we only satisfy ourselves by seeing that the right-hand side of the formula is indeed expressed as a topological invariant involving the items stated in the Gelfand problem. We note also that, when specifying to various concrete situations, the right-hand side of the above formula simplifies. For example, the afore-studied Chern–Gauss–Bonnet theorem and Hirzebruch–Riemann–Roch theorem can both be recovered as special cases.

Remarks on Proofs. There are many different proofs of the Atiyah–Singer index theorem. The first three classical proofs are (i) the original cobordism proof; (ii) the heat equation proof; (iii) the embedding proof. These proofs are concisely described and compared and can be consulted in the book of Booss and Bleecker.[17] Concerning the differences and similarities of these proofs, it may be interesting to quote the words of Atiyah:[17] "these different proofs differ only in their use and presentation of algebraic topology," but "the analysis is essentially the same in origin." It should be noted that, in recent years, there appeared several other proofs using modern ideas. These are (iv) the supersymmetric quantum mechanics proof of Windey,[101] Alvarez-Gaume,[2] Manes and Zumino,[67] Goodman,[46] Getzler,[44] based on some ideas of Witten;[103] (v) the probabilistic approach of Bismut;[15] (vi) the superspace formulation of Friedan and Windey.[43] A common feature of all these proofs is the use of K-theory[72] whose power is to allow one to reduce the proof of the general index theorem to special "twisted" cases.

Dimension of Moduli Space of Self-Dual Instantons

Interestingly, this dimension calculation was first carried out by Schwartz[79] using the Atiyah–Singer index theorem, and then by Atiyah, Hitchin, and Singer,[6,7] when the gauge group is $SU(2)$. See also.[21] Shortly after, Bernard, Christ, Guth, and Weinberg extended this work to arbitrary gauge groups and wrote a very readable article[12] on the subject.

Consider the self-dual equation

$$F_A = *F_A$$

where A is the connection 1-form of a principle G-bundle P over $M = S^4$, G is a compact Lie group, and F_A is the curvature 2-form induced from A with $F_A = \mathrm{d}A + A \wedge A$. We look at small fluctuations, say ω, around a solution of the above equation within the topological class

$$c_2(P) = -\frac{1}{8\pi^2} \int_M \mathrm{Tr}\,(F_A \wedge F_A) = k, \quad k \geq 1$$

Then ω gives rise to the linear fluctuations in F_A:

$$\mathrm{d}\omega + A \wedge \omega + \omega \wedge A = D\omega$$

Besides, in order to preserve self-duality, we must require $D\omega = *D\omega$ or

$$P_1 D\omega = 0$$

where $P_1 = 1 - *$ is the projection operator over 2-forms.

Recall also that we need to discount the gauge-transformed fluctuations which are characterized by

$$A \mapsto UAU^{-1} - U\mathrm{d}U^{-1}, \quad U \in G$$

With the exponential representation $U = \exp u$ where u is valued in the Lie algebra $\mathcal{G}$ of G and neglecting higher-order terms, the above corresponds to

$$A \mapsto A + \mathrm{d}u - [A, u]$$

Hence, "pure gauge" fluctuations are described by the relation

$$\omega = \mathrm{d}u - [A, u] = Du$$

which are physically nonmeasurable.

In summary, define

$$D_1 = P_1 D : \quad \Omega^1(\mathcal{G}) \to \Omega^2(\mathcal{G})$$
$$D_0 = D : \quad \Omega^0(\mathcal{G}) \to \Omega^1(\mathcal{G})$$

We see that the dimension of the moduli space $\mathcal{M}_k$ of self-dual instantons which are gauge-inequivalent is given by

$$\dim \mathcal{M}_k = \dim(\ker D_1 / \mathrm{image} D_0)$$

which happens to be the dimension number of a "cohomological" space.

The afore-going formulation suggests the following short complex

$$0 \xrightarrow{D_{-1}} \Omega^0(\mathcal{G}) \xrightarrow{D_0} \Omega^1(\mathcal{G}) \xrightarrow{D_1} \Omega^2_-(\mathcal{G}) \xrightarrow{D_2} 0$$

called the self-dual complex or the Atiyah–Hitchin–Singer complex in which the definitions of the operators D_{-1} and D_2 are self-evident and $\Omega^2_-(\mathcal{G})$ is the space of anti-self-dual $\mathcal{G}$-valued 2-forms.

Define the "Betty" numbers as the dimensions of the cohomological spaces of this short complex,

$$b_i = \dim(\ker D_i / \mathrm{image} D_{i-1}), \quad i = 0, 1, 2$$

which gives us the analytic index of the complex,

$$\mathrm{index}(D) = b_0 - b_1 + b_2$$

Note that b_1 is the desired number $\dim \mathcal{M}_k$.

b_0: If the connection is irreducible, then $\ker(D_0) = \{0\}$ (see[42]), which gives $b_0 = 0$. For $G = SU(n)$, irreducibility is ensured when

$$k \geq \frac{n-1}{2}$$

Hence, for the classical $SU(2)$ situation, nontriviality $k \geq 1$ guarantees $b_0 = 0$.

b_2: Since the kernel of D_2 is the entire $\Omega_-^2(\mathcal{G})$, b_2 is the dimension of the subspace of $\Omega_-^2(\mathcal{G})$ orthogonal to the image of D_1, which is just the kernel of D_1^* (D_1^* is the dual of D_1). Hence we may consider the dimension of

$$\ker D_1^* \subset \Omega_-^2(\mathcal{G})$$

If $T \in \ker D_1^*$, then $D_1 D_1^* T = 0$. In differential geometry, a useful device is called the Bochner formula or the Weitzenbock formula which relates two Laplacians by a zeroth-order curvature multiplicative operator. In our case, since the Weyl tensor vanishes on a conformally flat manifold, we have the relation[12,42]

$$D_1 D_1^* = D^* D - \frac{1}{3}R$$

Integration by parts and the condition $R > 0$ then lead to $T = 0$ which gives us $b_2 = 0$.

We now consider the right-hand-side quantity, say $I(D)$, in the Atiyah–Singer index formula. This is a topological invariant which may be calculated according to the specific situation here, i.e., gauge theory over S^4 housed in terms of a G-bundle, using a beautiful deformation theory approach.[13] We will not be able to get into this area but only list a special class of important results here for $G = SU(n)$, $n \geq 2$:

$$I(D) = \begin{cases} n^2 - 1 - 4nk, & k \geq \frac{n}{2}, \\ -4k^2 - 1, & k < \frac{n}{2} \end{cases}$$

In particular, for $n = 2$, we have $I(D) = 3 - 8k$. Inserting this into the Atiyah–Singer index formula and noting index$(D) = -b_1$, we obtain the classical result[6,79]

$$\dim \mathcal{M}_k = 8k - 3$$

For results involving other gauge groups such as $SO(n), Sp(2n), G, F, E$, see.[12]

5. Topological Classes and Instantons in All $4m$ Dimensions and Nonlinear Elliptic Equations

Physics is not restricted to four-dimensional spacetimes. In fact, modern theoretical physics thrives in higher dimensions as witnessed by the development of string theory. Other areas of applications of higher dimensional quantum field theory include cosmology and condensed-matter systems. Tchrakian showed in a series of papers[90–92] that one can systematically develop the Yang–Mills theory in $4m$ dimensions so that the $2m$-th order Chern–Pontryagin classes, c_{2m}, over S^{4m} (say) may be represented by self-dual or anti-self-dual Yang–Mills instantons. In order to obtain instantons representing arbitrarily prescribed Chern–Pontryagin classes in higher dimensions ($m > 1$), Chakrabarti, Sherry, and Tchrakian[26] extend Witten's axially symmetric instantons in 4 dimensions and find a system of self-dual or anti-self-dual equations over the Poincaré half plane unifying the problem in all $4m$ dimensions. We have seen that, when $m = 1$, the problem reduces to the integrable Liouville equation and Witten uses this fact to construct all possible solutions explicitly.[102] We shall see here that, when $m > 1$, the system reduces to a quasilinear elliptic equation and is no longer integrable.

We shall use PDE techniques to establish the general existence theorem that for any integer N one can realize the $2m$-th order Chern–Pontryagin class $c_{2m} = N$ by a self-dual or anti-self-dual instanton and in fact, for a given choice of the 'time' coordinate, the moduli space of these N-instantons has a dimension of at least $2|N|$ where the number $|N|$ corresponds to the number of 'vortices' or 'antivortices'.

This work was completed in two papers:

Existence of weak solutions – joint work with J. Spruck and D. H. Tchrakian.[84]

Regularity of solutions – joint work with L. Sibner and R. Sibner.[81]

The Yang–Mills Instantons and Characteristic Classes in Higher Dimensions and the (Main) Harmonic Representation Theorem

We take the base manifold $M = S^{4m}$. The most natural principal bundle to host the gauge fields over S^{4m} is the frame bundle associated with the tangent bundle. Hence, we are led to the largest possible structure group, $SO(4m)$. In 4 dimensions, we have $SO(4)$, which contains two copies of $SO(3)$. Since the Lie algebra of $SO(3)$ is the same as that of $SU(2)$, the $SU(2)$-gauge theory, which has been extensively studied by numerous people, is a special case of the $SO(4)$-gauge theory. Thus, we now formulate a general $SO(4m)$ pure Yang–Mills gauge theory over S^{4m}. The Lie algebra of $SO(4m)$ is conventionally denoted by $so(4m)$.

Let A be an $so(4m)$-valued connection 1-form over S^{4m} and F its induced curvature 2-form. Motivated from the Yang–Mills theory in 4 dimensions, Tchrakian introduces the following energy functional over M:

$$E = -\int \mathrm{Tr}\ (F(m) \wedge *F(m))$$

where

$$F(m) = \underbrace{F \wedge \cdots \wedge F}_{m}$$

is a $2m$-form generalizing the 2-form F.

Recall that for $so(4m)$-valued differential forms over M, the global inner product is given by

$$\langle \alpha, \beta \rangle = -\int \mathrm{Tr}\ (\alpha \wedge *\beta)$$

In view of this, the energy is nothing but the squared norm of the generalized curvature $F(m)$: $E = \|F(m)\|^2$.

We introduce the characteristic class

$$s_{2m}(F) = -\mathrm{Tr}\ (F(m) \wedge F(m)) = -\mathrm{Tr}\ (\underbrace{F \wedge \cdots \wedge F}_{2m})$$

Of course, $s_2(F)$ is proportional to the second Chern–Pontryagin form $c_2(F)$:
$s_2(F) = 8\pi^2 c_2(F)$. In general, $s_{2m}(F)$ is proportional to the $2m$th Chern–Pontryagin form $c_{2m}(F)$,

$$s_{2m}(F) = -(-1)^m (2\pi)^{2m} (2m)!\, c_{2m}(F)$$

The associated topological charge is then defined as

$$s_{2m} = \int_{S^{4m}} s_{2m}(F) = -\int_{S^{4m}} \mathrm{Tr}\,(F(m) \wedge F(m))$$
$$= -(-1)^m (2\pi)^{2m} (2m)!\, c_{2m}$$

We now decompose $F(m)$ into its self-dual and anti-self-dual parts,

$$F(m) = F^+(m) + F^-(m), \quad F^\pm(m) = \frac{1}{2}(F(m) \pm *F(m))$$

We see that $F^+(m)$ and $F^-(m)$ are orthogonal,

$$\langle F^+(m), F^-(m)\rangle = 0$$

Therefore, using the property

$$*F^\pm(m) = \pm F^\pm(m)$$

and the orthogonality of $F^+(m)$ and $F^-(m)$, we obtain

$$E = \langle F(m), F(m)\rangle$$
$$= \langle F^+(m) + F^-(m), F^+(m) + F^-(m)\rangle$$
$$= \|F^+(m)\|^2 + \|F^-(m)\|^2,$$
$$s_{2m} = \langle F(m), *F(m)\rangle$$
$$= \langle F^+(m) + F^-(m), F^+(m) - F^-(m)\rangle$$
$$= \|F^+(m)\|^2 - \|F^-(m)\|^2$$

Consequently, we arrive at

$$E = 2\|F^\mp(m)\|^2 \pm (\|F^+(m)\|^2 - \|F^-(m)\|^2)$$
$$= 2\|F^\mp(m)\|^2 + |s_{2m}|$$
$$\geq |s_{2m}|$$

The above topological lower bound is attained for $s_{2m} = \pm|s_{2m}|$ if and only if the curvature satisfies $F^\mp(m) = 0$; that is, $F(m)$ satisfies either the self-dual or anti-self-dual Yang–Mills equations

$$F(m) = \pm * F(m)$$

It will be instructive to consider the above equations in view of the Euler–Lagrange equations of the energy.

First we observe that we can derive the generalized Bianchi identity

$$DF(k) = 0, \quad \forall k \geq 1; \quad F(1) = F$$

Next we see that, after a straightforward calculation, we obtain the Euler–Lagrange equations of the energy

$$DF = 0,$$
$$\mathcal{F} = F(m-1) \wedge *F(m) + F(m-2) \wedge *F(m) \wedge F$$
$$+F(m-3) \wedge *F(m) \wedge F(2) + \cdots + *F(m) \wedge F(m-1)$$

which may be called the generalized Yang–Mills equations in $4m$ dimensions. When $m = 1$, it is the classical one,

$$D(*F) = 0$$

If $F(m)$ is self-dual or anti-self-dual, the generalized Yang–Mills equations is reduced to

$$DF(2m-1) = 0$$

which is automatically fulfilled because of the generalized Bianchi identity. This is a great comfort.

As in the classical 4-dimensional Yang–Mills theory case, we shall concentrate on the self-dual or anti-self-dual equation. Below is our main harmonic representation theorem.

Theorem 5.1. *For any integer $N \geq 1$, the self-dual equation on S^{4m} has a $8mN$-parameter family of N-instanton solutions representing the Chern–Pontryagin class $c_{2m} = N$ and carrying the minimum energy $E = (2\pi)^{2m}(2m)! \, N$.*

Note on notation: We are short of letters. So we now use N instead of k (before) to denote the value of the Chern–Pontryagin class.

Note on harmonicity: We consider the bundle Laplacian induced from the connection D:

$$\Delta_D = -(*D*D + D*D*)$$

Then the generalized Bianchi identity and the self-duality imply that $\Delta_D F(m) = 0$. In other words, the generalized "curvature" $2m$-form $F(m)$ is indeed harmonic.

The Witten–Tchrakian Vortex Equations and Governing Elliptic PDE

It will be convenient to work on the Euclidean space $\mathbb{R}^{4m}$ instead of the sphere S^{4m}. Such a reduction is possible because, through a stereographic projection, $\mathbb{R}^{4m}$ is conformal to a punctured sphere, say, $S^{4m} - \{P\}$. We also know that the Hodge dual $*F(m)$ is conformally invariant. Hence the Yang–Mills theory on $\mathbb{R}^{4m}$ is identical to that on $S^{4m} - \{P\}$. Finally, in analogy to Uhlenbeck's removable singularity theorem, the finite-energy condition implies that the solutions on $\mathbb{R}^{4m}$ behave well at infinity so that, when viewed on S^{4m}, they extend smoothly to the point P. Therefore, in this way, we have actually obtained a family of solutions on S^{4m}. Thus, from now on, we consider the Yang–Mills theory on $\mathbb{R}^{4m}$.

In order to obtain N-instanton solutions, Tchrakian uses the approach of Witten as described earlier and extends it to the general $SO(4m)$ setting over $\mathbb{R}^{4m}$. The algebra is quite involved.[90–92] Here we will only record the final form of the problem: A field configuration is represented by a complex scalar field ϕ and a real-valued vector field $\mathbf{a} = (a_1, a_2)$, both defined on the Poincaré half-plane

$$\mathbb{R}_+^2 = \{(r,t) \mid r > 0, \ -\infty < t < \infty\}$$

where $r = \sqrt{x_1^2 + x_2^2 + \cdots + x_{4m-1}^2}$ and $t = x_{4m}$; up to a positive numerical factor the energy functional is

$$E^{(m)} = \int_{-\infty}^{\infty} dt \int_0^{\infty} dr \, (1 - |\phi|^2)^{2(m-2)} e^{(m)}(\mathbf{a}, \phi)$$

$$e^{(m)}(\mathbf{a}, \phi) = r^2([1 - |\phi|^2] f_{12} - i[m-1][D_1\phi\overline{D_2\phi} - \overline{D_1\phi}D_2\phi])^2$$

$$+ 2m(2m-1)(1 - |\phi|^2)^2(|D_1\phi|^2 + |D_2\phi|^2) + \frac{(2m-1)^2}{r^2}(1 - |\phi|^2)^4$$

the topological charge is

$$s^{(m)} =$$

$$-\int_{-\infty}^{\infty} dt \int_0^{\infty} dr \left\{ (1 - |\phi|^2)f_{12} - i(2m-1)(D_1\phi\overline{D_2\phi} - \overline{D_1\phi}D_2\phi) \right\} (1 - |\phi|^2)^{2(m-1)}$$

and the self-dual equation becomes

$$D_1\phi = -iD_2\phi,$$

$$\frac{(2m-1)}{r^2}(1 - |\phi|^2)^2 = -(1 - |\phi|^2)f_{12} + i(m-1)(D_1\phi\overline{D_2\phi} - \overline{D_1\phi}D_2\phi),$$

$$x_1 = r, \quad x_2 = t, \quad x = (x_1, x_2) \in \mathbb{R}_+^2$$

where $f_{jk} = \partial_j a_k - \partial_k a_j$ and $D_j\phi = \partial_j\phi + ia_j\phi$ $(j, k = 1, 2)$. It is comforting to note that, when $m = 1$, we recover Witten's results. The above general equations for arbitrary $m = 1, 2, \cdots$ arising from the Yang–Mills theory in $4m$ dimensions were derived by Tchrakian[26] and may be called the Witten–Tchrakian equations.

It is direct to see the relation between the energy $E^{(m)}$ and the topological charge $s^{(m)}$. In fact, the integrand of $E^{(m)}$ can be rewritten as

$$\mathcal{H}^{(m)} =$$

$$(1 - |\phi|^2)^{2(m-2)} \left\{ \left(r([1 - |\phi|^2] f_{12} - i[m-1][D_1\phi\overline{D_2\phi} - \overline{D_1\phi}D_2\phi]) \right. \right.$$

$$\left. + \frac{(2m-1)}{r}(1 - |\phi|^2)^2 \right)^2 + 2m(2m-1)(1 - |\phi|^2)^2|D_1\phi + iD_2\phi|^2 \right\}$$

$$- 2(2m-1)(1 - |\phi|^2)^{2(m-1)} \left\{ (1 - |\phi|^2)f_{12} - i(2m-1)(D_1\phi\overline{D_2\phi} - \overline{D_1\phi}D_2\phi) \right\}$$

We obtain the following topological lower bound for the energy

$$E^{(m)} \geq 2(2m-1)s^{(m)}$$

This lower bound is saturated if and only if the field configuration satisfies the Witten–Tchrakian equations.

As in the case of Witten, our N-instanton solutions of the self-dual Yang–Mills equations on S^{4m} or $\mathbb{R}^{4m}$ stated in Theorem 5.1 will be obtained through a family of N-soliton solutions of the equations on the Poincaré half-plane $\mathbb{R}^2_+$ characterized by N zeros of the complex field ϕ.

Here is our main existence and uniqueness theorem for the N-soliton solutions of the elegant Witten–Tchrakian equations.

Theorem 5.2. *For any N points $p_1, p_2, \cdots, p_N$ in $\mathbb{R}^2_+$ and any integer $m = 1, 2, \cdots$, the Witten–Tchrakian equations have a unique solution so that ϕ vanishes exactly at these points, $|\phi| = 1$ at the boundary and the infinity of the Poincaré half-plane, the topological charge is given by $s^{(m)} = 2\pi N$, and the solution carries the quantized minimum energy $E^{(m)} = 4\pi(2m-1)N$.*

The N prescribed zeros stated in Theorem 5.2 give rise to $2N$ parameters stated in Theorem 5.1. There are $4m$ choices of the time variable. Hence we obtain a total of $8mN$ parameters as stated in Theorem 5.1.

With $p_1, p_2, \cdots, p_N \in \mathbb{R}^2_+$ (with possible multiplicities) being given as in Theorem 5.2, the substitution $u = \ln|\phi|$ transforms the problem into the following equivalent scalar equation,

$$(e^{2u} - 1)\Delta u = \frac{(2m-1)}{r^2}(e^{2u} - 1)^2 - 2(m-1)e^{2u}|\nabla u|^2 - 2\pi \sum_{j=1}^{N} \delta_{p_j},$$

$$x \in \mathbb{R}^2_+$$

where δ_p is the Dirac measure concentrated at p. We are to look for a solution u so that $u(x) \to 0$ (hence $|\phi(x)| \to 1$) as $x \to \partial\mathbb{R}^2_+$ or as $|x| \to \infty$.

It is clear that this is a quasilinear problem for $m \neq 1$. Since the maximum principle implies that $u(x) \leq 0$ everywhere, it will be more convenient to use the new variable

$$v = f(u) = 2(-1)^m \int_0^u (e^{2s} - 1)^{m-1} ds, \quad u \leq 0$$

It is easily seen that

$$f : (-\infty, 0] \to [0, \infty)$$

is strictly decreasing and convex.

Set

$$u = F(v) = f^{-1}(v), \quad v \geq 0$$

Then the equation is simplified into a semilinear one,

$$\Delta v = \frac{2(-1)^m(2m-1)}{r^2}(e^{2F(v)} - 1)^m - 4\pi \sum_{j=1}^{N} \delta_{p_j} \quad \text{in } \mathbb{R}^2_+$$

To approach this equation, we introduce its modification of the form

$$\Delta v = \frac{2(2m-1)}{r^2} R(v) - 4\pi \sum_{j=1}^{N} \delta_{p_j}$$

where the right-hand-side function $R(v)$ is defined by

$$R(v) = \begin{cases} (-1)^m (e^{2F(v)} - 1)^m, & v \geq 0, \\ mv, & v < 0 \end{cases}$$

Then it is straightforward to check that $R(\cdot) \in C^1$. In order to obtain a solution of the original equation, it suffices to get a solution satisfying $v(x) \geq 0$ in $\mathbb{R}_+^2$ and $v(x) \to 0$ as $x \to \partial\mathbb{R}_+^2$ or as $|x| \to \infty$.

The main technical difficulty is the singular boundary of $\mathbb{R}_+^2$. We will employ a limiting argument to overcome this difficulty. We first solve the equation on a given bounded domain away from $r = 0$ under the homogeneous Dirichlet boundary condition. It will be seen that the obtained solution is indeed nonnegative and thus the equation is recovered. Such a property also allows us to control its energy and pointwise bounds conveniently. We then choose a sequence of bounded domains to approximate the full $\mathbb{R}_+^2$. The corresponding sequence of solutions is shown to converge to a weak solution. This weak solution is actually a positive classical solution which necessarily vanishes asymptotically as desired. Then suitable decay rates are established by using certain comparison functions.

Existence of Weak Solution

To proceed, choose a function, say, v_0, satisfying the requirement that it is compactly supported in $\mathbb{R}_+^2$ and smooth everywhere except at $p_1, p_2, \cdots, p_N$ so that

$$\Delta v_0 + 4\pi \sum_{j=1}^{N} \delta_{p_j} = g(x) \in C_0^\infty(\mathbb{R}_+^2)$$

Let Ω be any given bounded domain containing the support of v_0 and $\overline{\Omega} \subset \mathbb{R}_+^2$ (where and in the sequel, all bounded domains have smooth boundaries). Then $v = v_0 + w$ changes the equation into a regular form without the Dirac measure right-hand-side source terms, which is the equation in the following boundary value problem,

$$\Delta w = \frac{2(2m-1)}{r^2} R(v_0 + w) - g \quad \text{in } \Omega,$$
$$w = 0 \quad \text{on } \partial\Omega$$

We first apply a variational method to prove the existence of a solution by using the functional

$$I(w) = \int_\Omega \left\{ \frac{1}{2}|\nabla w|^2 + \frac{2(2m-1)}{r^2} Q(v_0 + w) - gw \right\} dx,$$
$$w \in W_0^{1,2}(\Omega)$$

where $\mathrm{d}x = \mathrm{d}r\mathrm{d}t$ and the function $Q(s)$ is defined by

$$Q(s) = \int_0^s R(s')\,\mathrm{d}s' = \begin{cases} (-1)^m \int_0^s (e^{2F(s')} - 1)^m\,\mathrm{d}s', & s \geq 0, \\ \int_0^s ms'\,\mathrm{d}s' = \frac{m}{2}s^2, & s < 0 \end{cases}$$

which is positive except at $s = 0$. This property and the Poincaré inequality indicate that the functional is coercive and bounded from below on $W_0^{1,2}(\Omega)$. On the other hand, since $F(s) \leq 0$ for $s \geq 0$, we have

$$\left| \frac{\mathrm{d}}{\mathrm{d}s} Q(s) \right| = |R(s)| \leq \max\{1, m|s|\}.$$

This feature says that the functional is continuous on $W_0^{1,2}(\Omega)$ because $\overline{\Omega}$ is away from the boundary of $\mathbb{R}_+^2$ and, so, the weight $2(2m-1)/r^2$ is bounded. Besides, the definition of $F(s)$ gives us the result

$$\frac{\mathrm{d}^2}{\mathrm{d}s^2} Q(s) = \begin{cases} me^{2F(s)}, & s \geq 0, \\ m, & s < 0 \end{cases}$$

which says that the functional is also convex. Thus, by convex analysis, the functional is weakly lower semicontinuous on $W_0^{1,2}(\Omega)$ and the existence and uniqueness of a critical point is ensured. The standard elliptic theory then implies that such a critical point is a classical solution.

We observe that a simple application of the maximum principle proves that $v_0 + w > 0$ in Ω.

We now choose a sequence of bounded domains $\{\Omega_n\}$ satisfying

$$\text{supp}\,(v_0) \subset \Omega_1,\ \overline{\Omega_n} \subset \Omega_{n+1},\ \overline{\Omega_n} \subset \mathbb{R}_+^2,\ n = 1, 2, \cdots,\ \lim_{n \to \infty} \Omega_n = \mathbb{R}_+^2$$

Let w_n be the solution of obtained above for $\Omega = \Omega_n$ and $I(\cdot; \Omega_n)$ be the variational functional with $\Omega = \Omega_n$. Then, since w_n's are minimizers, we have the monotonicity

$$I(w_n; \Omega_n) \geq I(w_{n+1}; \Omega_{n+1}), \quad n = 1, 2, \cdots$$

To pass to the limit $n \to \infty$, we need to show that the $\{I(w_n; \Omega_n)\}$ is bounded from below. For this purpose, we need the following inequality:

For any $W_0^{1,2}(\mathbb{R}_+^2)$ function w, there holds the Poincaré inequality

$$\int_{\mathbb{R}_+^2} \frac{1}{r^2} w^2(x)\,\mathrm{d}x \leq 4 \int_{\mathbb{R}_+^2} |\nabla w(x)|^2\,\mathrm{d}x$$

The proof is a simple integration by parts: For $w \in C_0^1(\mathbb{R}_+^2)$ we have

$$\int_0^\infty \frac{1}{r^2} w^2(r, t)\,\mathrm{d}r = 2 \int_0^\infty \frac{1}{r} w(r, t) \frac{\mathrm{d}}{\mathrm{d}r} w(r, t)\,\mathrm{d}r$$

Thus the Schwartz inequality gives us

$$\int_{\mathbb{R}_+^2} \frac{1}{r^2} w^2(x)\,\mathrm{d}x \leq 4 \int_{\mathbb{R}_+^2} \left| \frac{\mathrm{d}w}{\mathrm{d}r}(x) \right|^2 \mathrm{d}x$$

which is actually stronger.

As a consequence, we obtain the lower estimate for the energy sequence,

$$I(w_n; \Omega_n) \geq \frac{1}{4} \int_{\mathbb{R}^2_+} |\nabla w_n|^2 \, dx - 4 \int_{\mathbb{R}^2_+} r^2 g^2 \, dx$$

We note that another consequence of the maximum principle is the pointwise monotonicity

$$w_n < w_{n+1} \quad \text{on } \Omega_n, \quad n = 1, 2, \cdots$$

We are now ready to pass to the limit.

We claim: For a given bounded subdomain Ω_0 with $\overline{\Omega_0} \subset \mathbb{R}^2_+$, the sequence $\{w_n\}$ is weakly convergent in $W^{1,2}(\Omega_0)$. The weak limit, say, w_{Ω_0}, is a solution of the equation with $\Omega = \Omega_0$ (neglecting the boundary condition) which satisfies $w_{\Omega_0}(x) > 0$.

In fact, from the above discussion, we see that there is a constant $C > 0$ such that

$$\sup_n \|\nabla w_n\|^2_{L^2(\mathbb{R}^2_+)} \leq C$$

which also gives us the boundedness of $\{w_n\}$ in $W^{1,2}(\Omega_0)$. Combining this with the monotonicity property, we conclude that $\{w_n\}$ in weakly convergent in $W^{1,2}(\Omega_0)$. It then follows from the compact embedding $W^{1,2}(\Omega_0) \to L^2(\Omega_0)$ that $R(v_0 + w_n)$ is convergent in $L^2(\Omega_0)$. On the other hand, since for sufficiently large n, we have $\Omega_0 \subset \Omega_n$, consequently

$$\int_{\mathbb{R}^2_+} \left\{ \nabla w_n \cdot \nabla \xi + \frac{2(2m-1)}{r^2} R(v_0 + w_n)\xi - g\xi \right\} dx = 0, \quad \forall \xi \in C^1_0(\Omega_0)$$

Letting $n \to \infty$, we see that w_{Ω_0} is a weak solution (without considering the boundary condition). The standard elliptic regularity theory then implies that it is also a classical (hence, smooth) solution. Since $w_n > 0$, we have $w_{\Omega_0} \geq 0$. The maximum principle then yields $w_{\Omega_0} > 0$ in Ω_0. Thus the claim follows.

Set $w(x) = w_{\Omega_0}(x)$ for $x \in \Omega_0$ for any given Ω_0 stated above. In this way we obtain a global solution of the equation over the full $\mathbb{R}^2_+$. Besides, we have seen that there is a constant $C > 0$ such that

$$I(w) \leq C, \quad \|\nabla w\|_{L^2(\mathbb{R}^2_+)} \leq C$$

Verification of Vanishing Boundary Condition

Note that the boundedness result above is not sufficient to ensure the decay of w at $r = 0$ and at infinity. We also need the pointwise boundedness of w over $\mathbb{R}^2_+$. This property will be assumed but not proved here. The problem appears to be similar to the multi-meron solution problem in the classical $SU(2)$ Yang–Mills theory.[24,45,55,76]

Claim: Let w be the solution of the equation. Then for $x = (r, t) \in \mathbb{R}^2_+$ we have the uniform limits

$$\lim_{r \to 0} w(x) = \lim_{|x| \to \infty} w(x) = 0.$$

Here is a proof adapted to our problem from:[55] Given $x = (r, t)$, let D be the disk centered at x with radius $r/2$. The Dirichlet Green's function $G(x', x'')$ of the Laplacian Δ on D (satisfying $G(x', x'') = 0$ for $|x'' - x| = r/2$) is defined by the expression

$$G(x', x'') = \frac{1}{2\pi} \ln \sqrt{|x' - x|^2 + |x'' - x|^2 - 2(x' - x) \cdot (x'' - x)}$$
$$- \frac{1}{2\pi} \ln \sqrt{\left(\frac{2|x' - x||x'' - x|}{r}\right)^2 + \left(\frac{r}{2}\right)^2 - 2(x' - x) \cdot (x'' - x)}$$
$$\text{where } x', x'' \in D \text{ but } x' \neq x''$$

Hence w at $x' \in D$ can be represented as

$$w(x') = \int_D dx'' \left\{ (-1)^m \frac{2(2m - 1)}{r''^2} (e^{2F(v_0 + w)} - 1)^m - gw \right\} (x'') G(x', x'')$$
$$+ \int_{\partial D} dS'' \left\{ \frac{\partial G}{\partial n''}(x', x'') \right\} w(x'')$$

where $x'' = (r'', t'')$ and $\partial/\partial n''$ denotes the outer normal derivative on D with respect to the variable x''. We need to first evaluate $|r(\nabla_x w)(x)|$. This can be done by differentiating the above and then setting $x = x'$. Note that

$$(\nabla_{x'} G(x', x''))_{x'=x} = \frac{1}{2\pi} \left(\frac{4}{r^2} - \frac{1}{|x'' - x|^2} \right) (x'' - x),$$
$$\left(\nabla_{x'} \frac{\partial G}{\partial n''}(x', x'') \right)_{x'=x} = \nabla_{x'} \left(\frac{x'' - x}{|x'' - x|} \cdot \nabla_{x''} G(x', x'') \right) \Big|_{x'=x}$$
$$= \frac{8}{\pi r^3}(x'' - x), \quad x'' \in \partial D$$

Now let

$$C_1 = \sup_{\mathbb{R}^2_+} \left\{ |2(2m - 1)(e^{2F(v_0 + w)} - 1)^m(x) - r^2 g(x) w(x)| \right\},$$
$$C_2 = \sup_{\mathbb{R}^2_+} \left\{ |w(x)| \right\}$$

Differentiate $w(x')$ again, set $x' = x$, apply the above results, and use $r'' \geq r/2$. We have

$$|\nabla w(x)| \leq \frac{2C_1}{\pi r^2} \int_D \frac{1}{|x'' - x|} dx'' + \frac{8C_2}{\pi r^3} \int_{\partial D} |x'' - x| dS''$$
$$\leq \frac{C}{r}$$

where C is a constant independent of $r > 0$. Thus the claimed bound for $|r\nabla w(x)|$ over $\mathbb{R}^2_+$ is established.

To show that w vanishes at $\partial\mathbb{R}^2_+$, we argue by contradiction. Let $x_n = (r_n, t_n)$ be a sequence in $\mathbb{R}^2_+$ satisfying either $r_n \to 0$ or $|x_n| \to \infty$ but $|w(x_n)| \geq$ some $\varepsilon > 0$. Without loss of generality we may also assume that the sequence is so chosen that the disks centered at x_n with radius $r_n/2$ are non-overlapping. Then set

$$D_n = \left\{ x \in \mathbb{R}^2_+ \;\middle|\; |x - x_n| < \varepsilon_0 r_n \right\}, \quad \varepsilon_0 = \min\left\{ \frac{1}{2}, \frac{\varepsilon}{4C} \right\}$$

where $C > 0$ is the constant given in $|r\nabla w(x)| \leq C$. For $x = (r, t) \in D_n$ we have $3r_n/2 \geq r \geq r_n/2$. Thus, integrating ∇w over the straight line L from x_n to $x \in D_n$ and using $|\nabla w(x')| < 2C/r_n$ ($\forall x' \in D_n$), we obtain the estimate

$$|w(x)| = \left| w(x_n) + \int_L \nabla w(x') \cdot dl' \right|$$
$$\geq \varepsilon - \frac{2C}{r_n} \frac{\varepsilon}{4C} r_n$$
$$= \frac{\varepsilon}{2} \qquad x \in D_n$$

Therefore we arrive at the contradiction

$$\int_{\mathbb{R}^2_+} \frac{w^2}{r^2} \, dx \geq \sum_{n=1}^{\infty} \int_{D_n} \frac{w^2}{r^2} dx$$
$$\geq \sum_{n=1}^{\infty} \left(\frac{2}{3r_n} \right)^2 \left(\frac{\varepsilon}{2} \right)^2 \pi (\varepsilon_0 r_n)^2$$
$$= \infty$$

because in view of the earlier discussion we have $w/r \in L^2(\mathbb{R}^2_+)$.

Remark on Parameter Count

It will be interesting to determine the maximal number of free parameters in the solutions or the dimensions of the moduli space of solutions of the $4m$-dimensional self-dual Yang–Mills equations.

Our study has shown that, when the Chern–Pontryagin number is N, the solutions contain at least $8mN$ parameters. In view of,[21,52] we may attempt an intuitive counting of the number of free parameters in the general solution in $4m$ dimensions as follows:

For an N-instanton solution, we need $4mN$ parameters and N parameters to determine the positions and sizes of the N localized instanton lumps. Besides, using the dimension formula $\dim(SO(n)) = n(n-1)/2$ and the fact that our generalized Yang–Mills theory has an $SO(4m)$ internal symmetry which is of the dimension $\dim(SO(4m)) = 2m(4m-1)$, we need $2m(4m-1)N$ extra parameters to determine the asymptotic orientations of these N instantons at infinity, from which the $2m(4m-1)$ parameters originated from the global $SO(4m)$ gauge equivalence is to be subtracted. Hence it appears that a plausible number-of-parameter count for the general N-instanton solution in $4m$ dimensions is given by

$$4mN + N + 2m(4m-1)N - 2m(4m-1) = (8m^2 + 2m + 1)N - 2m(4m-1)$$

For $m = 1$ (4 dimensions), this is $10N - 6$. Furthermore, if we make restriction of the Yang–Mills theory to one of the chiral representations, $SO(4m)_\pm$, of $SO(4m)$, we have only half of the dimension of $SO(4m)$: $\dim(SO(4m)_\pm) = m(4m-1)$. For example, the Witten–Tchrakian equations arise from such a restriction. Now the parameter count is instead

$$4mN + N + m(4m-1)N - m(4m-1) = (4m^2 + 3m + 1)N - m(4m-1)$$

For $m = 1$ (4 dimensions), the chiral representations of $SO(4)$ are simply two copies of $SO(3)$ which has the same Lie algebra as $SU(2)$ and the number count is $8N - 3$, which is the magic number obtained earlier for the classical $SU(2)$ Yang–Mills theory in 4 dimensions.[6,21,52,79] Recall that the proof of Atiyah, Hitchin, and Singer[6] in 4 dimensions is a study of the fluctuation modes around an N-instanton. Our theorem for the existence of N-instanton solutions in all $4m$ dimensions lays a foundational step for a general analysis of this type.

Remark on Stability of Higher Dimensional Yang–Mills Fields

For the classical Yang–Mills theory in 4 dimensions, Bourguignon and Lawson calculated the second variation of the energy functional and prove that stable solutions must be self-dual or anti-self-dual.[20] It will be interesting to establish this result in all $4m$ dimensions.

Remark on the Ω-Self-Dual Yang–Mills Fields

Consider a G-bundle ξ over an n-dimensional Riemannian manifold M. Let A be a $\mathcal{G}$-valued connection 1-form which gives rise to the connection D_A and the curvature 2-form F_A. Like before, the Yang–Mills energy is

$$E(A) = \int_M |F_A|^2$$

with the associated Yang–Mills equation

$$D_A * F_A = 0$$

It is clear that the Hodge dual of F_A is $*F_A$ which is an $(n-2)$-form so that the self-dual equation $F_A = \pm * F_A$ no longer makes sense. In order to overcome this, Tian[94,95] considers a new "self-dual" equation called the Ω-self-dual equation which is of the form

$$*(\Omega \wedge F_A) = \pm F_A$$

where Ω is a (scalar-valued) $(n-4)$-form. If the connection A satisfies this equation, then, using the usual Leibniz rule, we have

$$D_A * F_A = \pm d\Omega \wedge F_A \pm \Omega \wedge D_A F_A$$
$$= \pm d\Omega$$

Hence, we arrive at $D_A * F_A = 0$ if and only if Ω is closed. In other words, in this situation, the first-order Ω-self-dual equation is a first integral of the second-order Yang–Mills equation and we see an extension of the self-dual equation in higher dimensions.

The above extension of self-duality has several limitations.

1. Lack of exact nontrivial solutions.

2. Lack of conformal invariance. The physically most interesting noncompact space is $M = \mathbb{R}^n$. It is well known[32,53] from a simple rescaling argument on the energy functional that there is only the trivial solution of zero energy when $n > 4$. In order to achieve conformal invariance, one inevitably needs to consider instead an energy of the form

$$\int |F_A|^{n/2}$$

which has the same conformal property as the energy[84,90–92] we began with in this section.

3. Lack of higher-order nontrivial topological characterization. It is not hard to accept that a physically interesting compact space M would contain no "holes" in it. Topologically, this assumption could amount to assuming that

$$\pi_1(M) = 0, \cdots, \pi_4(M) = 0, \quad \dim M > 4$$

(Recall that a manifold M is called k-connected[50] if $\pi_1(M) = 0, \cdots, \pi_k(M) = 0$ for some $1 \le k < \dim M$. For example, the symmetric space $SU(2n)/Sp(n)$ arising in the soliton model of Witten[71,104,105] for baryons is 4-connected when $n \ge 3$.)

At this moment, we may recall the Hurewicz isomorphism theorem[41,50,71] which says that for a simply connected manifold the first nontrivial homotopy group and homology group appear at the same dimension and are isomorphic. Hence we conclude that $H_1(M, \mathbb{Z}) = 0, \cdots, H_4(M, \mathbb{Z}) = 0$. Using the Poincaré duality, we obtain

$$H^1(M, \mathbb{R}) = 0, \cdots, H^4(M, \mathbb{R}) = 0$$

We note that the higher dimensional Yang–Mills theory in the context of the Ω-self-duality[94,95] only involves the beginning two Chern classes, $c_1(\xi)$ and $c_2(\xi)$, which are known[73,83] to belong to the de Rham cohomology groups $H^2(M, \mathbb{R})$ and $H^4(M, \mathbb{R})$, which are all trivial when M satisfies the "no hole" condition stated above so that M is "like" a sphere.

In the extreme situation where all the homotopy groups up to the $(n-1)$th order vanish, then using the classical result that the nth integer homology $H_n(M, \mathbb{Z})$ of M (M is orientable) is $\mathbb{Z}$ itself and the Hurewicz theorem, we have

$$\pi_n(M) \cong H_n(M, \mathbb{Z}) = \mathbb{Z}$$

In other words, such a space M has the identical homotopy structure as that of S^n. According to the Poincaré conjecture (now the "Poincaré Theorem"), we have

$$M \cong S^n$$

Pause – A Brief History of the Poincaré Conjecture: The $n = 2$ case is classical and was known to the 19th century mathematicians, the $n = 3$ case, which is the original conjecture, appears to have been proved by recent work of G. Perelman (although the proof is yet to be fully verified), the $n = 4$ case was proved by Freedman (1982), the $n = 5$ case was proved by Zeeman (1961), the $n = 6$ case was proved by Stallings (1962), and the $n \geq 7$ case was proved by Smale (1961) (Smale subsequently extended his proof to include all the $n \geq 5$ cases).

Conclusion: The higher dimensional Yang–Mills theory of Tchrakian[90–92] presented here possesses rich classes of solutions representing all possible values of the top Chern–Pontryagin class and enjoys the same conformal invariance exactly as that in the classical 4 dimensional situation. However, the new technical difficulties we encounter are that we need to consider high tuples of the curvature 2-form and that we must confine ourselves to the correct "Pontryagin" dimensions, $n = 4m$, $m \geq 1$.

References

1. A. Actor, Classical solutions of $SU(2)$ Yang–Mills theories, *Rev. Mod. Phys.* **51**, 461–525 (1979).
2. L. Alvarez-Gaume, Supersymmetry and the Atiyah-Singer index theorem, *Commun. Math. Phys.* **90**, 161–173 (1983).
3. V. A. Andreev, Application of the inverse scattering method to the equation $\sigma_{xt} = e^\sigma$, *Theoret. Math. Phys.* **29**, 1027–1035 (1976).
4. M. M. Ansourian and F. R. Ore, Jr., Pseudoparticle solutions on the $O(5,1)$ light cone, *Phys. Rev. D* **16**, 2662–2665 (1977).
5. M. F. Atiyah, V. G. Drinfeld, N. J. Hitchin and Yu. I. Manin, Construction of instantons, *Phys. Lett. A* **65**, 185–187 (1978).
6. M. F. Atiyah, N. J. Hitchin, and I. M. Singer, Deformation of instantons, *Proc. Natl. Acad. Sci. USA* **74**, 2662–2663 (1977).
7. M. F. Atiyah, N. J. Hitchin, and I. M. Singer, Self-duality in four-dimensional Riemannian geometry, *Proc. Roy. Soc. A* **362**, 425–461 (1978).
8. M. F. Atiyah and I. M. Singer, The index of elliptic operators on compact manifolds, *Bull. Amer. Math. Soc.* **69**, 422–433 (1963).
9. M. F. Atiyah and I. M. Singer, The index of elliptic operators. I, *Ann. of Math.* **87**, 484–530 (1968); M. F. Atiyah and G. B. Segal, II, *ibid* **87**, 531–545 (1968); M. F. Atiyah and I. M. Singer, III, *ibid* **87**, 546–604 (1968); M. F. Atiyah and I. M. Singer, IV, *ibid* **93**, 119–138 (1971); V, *ibid* **93**, 139–149 (1971).
10. A. A. Belavin and A. M. Polyakov, Metastable states of two-dimensional isotropic ferromagnets, *JETP Lett.* **22** (1975) 245–247.
11. A. A. Belavin, A. M. Polyakov, A. S. Schwartz and Yu. S. Tyupkin, Pseudoparticle solutions of the Yang–Mills equations, *Phys. Lett. B* **59**, 85–87 (1975).

12. C. W. Bernard, N. H. Christ, A. H. Guth, and E. J. Weinberg, Pseudoparticle parameters for arbitrary gauge groups, *Phys. Rev. D* **16**, 2967–2977 (1977).

13. C. W. Bernard, A. H. Guth, and E. J. Weinberg Note on the Atiyah-Singer index theorem, *Phys. Rev. D* **17**, 10531055 (1978).

14. P. Bidal and G. de Rham, Les formes différentielles harmoniques, *Comm. Math. Helv.* **19**, 1–49 (1946).

15. J.-M. Bismut, The Atiyah–Singer theorems: a probabilistic approach, I. The index theorem, *J. Funct. Anal.* **57**, 56–99 (1984).

16. E. B. Bogomol'nyi, The stability of classical solutions, *Sov. J. Nucl. Phys.* **24**, 449–454 (1976).

17. B. Booss and D. D. Bleecker, *Topology and Analysis – The Atiyah–Singer Index Formula and Gauge-Theoretical Physics*, Springer, New York, 1985.

18. A. Borel and J.-P. Serre, Le thorme de Riemann–Roch, *Bull. Soc. Math. France* **36**, 97–136 (1958).

19. R. Bott, Lectures on Morse theory, old and new, *Bull. AMS* **7**, 331–358 (1982).

20. J. P. Bourguignon and H. B. Lawson, Jr., Stability and isolation phenomena for Yang–Mills fields, *Commun. Math. Phys.* **79**, 189–230 (1981).

21. L. S. Brown, R. D. Carlitz, and C. Lee, Massless excitations in pseudoparticle fields, *Phys. Rev. D* **16**, 417–422 (1977).

22. J. Burzlaff, A finite-energy $SU(3)$ solution which does not satisfy the Bogomol'nyi equation, *Czech. J. Phys. B* **32**, 624 (1982).

23. J. Burzlaff, D. O'Se, and D. H. Tchrakian, A finite-action solution to generalized Yang–Mills–Higgs theory, *Lett. Math. Phys.* **13**, 121–125 (1987).

24. L. Caffarelli, B. Gidas and J. Spruck, On multimeron solutions of the Yang–Mills equation, *Commun. Math. Phys.* **87**, 485–495 (1983).

25. D. Chae and O. Yu. Imanuvilov, The existence of nontopological multivortex solutions in the relativistic self-dual Chern–Simons theory, *Commun. Math. Phys.* **215**, 119–142 (2000).

26. A. Chakrabarti, T. N. Sherry, and D. H. Tchrakian, On axially symmetric self-dual field configurations in $4p$ dimensions, *Phys. Lett. B* **162**, 340–344 (1985).

27. H. Chan, C.-C. Fu, and C.-S. Lin, Non-topological multivortex solutions to the self-dual Chern–Simons–Higgs equation, *Commun. Math. Phys.* **231** (2002) 189–221.

28. X. Chen, S. Hastings, J. B. McLeod, and Y. Yang, A nonlinear elliptic equation arising from gauge field theory and cosmology, *Proc. Roy. Soc. A* **446**, 453–478 (1994).

29. S. S. Chern and J. Simons, Characteristic forms and geometric invariants, *Ann. of Math.* **99**, 48–69 (1974).

30. A. Comtet and G. W. Gibbons, Bogomol'nyi bounds for cosmic strings, *Nucl. Phys. B* **299**, 719–733 (1988).

31. G. de Rham, Sur la théorie des formes différentielles harmoniques, *Ann. Grenoble (Nouv. sér.)* **22**, 135–152 (1946).

32. G. H. Derrick, Comments on nonlinear wave equations as models for elementary particles, *J. Math. Phys.* **5**, 1252–1254 (1964).

33. J. Dziarmaga, Low energy dynamics of $[U(1)]^N$ Chern–Simons solitons, *Phys. Rev. D* **49** (1994) 5469–5479.

34. J. Dziarmaga, Only hybrid anyons can exist in broken symmetry phase of nonrelativistic $U(1)^2$ Chern-Simons theory, *Phys. Rev. D* **50**, R2376–R2380 (1994).

35. J. Eells and M. J. Ferreira, On representing homotopy classes by harmonic maps, *Bull. London Math. Soc.* **23**, 160–162 (1991).

36. J. Eells and J. H. Sampson, Harmonic mappings of Riemannian manifolds, *Amer. J. Math.* **86**, 109–160 (1964).

37. J. Eells and L. Lemaire, A report on harmonic maps, *Bull. London Math. Soc.* **10**, 1–68 (1978).

38. J. Eells and J. C. Wood, Restrictions on harmonic maps of surfaces, *Topology* **15**, 263–266 (1976).

39. T. Eguchi, P. Gilkey, and A. Hanson, Gravitation, gauge theories, and differential geometry, *Phys. Rep.* **66**, 213–393 (1980).

40. M. J. Esteban, Existence of symmetric solutions for the Skyrme's problem, *Ann. Math. Pura Appl.* **147** (1987) 187–195.

41. T. Frankel, *The Geometry of Physics, an Introduction*, Cambridge U. Press, Cambridge, 1997.

42. D. S. Freed and K. Uhlenbeck, *Instantons and Four-Manifolds*, Springer, New York, 1984.

43. D. Friedan and P. Windey, Supersymmetric derivation of the Atiyah–Singer index and the chiral anomaly, *Nucl. Phys.* B **235**, 395–416 (1984).

44. E. Getzler, Pseudodifferential operators on supermanifolds and the Atiyah–Singer index theorem, *Commun. Math. Phys.* **92**, 163–178 (1983).

45. J. Glimm and A. Jaffe, Multiple meron solution of the classical Yang–Mills equation, *Phys. Lett.* B **73**, 167–170 (1978).

46. M. W. Goodman, Proof of character-valued index theorems, *Commun. Math. Phys.* **107**, 391–409 (1986).

47. C. H. Gu, Conformally flat spaces and solutions to Yang-Mills equations, *Phys. Rev.* D **21**, 970–971 (1980).

48. P. Hartman, On homotopic harmonic maps, *Cand. J. Math.* **19**, 673–687 (1967).

49. W. V. D. Hodge, *The Theory and Applications of Harmonic Integrals*, Cambridge U. Press, 1941.

50. B. Y. Hou and B. Y. Hou, *Differential Geometry for Physicists*, World Scientific, Singapore, 1997.

51. R. Jackiw, C. Nohl, and C. Rebbi, Conformal properties of pseudoparticle configurations, *Phys. Rev.* D **15**, 1642–1646 (1977).

52. R. Jackiw and C. Rebbi, Degrees of freedom in pseudoparticle systems, *Phys. Lett.* B **67**, 189–192 (1977).

53. A. Jaffe and C. H. Taubes, *Vortices and Monopoles*, Birkhäuser, Boston, 1980.

54. A. Jaffe and E. Witten, Quantum Yang–Mills theory, *Preprint* (available at http: www.claymath.org).

55. T. Jonsson, O. McBryan, F. Zirilli and J. Hubbard, An existence theorem for multimeron solutions to classical Yang–Mills field equations, *Commun. Math. Phys.* **68**, 259–273 (1979).

56. B. Julia and A. Zee, Poles with both magnetic and electric charges in non-Abelian gauge theory, *Phys. Rev.* D **11**, 2227–1232 (1975).

57. C. Kim, C. Lee, P. Ko, B.-H. Lee, and H. Min, Schrödinger fields on the plane with $[U(1)]^N$ Chern–Simons interactions and generalized self-dual solitons, *Phys. Rev.* D **48** (1993) 1821–1840.

58. F. R. Klinkhamer and N. S. Manton, A saddle-point solution in the Weinberg–Salam theory, *Phys. Rev.* D **10**, 2212–2220 (1984).

59. K. Kodaira, Harmonic fields in Riemannian manifolds (generalized potential theory), *Ann. of Math.* **50**, 581–665 (1949).

60. P. F. Leung, On the stability of harmonic maps, *Lect. Notes in Math.* **949**, 122–129 (1982).

61. A. N. Leznov, On the complete integrability of a nonlinear system of partial differential equations in two-dimensional space, *Theoret. Math. Phys.* **42**, 225–229 (1980).

62. C. S. Lin, A. C. Ponce, and Y. Yang, A system of elliptic equations arising in Chern–Simons field theory, *J. Funct. Anal.* **247**, 289–350 (2007).

63. F. Lin and Y. Yang, Existence of energy minimizers as stable knotted solitons in the Faddeev model, *Commun. Math. Phys.* **249**, 273–303 (2004).

64. F. Lin and Y. Yang, Energy splitting, substantial inequality, and minimization for the Faddeev and Skyrme models, *Commun. Math. Phys.* **269**, 137–152 (2007).

65. F. Lin and Y. Yang, Analysis on Faddeev knots and Skyrme solitons: recent progress and open problems, *Contemporary Mathematics* **446**, pp. 319–344, American Mathematical Society, 2007.

66. J. Liouville, Sur l'équation aux différences partielles $\frac{\mathrm{d}^2 \log \lambda}{\mathrm{d}u \mathrm{d}v} \pm \frac{\lambda}{2a^2} = 0$, *Journal de Mathématiques Pures et Appl.* **18**, 71–72 (1853).

67. J. Manes and B. Zumino, WKB method, SUSY quantum mechanics and the index theorem, *Nucl. Phys.* B **270**, 651–686 (1986).

68. N. S. Manton, Topology in the Weinberg–Salam theory, *Phys. Rev.* D **8**, 2019–2026 (1983).

69. P. J. McCarthy, Bäcklund transformations as nonlinear Dirac equations, *Lett. Math. Phys.* **2**, 167–170 (1977).

70. A. N. Milgram and P. C. Rosenbloom, Harmonic forms and heat conduction I: closed Riemannian manifolds, *Proc. Nat. Acad. Sci. USA* **37**, 180–184 (1951).

71. M. Monastyrsky, *Topology of Gauge Fields and Condensed Matter*, Plenum, New York and London, 1993.

72. A. Mostafazadeh, *Supersymmetry, Path Integration, and the Atiyah–Singer Index Theorem*, Thesis, U. Texas at Austin, 1994.

73. C. Nash and S. Sen, *Topology and Geometry for Physicists*, Academic, London and New York, 1983.

74. R. S. Palais, *Seminar on the Atiyah-Singer Index Theorem*, Annals of Mathematics Studies, Princeton U. Press, Princeton.

75. M. K. Prasad and C. M. Sommerfield, Exact classical solutions for the 't Hooft monopole and the Julia–Zee dyon, *Phys. Rev. Lett.* **35**, 760–762 (1975).

76. M. Renardy, On bounded solutions of a classical Yang–Mills equation, *Commun. Math. Phys.* **76**, 277–287 (1980).

77. L. Sadun and J. Segert, Non-self-dual Yang–Mills connections with quadrupole symmetry, *Commun. Math. Phys.* **145**, 362–391 (1992).

78. J. Schiff, Integrability of Chern–Simons Higgs and Abelian Higgs vortex equations in a background metric, *J. Math. Phys.* **32**, 753 (1991).

79. A. S. Schwartz, On regular solutions of Euclidean Yang–Mills equations, *Phys. Lett.* B **67**, 172–174 (1977).

80. L. M. Sibner, R. J. Sibner, and K. Uhlenbeck, Solutions to Yang–Mills equations that are not self-dual, *Proc. Nat. Acad. Sci. USA* **86**, 8610–8613 (1989).

81. L. Sibner, R. Sibner, and Y. Yang, Multiple instantons representing higher-order Chern–Pontryagin classes. Part II, *Commun. Math. Phys.* **241**, 47–67 (2003).

82. R. T. Smith, Harmonic maps of spheres, *Amer. J. Math.* **97**, 364–385 (1975).

83. M. Spivak, *Differential Geometry*, Vol. 5, Publish or Perish, Boston, 1975.

84. J. Spruck, D. H. Tchrakian, and Y. Yang, Multiple instantons representing higher-order Chern–Pontryagin classes. Part I, *Commun. Math. Phys.* **188**, 737–751 (1997).

85. J. Spruck and Y. Yang, Topological solutions in the self-dual Chern–Simons theory: existence and approximation, *Ann. Inst. H. Poincaré – Anal. non linéaire*, **12**, 75–97 (1995).

86. J. Spruck and Y. Yang, The existence of non-topological solitons in the self-dual Chern–Simons theory, *Commun. Math. Phys.* **149**, 361–376 (1992).

87. J. L. Synge, On the connectivity of spaces of positive curvature, *Quart. J. Math. Oxford* **7**, 316–320 (1936).

88. G. Tarantello, Multiple condensate solutions for the Chern–Simons–Higgs theory, *J. Math. Phys.* **37**, 3769–3796 (1996).

89. C. H. Taubes, The existence of a non-minimal solution to the $SU(2)$ Yang–Mills–Higgs equations on $\mathbb{R}^3$, Parts I, II, *Commun. Math. Phys.* **86**, 257–320 (1982).

90. D. H. Tchrakian, N-dimensional instantons and monopoles, *J. Math. Phys.* **21**, 166–169 (1980).

91. D. H. Tchrakian, Spherically symmetric gauge field configurations in $4p$ dimensions, *Phys. Lett. B* **150**, 360–362 (1985).

92. D. H. Tchrakian, Yang–Mills hierarchy, *Int. J. Mod. Phys.* (Proc. Suppl.) A **3**, 584–587 (1993).

93. G. 't Hooft, Computation of the quantum effects due to a four-dimensional pseudoparticle, *Phys. Rev. D* **14**, 3432–3450 (1976).

94. G. Tian, Gauge theory and calibrated geometry. I, *Ann. of Math.* **151**, 193–268 (2000).

95. G. Tian, Elliptic Yang–Mills equations, *Proc. Nat. Acad. Sci. USA* **99**, 15281–15286 (2002).

96. K. Uhlenbeck, Harmonic maps: a direct method in the calculus of variations, *Bull. Amer. Math. Soc.* **76**, 1082–1087 (1970).

97. K. Uhlenbeck, Removable singularities in Yang–Mills fields, *Commun. Math. Phys.* **83**, 11-29 (1982).

98. A. Vilenkin, Cosmic strings and domain walls, *Phys. Rep.* **121**, 263-315 (1985).

99. A. Vilenkin and E. P. S. Shellard, *Cosmic Strings and Other Topological Defects*, Cambridge University Press, 1994.

100. H. Weyl, On Hodge's theory of harmonic integrals, *Ann. of Math.* **44**, 1–6 (1943).

101. P. Windey, Supersymmetric quantum mechanics and the Atiyah–Singer index theorem, *Acta Phys. Polonica B* **15**, 435–452 (1984).

102. E. Witten, Some exact multipseudoparticle solutions of classical Yang–Mills theory, *Phys. Rev. Lett.* **38**, 121–124 (1977).

103. E. Witten, Constraints on supersymmetric breaking, *Nucl. Phys. B* **202**, 253–316 (1982).

104. E. Witten, Global aspects of current algebra, *Nucl. Phys. B* **223**, 422–432 (1983).

105. E. Witten, Current algebra, baryons, and quark confinement, *Nucl. Phys. B* **223**, 433–444 (1983).

106. J. Wood, Singularities of harmonic maps and applications of the Gauss–Bonnet formula, *Amer. J. Math.* **99**, 1329–1344 (1977).

107. Y. L. Xin, Some results on stable harmonic maps, *Duke Math. J.* **47**, 609–613 (1980).

108. Y. L. Xin, *Geometry of Harmonic Maps*, Birkhäuser, Boston, 1996.

109. X. Xu and P. C. Yang, A construction of m-harmonic maps of spheres, *Internat. J. Math.* **4**, 521–533 (1993).

110. C. N. Yang and R. Mills, Conservation of isotopic spin and isotopic gauge invariance, *Phys. Rev.* **96**, 191–195 (1954).

111. Y. Yang, On the global behavior of symmetric Skyrmions, *Lett. Math. Phys.* **19** (1990) 25–33.

112. Y. Yang, Obstructions to the existence of static cosmic strings in an Abelian Higgs model, *Phys. Rev. Lett.* **73**, 10–13 (1994).

113. Y. Yang, Prescribing topological defects for the coupled Einstein and Abelian Higgs equations, *Commun. Math. Phys.* **170**, 541–582 (1995).

114. Y. Yang, *Solitons in Field Theory and Nonlinear Analysis*, Springer-Verlag, New York, 2001.

115. Y. L. Yu, *The Index Theorem and the Heat Equation Method*, Nankai Tracts in Mathematics, World Scientific, Singapore, 2001.

Postscript

The publication of this volume, dedicated to Professor Youzhong Guo on the occasion of his 75th birthday and 55 years of research in mathematical sciences, is a suitable honor for a legendary scientist and wonderful human being.

At the early stage of his career, Guo was known for furthering the work of the late Professor Guoping Li. The intensive collaboration of Guo and Li completed the theory of the Minkowski–Denjoy functions, a field of mathematics which provides a notable example of the Sino-Soviet collaboration in science.

During the chaotic period of the so-called 'Great Cultural Revolution,' the draft manuscript by Guo and Li on the theory of the Minkowski–Denjoy functions was destroyed, and Guo was unjustifiably imprisoned. His passion for research was undiminished by the ten years of his imprisonment, from 1968 to 1978. Fortunately, after the Great Cultural Revolution had ended a reconstructed version of their draft manuscript — the first monograph on the subject — was published.

Upon his rehabilitation, Guo compensated for his years of imprisonment by the pace of his research activity, initiating new projects with younger colleagues. He published more than 198 theses, treatises, and translations, 11 of which were awarded prizes. They are highly regarded by colleagues in the fields of mathematical physics, mechanics and system science.

During the new era of reform and openness in China, Guo has augmented his leadership in diverse fields of research by serving as Vice Mayor of Wuhan City, a metropolis of 9.73 million people and one of several politico-economic centers in China, comparable to Chicago in the United States. Even after he retired from his government posts, Guo has remained active in strategic studies for the development of the Middle China Economic Region along the Yangtze River. He has also advised many graduate students, some of whom have emerged as leading scientists.

My own appreciation of Guo's published works is increased by their personal cost to him in the difficult times in which he created them. While celebrating his achievements, I often cannot help but imagine how much more he could have accomplished had he the good fortune that our younger generation enjoys now. He personifies a tradition of faith and fortitude in the presence of adversity, a tradition that stretches back to our greatest ancient authors.

Here I mention only two who exemplify our great culture. After suffering humiliating punishment for pleading on behalf of the defeated General Li Guang, Sima Qian (145–87 B.C.) completed his masterpiece Shiji (*Records of the Historian*) in his remaining years, hoping "for justification only after my death," as he explained in a letter to his friend Ren An. His giant book covers about 3000 years in the history of China up to Emperor Wu of the Han Dynasty, providing a model of biographical prose even today. The second, a millennium later, was the woman poet of the Song Dynasty, Li Qingzhao (1084–1151), who created resonant verses while living as a widow and refugee from foreign invasion. When I was young, Sima Qian's Letter to Ren An, and the Postscript Li Qingzhao wrote for her husband's archaeological book *Bronze and Stone Inscriptions,* impressed me with the authors' dedication to history and literature, oblivious to their personal disasters. Decades later, tales about Guo's misfortune reminded me of these two great masters in our history.

I first met Youzhong in 1958, during my brief stay with the Three-Gorge Rock Foundation Research Group. In all these years, he has never expressed concern over personal gain or loss nor complained about his misfortune or psychological trauma. Rather, I have found his enthusiasm and sagacity about the development of mathematical physics, as well as his casual conversation, to be consistently enlightening, wise, and optimistic.

Youzhong is noted not only for his rigorous scholarship, but especially for his warmth and honesty. During his career, spanning half a century and involving extensive collaboration in different fields, he has established effective and harmonious relationships with his colleagues and earned support and affection from three generations of scientists. In my mind he is a rare role model for our times, one who has attained the high realm advocated by Confucius, who said "a gentleman's deeds should make the elderly feel at ease about him, juveniles think fondly of him, and friends trust him." That realm is difficult to attain, but clear and beautiful to imagine. Once the goal is reached, one can be regarded as successful. Sima Qian wrote a biography of Li Guang, who fought and won more than seventy battles in defense of the empire but never boasted about his military victories. To conclude the story about Li Guang, Sima Qian used an eight-character metaphor which, in my loose vernacular adaptation, says "though peach and plum trees are silent, the foot paths underneath have been trudged out by admiring viewers." This phrase has become an inspiration to all successful and unassuming people, and I take the liberty of quoting it here to epitomize Youzhong Guo, my instructive teacher and sincere friend.

金 汉平

Han-ping Chin
Albany, California, USA
March 8, 2009